Architecting Robust Co-Design of Materials, Products, and Manufacturing Processes

Anand Balu Nellippallil · Janet K. Allen ·
B. P. Gautham · Amarendra K. Singh ·
Farrokh Mistree

Architecting Robust Co-Design of Materials, Products, and Manufacturing Processes

Springer

Anand Balu Nellippallil
Department of Mechanical and Civil Engineering
Florida Institute of Technology
Melbourne, FL, USA

Janet K. Allen
School of Industrial and Systems Engineering
University of Oklahoma
Norman, OK, USA

B. P. Gautham
TCS Research, TRDDC
Tata Consultancy Services
Pune, India

Amarendra K. Singh
Department of Materials Science and Engineering
Indian Institute of Technology Kanpur
Kanpur, India

Farrokh Mistree
School of Aerospace and Mechanical Engineering
University of Oklahoma
Norman, OK, USA

ISBN 978-3-030-45323-7 ISBN 978-3-030-45324-4 (eBook)
https://doi.org/10.1007/978-3-030-45324-4

This Springer imprint is published by the registered company Springer Nature Switzerland AG
The registered company address is: Gewerbestrasse 11, 6330 Cham, Switzerland

Foreword

Since the establishment of the Integrated Computational Materials Engineering (ICME) initiative in 2008 and the Materials Genome Initiative in 2011, focus has been increasingly drawn toward the use of computational materials science and mechanics tools and methods to predict material process–structure–property relations. One branch of inquiry has focused on predictive materials computation to discover new materials, accelerated using combinatorial strategies. Another pursuit involves enhancement of materials selection for design applications, which had been the overarching goal of "materials design" in the late twentieth century.

Sometime around 2005, Professors Farrokh Mistree, Janet K. Allen, and I envisioned the pressing need to extend formal engineering systems design, namely concepts of decision-making in design and multidisciplinary design optimization, to address the hierarchical structure of materials, along with process path dependence of this structure and resulting properties. This collaboration at Georgia Tech ultimately led to the first monograph in this area, *Integrated Design of Multiscale, Multifunctional Materials and Products*, published by Butterworth-Heinemann in 2010 (ISBN 978-1-85617-662-0) and co-authored with some of our joint Ph.D. students (Carolyn C. Seepersad, Hae-Jin Choi, and Jitesh H. Panchal). That initial body of work focused on the integration of product design specifications with the top-down requirements on material properties/responses. An inductive design exploration framework was introduced to pursue multilevel robust inverse design across length scales of material structure comprising physical products. A fundamental assertion was that materials design is a process that necessitates an engineering systems approach, and cannot be addressed in a decoupled way as a materials selection problem without greatly compromising the breadth of available solutions. Moreover, uncertainty plays a key role in materials design, necessitating the notion of robust design across a set of objectives or goals rather than single point/goal optimization. That initial book ended with a discussion of future directions in distributed, collaborative concurrent design of materials and products and distributed product realization. The resulting concepts of decision-based materials design are central to unleashing the real power of computation and experiment in accelerating the rate of development of new and improved materials for applications.

It has become apparent in the years following our 2010 book that one of the most underdeveloped aspects of this vision of multilevel integrated design of materials and products has to do with the consideration of manufacturing processes. In particular, materials processing/synthesis is extremely complex to model, often involving highly nonlinear and non-equilibrium evolution of the material structure. Evolution of material structure strongly affects properties, and is tightly coupled to the selected product manufacturing route. Often, only heuristics are available to provide guidance for material process–structure relations, which carry high uncertainty. Though part of the vision of our earlier monograph, the relation of materials process path to structure and properties was not emphasized, the link to manufacturing processes was evident, but not elaborated to a significant extent. In this new work with their former student Dr. Nellippallil and colleagues at Tata and IIT Kanpur, Professors Allen and Mistree have focused on the coupling of material process–structure–property relations with product design *and manufacturing*. Doing so requires more formal treatment of identification of material structures and processing paths that are constrained by available manufacturing processes and the satisfaction of certain required product and manufacturing process-level properties and performance. This requires detailed mappings of steps involved in materials processing that precedes or associates with product manufacture. Moreover, models framed at various levels of fidelity and uncertainty quantification must be employed. To this end, the work in this volume pursues the concepts of vertical and horizontal integration of information from models and experiments/sensors; vertical integration addresses multiscale modeling over relevant material length and time scales, while horizontal integration relates to integration of manufacturing processes. This advancement of the inverse decision-based design paradigm is necessary to realize true integration of various manufacturing processes in concurrent materials and product design. Design variables lie within the triad of material, product, and manufacturing process.

This is truly rich and fertile ground for materials design research and points to the need to extend next-generation CAD systems in our increasingly digital manufacturing-materials world. Broadly speaking, I think this is the way engineering design will be taught sometime in the (hopefully not too distant) future. In the meantime, this new foundational work identifies and addresses research gaps and challenges in robust concept exploration as a precursor to detail design, which is an important first step in aligning investment strategies toward feasible solutions that include available manufacturing routes. The networking of stakeholders involved in co-creation of value supporting open innovation is defined by the authors as enabling the capability to carry out *co-design*. Linking co-design to achieving goals of integrated materials and product design as required by ICME is a critical aspect of this volume. Accordingly, the authors address future cloud-based design support systems for networked workflows and collaboration among widely distributed experts and stakeholders. Moreover, they include the role of emergent data science methods, including machine learning.

I am pleased to recommend this stimulating and insightful book on integrated design material, product, and manufacturing process to your personal library, organization, university, or company (whether Fortune 500 or a start-up). I would also encourage the ideas presented here to become part of a broader conversation regarding design education at the interface of various engineering disciplines, including materials science and engineering, as well as materials physics and chemistry. Their approach to robust co-design is systematic and well-conceived, helping the reader organize the relevant challenges and opportunities for transformational change. I trust its value will endure.

David L. McDowell
Regents' Professor and Carter N. Paden
Jr. Distinguished Chair in Metals Processing
Georgia Institute of Technology
Atlanta, GA, USA

Preface

In the context of the emerging Integrated Computational Materials Engineering (ICME) and Digital Manufacturing and Design Innovation (DMDI) paradigms, we present in this monograph our vision, that is, to collaboratively (with academic and industrial partners) define the emerging frontiers for architecting robust co-design of materials, products, and associated manufacturing processes. We recognize in this monograph that in order to foster the integrated realization of materials, products, and manufacturing processes, there is a need to not just design, but facilitate *co-design*. We define co-design as the ability of a network of participants, which includes material scientists, systems designers, software developers, and end customers to come together and share material/product/manufacturing process/market data, information, knowledge, and resources instantly and thereby collaborate to facilitate cost-effective co-creation of value supporting open innovation.

Our foundational premise is that systems-based co-design makes it possible for tailoring materials, their processing paths, and the end products constituted by these materials in an integrated fashion for challenging applications to satisfy conflicting product and process-level property and performance requirements. Accordingly, we establish a systems-based design architecture that includes systems-level synthesis methods and tools for the robust co-design of complex materials, products, and associated manufacturing processes starting from the end requirements. The core question that we address is:

> *What are the theoretical, mathematical, and computational foundations needed for establishing a comprehensive systems-based design architecture to realize the robust co-design of the product, its environment, manufacturing processes and material as a system?*

The major research challenges addressed in this monograph are:

- Integration of models of different fidelities (material, process, and product) to establish processing–structure–property–performance relationships.
- Goal-oriented, inverse design of material microstructures and processing paths to meet multiple conflicting performance/property requirements.

- Robust concept exploration by managing uncertainty across process chains.
- Systematic, domain-independent, modular, reconfigurable, reusable, computer interpretable, archivable, and multiobjective decision support in the early stages of design to different users.

The components of the systems-based design architecture proposed in this monograph to address these research challenges are shown in Fig. 1. The components include:

- Systematic model integration and information flow.
- Systematic concept exploration.
- Systematic inverse design exploration.
- Systematic robust concept exploration by managing uncertainty.
- Knowledge-based platform for decision support.
- Cloud-based decision support.

The key features of the systems-based design architecture adding value to the emerging ICME and DMDI paradigms are:

- Systematic function structure-based identification and integration of material, process, and product models and workflows to define processing–structure–property–performance mapping and information workflow.
- A framework supporting systematic design and solution space exploration.
- A generic method for inverse design of materials and products across process chains.
- Metrics, robust design constraints, and goals for facilitating robust design by managing uncertainty across process chains for multiple conflicting goals.

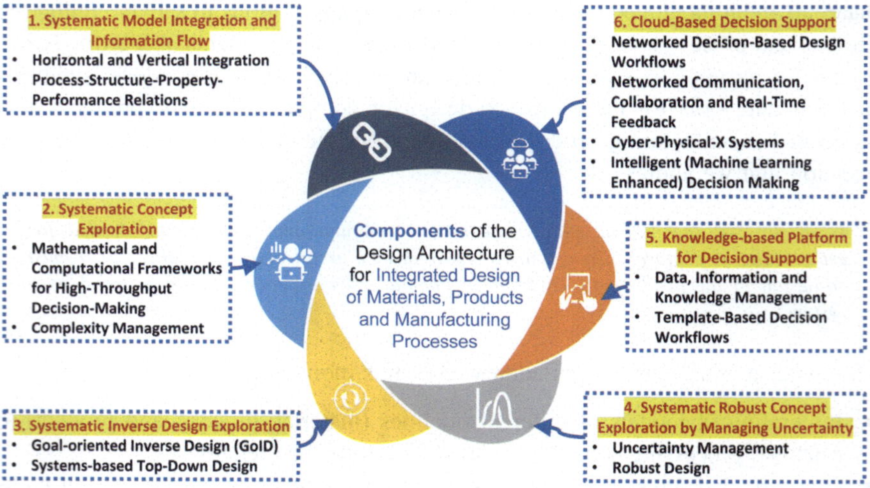

Fig. 1 Components of the systems-based design architecture for the robust co-design of materials, products, and manufacturing processes as proposed in this monograph

- Capture knowledge using original design, maintain consistency using adaptive design, and provide a package of documented knowledge using variant design.
- Cloud-based decision support platform that embodies the above-said features and functionalities enabling networked decision-based design workflows, design collaborations, instant communication, and feedback.

This monograph is comprised of eight chapters; see Fig. 2. While we recommend reading the chapters sequentially, each chapter is self-contained, and chapters can be read independently or in any preferred sequence. Following the introductory chapter, in Chap. 2, a review of the existing efforts associated with materials and

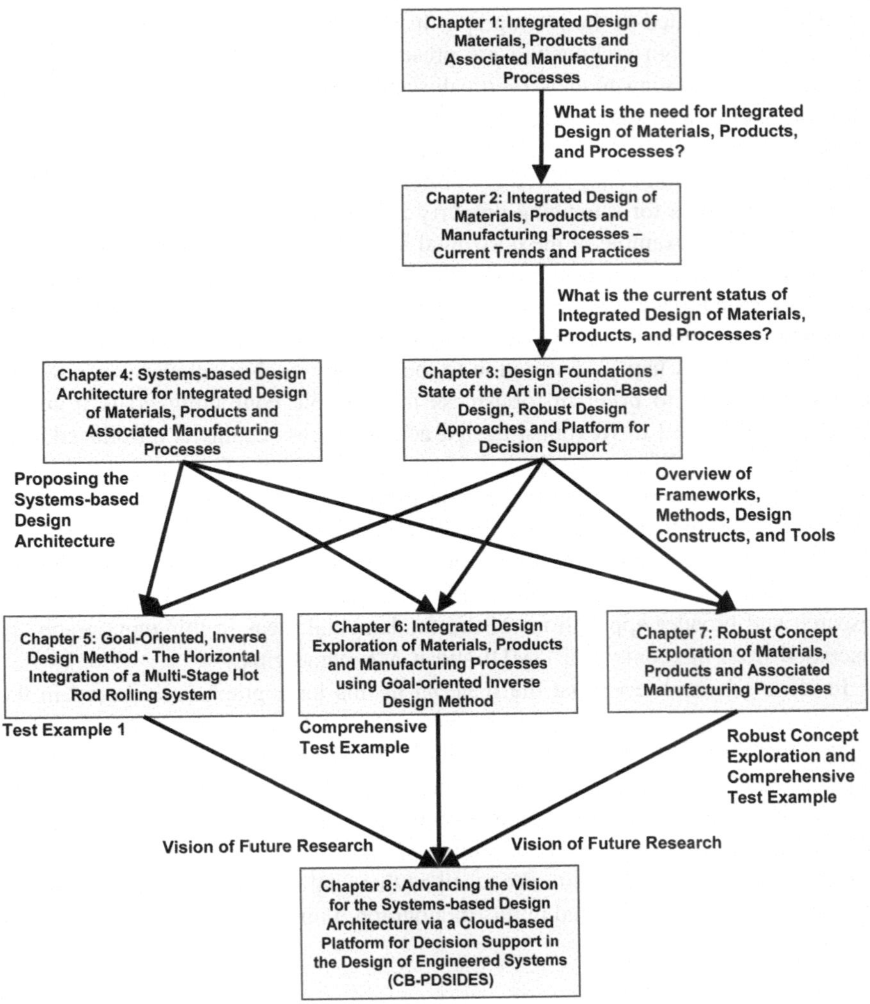

Fig. 2 Organization of the monograph

product design is given. Some of the major elements of modern materials design like material informatics, uncertainty management, verification and validation, multiscale modeling, systems design, and so on are discussed in this chapter. In Chap. 3, the theoretical foundations for designing simulation-based design processes are discussed. These foundations include existing design construct related to the decision-based design paradigm, meta-design and decision support problem technique, robust design, compromise decision support problem, robust concept exploration method, robust design metrics, indices for robust design, inductive design exploration method, template-based decision-centric design, design concept flexibility, function-based systematic design, and solution generation approaches. Relevant literatures for each of these areas are referenced, discussed, and critically evaluated to show the appropriateness of use of these constructs for the design architecture presented in the monograph. In Chap. 4, the major components of the systems-based design architecture are presented: (1) a systematic function-based approach of integrating models (vertically and horizontally) to formulate forward material workflows establishing processing–structure–property–performance relations, (2) a concept exploration framework for systematic design problem formulation and exploration along with a goal-oriented, inverse design (GoID) method for designing materials for satisficing property and performance goals.

In Chap. 5, an example of the horizontal integration of the multistage hot rolling process and product design is presented. In Chap. 6, the design architecture presented is applied to a multiscale, multistage materials design problem—vertical and horizontal integration and integrated design of hot rod rolling process chain, steel, and rolled rod. In Chap. 7, a variation to the existing goal-oriented, inverse design method (Chap. 6) to bring in robustness for multiple conflicting goals from the standpoint of Type I to III robust design across process chains is introduced. The design of a hot rolling process chain for the production of a rod is used as an example to verify and validate the approach proposed. In Chap. 8, we present the architecture and functionalities of a cloud-based computational platform to facilitate mass collaboration and open innovation, thereby supporting the co-design of materials, products, and manufacturing processes. Finally, avenues for future research and broader applications of the fundamental ideas in this monograph are discussed from the context of DMDI and ICME paradigms.

In closing, we observe that the material in this monograph is anchored in the doctoral dissertation of Anand Balu Nellippallil whom we were privileged to mentor together with our colleagues Dr. B. P. Gautham (TCS Research, Pune, India) and Prof. Amarendra Kumar Singh (IIT Kanpur, India). We recognize Zhenjun Ming, Ru Wang, Vignesh Rangaraj, Kevin Song, and Pranav Mohan, who collaborated closely with Nellippallil during his doctoral research and added value to the material presented herein. We received financial support from several sources to pursue the concepts outlined in this monograph. Funds from the LA Comp Chair account, the John and Mary Moore Chair account at the University of Oklahoma, and the Tata Consultancy Services, India were used to support Nellippallil through his doctoral studies in the School of Aerospace and Mechanical Engineering at the University of Oklahoma, Norman, USA.

Norman, USA
February 2020

Janet K. Allen
Farrokh Mistree

Contents

List of Figures

Chapter 1
Integrated Design of Materials, Products, and Associated Manufacturing Processes

1.1 The Motivation for Integrated Design of Materials, Products, and Associated Manufacturing Processes

In practice, "*design*" involved the selection of a suitable material for a given application (Norton 1996; Shigley 1972; Ashby and Cebon 1993; Pahl and Beitz 2013). The performance of many engineered systems involving materials and products is limited by the available properties of the constituent materials. The difficulty here with material selection is the inherent inability to tailor a material microstructure and constituents for satisfying application-specific requirements. These requirements tend to conflict with the actual achievable performance from the material microstructure and properties. The discovery of new materials has always been arduous, fortuitous, and instinctive for the people in this domain. The lead times for developing new materials have remained almost constant and unacceptably long when compared to the development cycle of the desired product. This has thus resulted in increased cost and time in the development of new materials and products which is partly due to the predominantly empirical, trial-and-error approach adopted by materials engineers and designers till now (McDowell and Story 1998; Olson 2000).

The U.S. National Science Foundation (NSF) defines *design* as a process by which products, processes, and systems are created to perform the desired functions through specification. In a design process, requirements, also termed as "functions", are transformed into design descriptions (Gero 1990a). The "functions" embody the expectations of the purposes of the resulting artifact. Gero (1990b) describes "design" as a goal-oriented, constrained, decision-making exploration, and learning activity that operates within a context that depends on the designer's perception of the context.

A materials design transformation is underway in the recent past where the focus is to ***design*** (not select) the material microstructure and processing paths to achieve multiple property or performance requirements that are often in conflict. Recent advancements in computational modeling tools and frameworks that support simulation-based, integrated design exploration of materials, products, and the manufacturing

A. B. Nellippallil et al., *Architecting Robust Co-Design of Materials, Products, and Manufacturing Processes*, https://doi.org/10.1007/978-3-030-45324-4_1

processes through which they are made have resulted in the speeding up of the process of discovering new materials and has paved the way for rapid assessment of process–structure–property–performance relationships of materials, products, and processes. This has led to the simulation-based design of material microstructure (microstructure-mediated design) to satisfy multiple property or performance goals of the product/process/system, thereby replacing the classical material design and selection approaches (see Olson 2000; Panchal et al. 2005; Pollock et al. 2008, McDowell and Olson 2008; Horstemeyer 2012). *The foundational premise therefore for this monograph and the emerging field of materials design, in general, is that systems-based materials design techniques offer the potential for tailoring materials, their processing paths, and the end products that employ these materials in an integrated fashion for challenging applications to satisfy conflicting product- and process-level property and performance requirements.*

At this point, it is important to recognize the distinction between multiscale modeling efforts, which is the focus of materials scientists and engineers, and systems-based materials design, which is the focus of systems designers to be addressed in this monograph. The focus of materials scientists and engineers is on creating increasingly sophisticated, realistic, physics-based, and history-dependent models that accurately predict the material microstructure and properties which can then be used to support a design process to satisfy a ranged set of performance requirements. Systems designers recognize the potential of integrated design of materials, products, and processes and focus on designing material microstructures that satisfy system-level design objectives. The distinction is captured in Fig. 1.1. The major challenge arising in systems-based materials design is the management of uncertainty and complexity of design problems. Multiscale modeling approaches are usually domain-specific

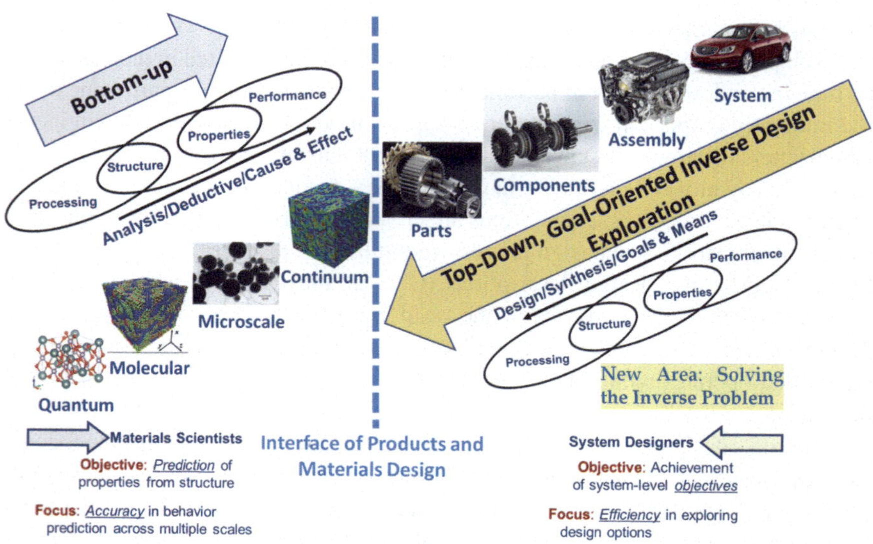

Fig. 1.1 Distinction between multiscale modeling and systems-based materials design

demanding considerable knowledge and insight in mechanisms, material hierarchy, and information flow and thus correspond to detailed design. Systems-based design approaches are mostly domain-independent and facilitate design "exploration" rather than detailed design.

The conventional way of modeling hierarchical processes and systems is a "bottom-up", cause and effect (deductive) approach of modeling the material's processing paths, microstructures, resulting properties, and then mapping the property relations to performance functions (see Fig. 1.1). Such deductive links are necessary but not sufficient for materials design. Systems designers, as discussed by Olson (Olson 1997) and illustrated in Fig. 1.1, seek a "top-down", goals/means, inductive, or inverse methods to explore the design space of processing paths and resulting microstructures of a material *satisfying* a set of specified performance requirements that could be conflicting in nature.

In this monograph, design is defined as a *top-down, simulation-supported, integrated, decision-based process to satisfy a ranged set of product-level performance requirements* (McDowell et al. 2009; Allen et al. 2015). Keeping with this, the integrated design of materials, products, and manufacturing processes is defined in this monograph as ***fundamentally an inverse, goal-oriented synthesis activity in which the designer (decision-maker) aims at identifying material structures and processing paths that achieve/satisfy certain required product and manufacturing process-level properties and performances***. Further, we recognize in this monograph that in order to foster the integrated realization of materials, products, and manufacturing processes, there is a need to not just design, but facilitate *co-design*. ***We define co-design as the capability of a network of participants, that includes material scientists, systems designers, software developers, end customers to come together and share material/product/manufacturing process/market data, information, knowledge, and resources instantly and in an integrated fashion thereby collaborate so as to facilitate a cost-effective co-creation of value supporting open innovation***. We bring-in the idea of co-design throughout the monograph and it will be addressed in greater detail in Chap. 8.

1.2 The Integrated Material, Products, and Manufacturing Process Design Focus in This Monograph: Integrated Design of Steel (Material), Manufacturing Processes (Rolling and Cooling), and Hot Rolled Rods (Product) for Automotive Gears

Steel mills are involved in the production of semi-products like sheets or rods with a certain grade of steel. Steel manufacturers are focused on developing newer grades of steels with improved properties and performances due to the increasing competition arising from new engineering materials. However, there has not been a decline in the popularity of steel as an engineering material in manufacturing industries as

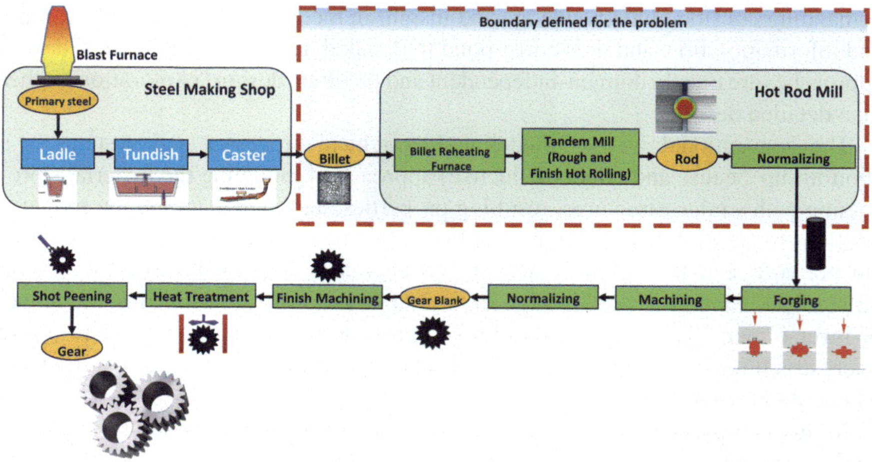

Fig. 1.2 Steel Manufacturing Process Chain for Automotive Gear Production

careful managing of material processing during steel manufacturing will lead to the development of diverse ranged sets of mechanical properties and microstructures resulting in improved performances of products.

Modern steelmaking for the production of automotive gears typically involves the following processes listed in sequential order as depicted in Fig. 1.2.

Ironmaking: In the first stage, known as ironmaking, the input raw materials such as iron ore/sinter/pellets (iron-bearing charge materials), coke (fuel), and lime (flux) are smelted in a blast furnace. The blast furnace is a high temperature metallurgical furnace used for smelting of iron ore to produce pig iron. The resulting molten iron, also known as hot metal, contains 4–4.5% carbon and other impurities that make it brittle upon solidification.

Primary Steelmaking: Primary steelmaking is carried out in two ways. The first involves the basic oxygen furnace (BOF) and the second involves the electric arc furnace (EAF) method. In BOF, the molten iron is converted to steel. Oxygen is blown into the furnace through lance at supersonic speed resulting in oxidation of carbon and other impurities. The heat generated due to reactions results in a very high temperature of the metal. Scrap is added to utilize excessive heat. The carbon content is reduced below 0.05%. In EAF, recycled steel scrap is fed through high power electric arcs that melts the metal converting it to molten steel.

Secondary Steelmaking: In secondary steelmaking (steelmaking shop in Fig. 1.2), the molten steel produced from BOF or EAF is further treated in ladle furnace (one of the secondary steelmaking unit operations) to refine the composition to the desired steel quality. This is carried out by adding or removing certain elements and/or manipulating the temperature and production environment. The count and nature of inclusions present and the levels of tramp elements such as sulfur, phosphorus, and total oxygen present in the liquid steel are factors assessed for checking the quality of steel. The desired composition is maintained with respect to alloying elements (Ni, Cr, Mn, etc.) that are added to impart certain properties to the steel.

The ladle furnace (in steelmaking shop, Fig. 1.2) is one of the key unit operations for carrying out deoxidation and desulfurization to maintain the levels of oxygen and sulfur within a tolerable limit. The steel from basic oxygen furnace (BOF) or electric arc furnace (EAF) is tapped into the ladle where several unit operations such as addition of deoxidizers (e.g. aluminum) to reduce oxygen level, addition of ferro-alloys and other alloying agents to obtain the desired composition, addition of slag forming compounds to maintain the basicity of slag and promote desulfurization and refining, argon stirring to promote dissolution of additives, bath homogenization, desulfurization, inclusion floatation, degassing, and so on, and arcing to maintain the heat content in steel required for subsequent casting are carried out to meet the compositional and cleanliness requirements of steel (Shukla et al. 2015). The molten steel from the ladle is transferred to the caster via tundish (see Fig. 1.2). A modern steelmaking tundish is used to facilitate inclusion removal, to maintain chemical and thermal homogeneity, and to provide molten steel to the next unit operation, the continuous caster with the required amount of superheat (Anapagaddi et al. 2013, 2014).

Continuous Casting **(Molten steel to Billet)**: The molten steel from the ladle is passed to the continuous casting unit via tundish (see Fig. 1.2). The continuous casting unit includes a mold usually made of copper. The hot melt as it comes in contact with the mold gets cooled as heat is extracted by the mold (primary cooling) and solidifying of the metal begins at the mold–melt interface. Steel is withdrawn from the mold by a dummy bar to start the casting operation. The mold oscillates in the vertical direction in order to avoid sticking and to ensure the separation of the solidified steel from the copper mold. The thickness of the solidified layer (shell thickness) continuously increases as the melt moves down the mold. The movement of the solidified block is guided and supported by rolls beyond the mold exit. The solidified metal block is cooled by water with the help of spray nozzles. The block is cut into desired lengths depending on the application. Different mold cross-sections are used to produce products for various applications: slabs for flat products (plate and strip), blooms for sections (beams), billets for long products (wires), or thin strips. For automotive gear production, usually solidified billets of square/rectangular cross-section, often in the range of 80–250 mm side, are used. During the solidification process, the compositional elements segregate leading to a variation across the cross-section. This segregation at the macroscopic scale, that is the width of the billet, is called macrosegregation. Besides the macrosegregation, the alloying elements also segregate at the level of dendritic arm spacing leading to microsegregation. Besides the spatial variation in chemistry, due to the different cooling rates seen by different portions of the cross-section, the macrostructure across the cross-section will also be inhomogeneous. Predominantly, the cross-section will have three zones consisting of central equiaxed morphology, surrounded by columnar region and finally a fine-grained chill zone on the surface. The variation of chemistry, as well as the cooling rate will impact the overall microstructure of the cast billet and also lead to the formation of precipitates and so on. Another aspect that is of importance is the distribution of inclusions present in the incoming steel melt across the cross-section of the billet. All these aspects lead to a significant impact on the final product as some of these signatures stay till the end, even though modified to some extent by reheating and rolling operations.

***Rolling Mill* (Billet to Rod)**: The steel billet that is cast is then sent to the rolling mill. The rolling mill includes a reheating furnace to heat the billet before rolling to ensure thermo-mechanical deformation and refinement of microstructure during rolling. The billet is formed into various shapes depending on the end application. The process is a high temperature, high strain rate process. The process eliminates cast defects and achieves the required shape, surface quality, and microstructure for the semi-product after rolling. Microstructural phenomena like dynamic, meta-dynamic and static recrystallizations, and grain growth occur during rolling, resulting in a change in the microstructure of the steel. The rolled rods are further sent to a cooling unit, where the phase transformation of the steel takes place. Hot rolled products are divided into flat products, long products, seamless tubes, and specialty products. Rods are one of the products from a rolling mill.

***Forging* (Rod to Blank)**: Rolled rods of desired grade and diameter are the raw materials for the forging industry. The rods are cut to required length, forged in one or multiple steps for obtaining the desired shape, and finally heat-treated to relieve of the stresses and are known as blanks. For automotive gears, the rods are forged to gear blanks. These blanks forged are machined to obtain the desired final blank shape. One of the key factors influencing the process is the distortion of the forged piece subsequent after to forging and subsequent heat treatment. Higher distortion leads to providing for higher machining allowances. The magnitude of distortion depends on the incoming material state, forging process sequence (die design and operations), and heat treatment and can be controlled by appropriate die design. However, the segregation and microstructural non-uniformity of the material will also be an important aspect in the final distortion as this signature will stay till the end and may cause distortion. One needs to look at all these aspects in tandem and recommend the best status for incoming material (rod) state without adding to the cost.

Machining, heat treatment, and finish machining: Finally, secondary forming techniques give the steel its final shape and properties. These techniques include machining (e.g. drilling), joining (e.g. welding), coating (e.g. galvanizing), heat treatment (e.g. tempering), surface treatment (e.g. carburizing), shot peening, and so on, to finally produce the product—a gear in this case.

A boundary is defined for the problem addressed in this monograph within the hot rod rolling process with the billet coming from the casting unit as the input and the hot rolled rod as the output (see Fig. 1.2).

1.3 Defining Boundary—Hot Rolling Process Chain

Typical steel mills produce intermediate products, such as slabs, billets, blooms, and finished products such as sheets and rods/bars. The round rod produced in steel mills after passing the raw steel material through several manufacturing processes like casting, reheating, rolling, and cooling forms the input material for the production of gears. The chemical composition of material, including the segregation of alloying elements, the deformation history during rolling, the cooling after rolling, and the microstructure generated, defines the end properties of the steel product that is rolled. The presence of a large number of design variables, constraints and bounds, conflicting goals, and sequential information/material flow during material processing makes the steel rod making process chain to be highly complex in nature. A large number of plant trials are therefore required to produce a product from a new steel grade having desired properties and performances. These trials are usually expensive and time-consuming. Process designers are very much aware of the operating constraints and process requirements for each of the operations as they are involved in the whole process day-in and day-out. Due to the advancements in material technology, new improved materials with enhanced properties are introduced to the market and this has posed a serious challenge to steel manufacturers. Suppose, a situation happens that owing to the changed properties and performance requirements, manufacturers are asked to produce a semi-product like a rod with a newer grade of steel with enhanced properties. This new steel grade is used at laboratory scale to produce a rod, but the current challenge posed to a steel manufacturer is to scale-up the production of this rod from laboratory scale to industrial scale. This has created a requirement to explore the design setpoints of each unit operation involved in the production of the rod with some target properties at the plant scale. Experimentations and plant trials are one way of achieving this requirement, which is typically expensive and time-consuming. Usually, for automotive applications, the materials research may take up to 8–10 years when this option is adopted. As per the information from Tata Steels, India, it takes around 20–25 plant trials to come up with these design setpoints and each such trial will cost a $100,000. Thus this is a huge challenge to the industry.

To address this issue, there is a huge drive by industry, especially in the past decade to use computational models for exploring the design setpoints for these operations and thereby reduce the time and cost involved in the development. The focus is to carry out simulation-based, integrated design exploration of the different manufacturing processes involved by exploiting the advancements in computational modeling tools and frameworks. The fundamental question addressed in this monograph from the problem perspective for the integrated realization of steel (material), rolled rod (product), and hot rolling process chain (manufacturing process chain) is:

How to realize this complex system involving the material, product, and manufacturing processes using simulation models that are typically incomplete, inaccurate, and not of equal fidelity?

Box (1979) is reputed for his aphorism that "all models are wrong but some are useful". In keeping with George Box's observation, the challenge here is to determine the design and operating setpoints for the hot rolling process chain involving the material steel and end product rod using computational models and simulations that at best capture the essence of reality but not reality itself. Therefore, there is a need to explore solutions that are relatively insensitive to the inherent uncertainties embodied in the computational models and simulations while satisficing the conflicting goals associated with the material, product, and process. An integrated design exploration approach is needed, where a ranged set of solutions are sought that satisfy the requirements identified for the steel manufacturing process as well as the end rod product. However, these models and simulations are specific for specific phenomena that happen during a process, and an isolated design using individual models will not be a true representation of the whole system and the solution desired. Thus, there requires an integration of these models to allow information flow so as to explore the design and operating setpoints for the production of the rod. For exploring the design and operating setpoints, knowledge of the operational constraints and requirements are necessary and for a newer grade of steel, this information is not readily available for a manufacturer. Therefore, the first task is to identify operating constraints and requirements for each operation, which is imposed by the subsequent unit operations as each process step is connected and information flows from one operation to another. To identify these operating constraints, there needs to be detailed information regarding each operation. This needs integration of different models that are at multiscale for a unit operation so as to obtain information in much greater detail which can then be passed to other unit operations. This is termed as ***vertical integration*** of models for one particular unit operation. The integrated study possible by the flow of information from one unit operation to others is termed as ***horizontal integration***.

Thus, in order to effectively couple the material processing–structure–property–performance spaces, there needs to be an interplay of the systems-based design of materials with an enhancement of models of various manufacturing processes through multiscale modeling methodologies and integration of these models at different length scales (vertical integration). This ensures the flow of information from one process to another, thereby establishing the integration of manufacturing processes (horizontal integration). Together these types of integration will support the decision-based design of the manufacturing process chain so as to realize the end product.

To achieve the vertical and horizontal integration of models, there must be analysis models and simulations that can link the different manufacturing processes by predicting the different material phenomena associated, thereby ensuring the proper forward flow of information. To predict the properties and performances of hot rolled steel product, there needs to be a modeling of the material flow behavior during hot rolling followed by the microstructural changes that happen during hot deformation and at the interstand region. This is followed by the modeling of the phase transformation phenomena that happens during the cooling process after rolling. Mathematical models and simulation programs for the different metallurgical events that happen during rolling and cooling when integrated into the right sequence will be able to help a designer predict the microstructure evolution as a function of process parameters. The integration of these models enables the designer to identify new processing routes, composition maps, and mill sequences that will provide a microstructure and to track their impacts on the end mechanical properties of the product. Thus, using these analysis models, the designer will be able to solve the standard forward problem: given the input parameters related to the processing and microstructure models, what are the properties and performances of the end product? These standard forward problems are characterized by the availability of a single-point solution and the designer iterates the analysis several times to identify a solution that meets the end property/performance requirements. Again, such an approach takes a huge amount of time and cost for the designer to make design decisions and is not top-down or goal-oriented.

Thus, the questions to be asked here are:

Given these analysis models and simulations that establish the forward material workflow for the system, how do a designer design the system from a systems perspective taking into account the end goals and requirements and make design decisions that are critical for the integrated simulation-based realization of materials, products, and processes?

From a systems perspective, the interest, therefore, lies in formulating and solving the inverse design problem: given the required end properties/performances, what should be the input parameters and variables in terms of material microstructures and processing paths so as to achieve the model-based realization of the material, product, and the manufacturing processes?

There are several challenges in addressing these questions. The challenges are discussed in the next section.

1.4 Challenges in Systems-Based Integrated Design of Materials, Products, and Processes

The philosophical underpinning of the goal-oriented approach to materials design has been provided by Olson (1997) and reiterated by many others (McDowell et al. 2009; Horstemeyer 2012, 2018; McDowell 2018). Several challenges associated with the top-down, goal-oriented approach of materials design have been highlighted (see Panchal 2005; McDowell et al. 2007, 2009; McDowell 2018). The challenges that need to be addressed for the integrated realization of materials, products, and manufacturing processes from a systems-based, simulation-assisted, top-down design perspective are:

i. **Managing uncertainty** associated with material microstructure and behavior;
ii. **Managing uncertainty** associated with complex manufacturing processes resulting from its environment and the factors affecting the processes;
iii. Material, product, and process models and simulations of complex systems are typically **incomplete, inaccurate, and not of equal fidelity,** and thus the models and simulations are **uncertain**—need to be managed;
iv. **Propagation of uncertainty** across multiple scales and across processing, microstructure, property, and performance spaces – need to be managed;
v. Modeling different phenomenon related to a manufacturing process and integration of models for these different phenomena (across scales mostly) to generate information specific to the manufacturing process—achieve **vertical integration**;
vi. Integration of different manufacturing processes and ensuring information (generated through vertical integration) flow across processing, microstructure, property, and performance spaces to come up with the end product—achieve **horizontal integration**;
vii. **Nonlinear, history-dependent behavior** of metals and alloys limits the extent of parametric studies and imparts dependence on initial conditions;
viii. **Non-unique and large number of solutions** possible for a given property or performance requirement—need exploration of design and solution space;
ix. **Computational tractability of microstructure** (new) that satisfy conflicting property and performance requirements;
x. Availability of bottom-up models and deductive links that establish forward information flow while **lacking generic and reliable inverse design techniques and tools for top-down materials design "exploration"**;
xi. Dynamic design scenarios where the design goals change with time, thereby establishing the **need for reusable, flexible, and adaptable design processes**;
xii. Balancing **model accuracy and computational cost**;
xiii. Modeling the physics associated with the materials system—ensuring the **capture of relevant information via modeling the appropriate number** of subsystems, components, parts, and material phenomenon;
xiv. Identification and selection of **appropriate models, model parameters, and the associated variabilities** at each scale/space of a complex material system;
xv. Managing **large quantities of information,** especially related to material structure and properties at different levels of abstraction;
xvi. Capture, storage, reuse, and updating of the **material, process, and product knowledge and database**;
xvii. **Verification and validation** of algorithms, models, and design results;

xviii. Systems designers and materials engineers have different **backgrounds and expertise** and may not share the same "language" in terms of materials design;

Systems-based robust design methods are needed to address the major challenges arising due to (i) uncertain material models (that includes input factors, parameters, responses, etc.) due to simplification/idealization or a lack of complete knowledge and (ii) the propagation of uncertainty due to hierarchical information dependence in a multiscale model chain or in Olson's processing–structure–property–performance relations. An effective top-down, goal-oriented systems approach for materials design must be able to manage the uncertainty with regard to all relevant information ensuring feasible designs that meet specified ranges with high confidence. McDowell (2018) asserts that such an approach must address the ***uncertainty of models and experiments at each scale, as well as uncertainty propagation through a chain of models and/or experiments at different levels of hierarchy with the ability to provide decision support through rapid design space exploration***. The report by the U.S. National Academy of Engineering National Materials Advisory Board on Integrated Computational Materials Engineering (ICME) (Pollock et al. 2008) outlines a broad set of challenges and opportunities for the integrated realization of materials, products, and processes from the emerging ICME perspective. In this monograph, the systems-level strategies and their implications are explored for the integrated realization of materials, products, and associated manufacturing processes, building on the foundational ideas laid by Olson on Materials-by-Design and the ICME community.

1.5 Addressing the Systems-Based Material, Product, and Process Design Challenges

To address many of the challenges associated with the integrated realization of materials, products, and manufacturing processes, **a systems-based approach** is necessitated. Designing the materials cannot be done in an isolated fashion. Materials are subsystems of a larger system that includes parts, assemblies, products, and physical systems (see Fig. 1.1). Engineering applications demand materials that satisfy multiple performance functions that require a system-level analysis to be defined properly. A hierarchical structure can be assigned to materials themselves where information passes from one scale to another and the desired end material properties and product performances often depend on the material phenomena that occur at these different length and time scales. The challenge here is in developing/formulating a single model that predicts the material properties at macroscale by unifying information from all the length scales (McDowell and Story 1998). Atomistic and molecular-level simulations based on first-principles predict the structure and properties of materials. However, these simulations are computationally too expensive and often too idealized to model materials having heterogeneous structures (McDowell et al.

2009). Similarly, continuum mechanics models and simulations predict the properties of materials and products at the macroscopic level but are inappropriate for incorporating lower scale information that involves atoms, dislocations, defects, and so on. Horstemeyer (2012) addresses the bridging between scales from the perspective of different disciplines/approaches like solid mechanics, numerical/concurrent methods, materials science, physics, mathematics, and design. Horstemeyer coins the term "upscaling" for the bottom-up approach of modeling and running simulations at each scale and averaging the results in some sense to be passed to the next scale and the term "downscaling" for the top-down approach of requirements-driven simulation at macrostructural level taking into account lower length scale features.

Developing physics-based models that capture process–structure–property–performance relations at different length and time scales is challenging. The integration of these models across length and time scales is difficult owing to the complexity and limitations involved due to the different domains of application. The requirement here is to link these models in an integrated fashion that facilitates the "***exploration***" of the systems-level design and solution space. The design and solution space exploration need to be carried out by distributing analysis and synthesis activities in a collaborative manner by a team of experts involving material scientists, product designers, and engineers—referred to as co-design in this monograph. This allows leveraging the different domain-specific knowledge and expertise associated with these experts related to maybe length and time scales, multiscale modeling, material classes, and material functionalities, and so on. The fundamental role of each of these domain-specific design experts is to make decisions given the information available. These decisions include synthesizing, analyzing, and identifying design alternatives that satisfy conflicting material property and performance goals by carrying out trade-offs. Decisions taken for material systems depend on the information available from different levels of hierarchy as the system is highly interconnected and interdependent. This demands the need for ***multi-objective decision protocols and workflows*** that allow the interfacing of individual decisions and decision-making experts so that information flows across the material hierarchy. A systems-based approach is sought for making these decisions by identifying solutions that satisfy ***systems-level*** objectives.

A deterministic approach to materials design is not recommended as there is a certain degree of randomness in material systems. There is also uncertainty associated with the model-based realization of complex material systems. Model parameters are subject to variations associated with the variability of microstructure and variability due to processing. There is also uncertainty associated with model predictions due to various sources. The assumptions and approximations incorporated in a model also contribute to the uncertainty associated. The uncertainty associated can be magnified if a model is used in its boundary of applicability and when information is passed from one model to another resulting in propagation of uncertainty. Surrogate models developed to facilitate design space exploration of broad space also contribute to the uncertainty as fidelity is compromised for computational efficiency in such models. The experimental data available to verify and validate the model predictions may be sparse and may be affected by errors associated with measurements. Removing or mitigating these sources of uncertainty is expensive or impossible in most of the cases. The impact of these uncertainty sources, however, could be profound on the model predictions and final system performance. The need, therefore, is for system-level design methods that take into account these sources of uncertainty without removing or eliminating them. The method should support a designer to manage uncertainty and facilitate the identification of robust design solutions that are relatively insensitive to these sources.

One of the foundational elements for material, product, and process design that is often neglected is databases for material structures, processing paths, and properties based on both experiments and simulations. Databases serve the same purpose as models and simulations and are considered as instruments of informing design decisions. Databases should also convey uncertainty associated with the material–product–process system to facilitate the systems-based robust design.

Design space exploration will be much more efficient if knowledge-guided assistance can be provided to designers at various decision points in the material system design process. Knowledge engineering plays a key role in enabling this, with learning being an integral part of it. Previous design problems, methods, results, and so on can be systematically evaluated to gain knowledge that can thus make future design exploration process more efficient. The ability to capture knowledge from various sources is also essential. A knowledge-based platform for decision support is therefore essential for the integrated design of materials, products, and processes. The platform should serve as a computing infrastructure for a simulation-based design process supporting decision making. The computing infrastructure should be easily extensible and platform-independent serving systems-based design. The process of executing, linking models, exploring solution spaces, and so on should be automated by the computing infrastructure. It should also have the capability to archive and organize large amounts of data and should be able to capture the relevant information and knowledge associated. Sharing of data, information, and knowledge with different types of users depending on their levels in the design process is necessary and the platform should facilitate real-time sharing, collaboration, communication, visualization, and search-based retrieval of design information and knowledge.

Based on the challenges identified in Sect. 1.4 and the needs identified to address the challenges in Sect. 1.5, key challenges to be addressed in this monograph and the associated research gaps are identified to achieve the model-based integrated realization of materials, products, and associated manufacturing processes. The key challenges identified are listed below and summarized as research gaps in Tables 1.1, 1.2 and 1.3.

(1) ***Integration of Models Across Process Chain***.

- *Integrated design of materials and systems*: Design of the system/components and the design of the materials need to be connected enabling the integrated design of materials, product systems, and manufacturing processes. To achieve this, models across scales for a specific process need to be integrated (vertical integration) for generating information that can be passed to subsequent manufacturing processes or material design space. Through this, designers will be able to incorporate model-based material microstructure and process design along with system and component design in a single, concurrent design process at the early stages of design, thereby reducing the lead time and cost involved with product development.

- *Integration of manufacturing process chains to ensure a joined product development lifecycle*: The model-based integration (horizontal integration) of several manufacturing processes enabling communication of relevant design information across the different product processing stages allows for a faster, more efficient iterative product development process. The integration of the manufacturing process stages allows for a unified representation of the relevant data and knowledge with uncertainty managed to be shared across the manufacturing process chain to facilitate the flexible, end-user specific product development process.

(2) ***Processing–Structure–Property–Performance Relationships***

- *Simulation-based definitions of material properties and performances at systems level using process–structure–property–performance relations – and exploiting them as system variables for designing materials and products with tailored performance characteristics*: Current practices use experimental and plant trials to define material properties and performances. These are often costly and time-consuming. Such material property definitions are static and are unable to change with the design of the product at the systems level. The material data is generated/available often in the past and is not updated with the current trends and practices, resulting in a gap between research and design. Simulation/model-based definitions of material property do not rely on such plant trials and are dynamic in nature that responds to changes in the design process at the systems level in the early stages of design. Verification and validation of individual models and models at different levels are critical, which further allows us to define the "design space" for material definitions and consider the management of uncertainty across the process–structure–property–performance spaces. The variables associated with the models are used as system variables to carry out the design of the material to tailored properties and performances (Table 1.1).

(3) ***Domain-Independent Design Methods and Tools***

- *Design methods, tools, mathematical constructs and frameworks, and ontologies that are domain-independent*: Currently, materials design methods,

Table 1.1 Research gaps in systematic model integration and information flow

Research gaps	
G1	Systematic approaches to identify and integrate material, process, and product models based on their function structures to frame system-level structure
G2	Systematic approaches to define the forward processing–structure–property–performance relationships and develop material and product workflows

tools, constructs, frameworks, and ontologies that describe material and product-related data, information, and knowledge are domain-specific and are not favorable for design exploration at early stages of design. There is a need for domain-independent design methods, tools, constructs, frameworks, and ontologies that are standard and interoperable ensuring easy storage, accessibility, and reusability of information and knowledge among designers, tools, domains, and different communities with different design school of thoughts.

(4) ***High-Throughput Decision Support and Inverse Methods Supporting Exploration***

- *High-throughput decision support and goal-oriented, inverse methods for materials design exploration*: High-throughput decision support is critical in all stages of material, product, and process design. There is a need for decision-based design exploration protocols, workflows, and inverse design methods that provide robust decision support for the top-down design for process–structure–property–performance relations using bottom-up simulations, models, and experiments. Uncertainty management, verification, and validation are foundational to address this requirement.

(5) ***Managing Uncertainty***

- *Robust Design of Materials, Products, and Processes under Uncertainty*: Robustness of materials, products, and processes with respect to variations in the boundary or operating conditions, material properties, or material microstructures or processing paths are important to consider during the design process because they can have significant impacts on the final performance of the end product. Typically, robustness is not considered during a material and product design process, and the design problems are usually formulated as deterministic optimal design problems. The need, therefore, is for system-level design methods that take into account these sources of uncertainty without removing or eliminating them. The method should take into account uncertainty of models and experiments at each scale, as well as uncertainty propagation through a chain of models and/or experiments at different levels of hierarchy with the ability to provide decision support through rapid design space exploration (Table 1.2).

(6) ***Reusable, Flexible, Adaptive Design Processes***

- Product and process requirements are subject to dynamic changes due to changing market requirements. This necessitates the need to design products by reusing existing knowledge for other products; flexible enough to dynamic market changes and accommodate them; adaptive enough to work satisfactorily in the changing environment that is prone to uncertainty.

Table 1.2 Research gaps in systematic "robust" concept exploration

Research gaps	
G3	Support systematic and rapid concept exploration of materials, products, and processes to generate satisficing design specifications
G4	Systematic design methods to carry out inverse design exploration of materials, products, and processes meeting end goals
G5	Systematic strategies to carry robust concept exploration of material, product, and process system in an inverse manner by managing uncertainty

Table 1.3 Research gaps in knowledge-based decision support

Research gaps	
G6	Constructs and tools to capture and reuse the knowledge associated with material and product systems design
G7	Facilitation of original, adaptive, and variant design decision support
G8	Facilitation of systematic design exploration through decisions that are robust, flexible, and modifiable, particularly in the early stages of design

(7) ***Capture, Storage, and Reuse the Knowledge***

- *Knowledge-Based Platforms for Decision Support in the Design of Complex Material and Product Systems*: Decision-making is a knowledge-intensive process, with knowledge playing a significant role in speeding up and effecting decisions. Capturing, managing, and reusing of decision-related knowledge such as alternatives, parameters, constraints, goals, dependencies, and the design process in the design of complex material and product systems is an effective way of providing decision support. Hence there is a need for a decision-based computational framework to manage (off-line and in real time) complexity and risk associated with the realization of complex (cyber-physical-social) material and product systems that necessitate the integration of information technology and operational technology (Table 1.3).

1.6 Research Gaps Identified

The focus in this monograph is on the integrated model-based realization of materials, products, and associated manufacturing processes from a systems perspective. Knowledge-based decision support during the early stages of design by managing uncertainty has not attracted adequate research attention. How to define the product, material, and manufacturing process or design process amidst customer, engineering, and production uncertainty remains largely an open question. In the preceding sections, the challenges in achieving the integrated model-based realization of materials,

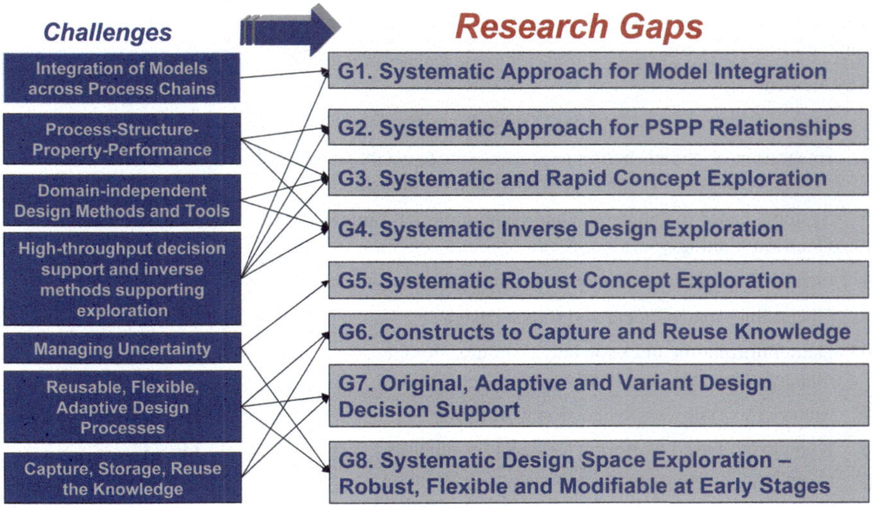

Fig. 1.3 Mapping between research challenges and research gaps

products, and associated manufacturing processes are reviewed and research gaps are identified. The mapping between the challenges and the research gaps identified is shown in Fig. 1.3. Research gaps thus identified are summarized as follows:

G1. Systematic approaches to identify and integrate material, process, and product models based on their function structures to frame system-level structure
G2. Systematic approaches defining the forward processing–structure–property–performance relationships and develop material and product workflows
G3. Support systematic and rapid concept exploration of materials, products, and processes to generate satisficing design specifications
G4. Systematic design methods to carry out inverse design exploration of materials, products, and processes meeting end goals
G5. Systematic strategies to carry robust concept exploration of material, product, and process system in an inverse manner by managing uncertainty
G6. Constructs and tools to capture and reuse the knowledge associated with material and product systems design
G7. Facilitation of original, adaptive, and variant design decision support
G8. Facilitation of systematic design exploration through decisions that are robust, flexible, and modifiable, particularly in the early stages of design.

Based on the research gaps identified, nine key elements for the integrated realization of materials, products, and manufacturing processes are identified and listed in Table 1.4. In this monograph, these nine key elements will be discussed in the context of the materials design problem of focus in this monograph, namely the integrated design of steel (material), manufacturing process (rolling and cooling), and hot rolled rods (product).

Table 1.4 Key elements for the integrated model-based realization of materials, products, and manufacturing processes

Key elements		Ties to research gaps
1	Individual models and multiscale models and modeling methods	G1
2	Integration of models across process chains to establish forward Process–Structure–Property–Performance relationships	G2
3	Systems-based mathematical and computational frameworks for Decision-making	G3, G4, G5
4	Generic Goal-Oriented, inverse design exploration methods	G3, G4
5	Uncertainty management and robust design	G5
6	Data, information and knowledge capture, storage, access, reuse, management, and visualization	G6, G7, G8
7	Material workflows and decision workflows	G6, G7
8	Knowledge-based platforms for decision support	G6, G7, G8
9	Verification and validation	G1 to G8

1.7 Components of the Systems-Based Design Architecture as Included in This Monograph

As discussed in the previous section, the primary goal in this monograph is to establish some of the scientific foundations and tools that are needed for the integrated realization of materials, products, and manufacturing processes using simulation models that are typically incomplete, inaccurate, and not of equal fidelity by managing the uncertainty associated. To realize the design of such complex engineering systems and make design decisions requires domain-specific knowledge, expertise, and multiscale (length and time scales) models with the consideration of not only the products themselves but also the manufacturing processes, the constituent materials, and the operating environments as a system.

Accordingly, the interest in this monograph lies in establishing a ***systems-based design architecture*** that includes system-level synthesis methods and tools that are required for the integrated design of complex materials, products, and associated manufacturing processes starting from the end requirements. Keeping with this interest, focus is put on this monograph to establish synthesis techniques that facilitate exploration of the design and solution spaces starting from the end performance to identify satisficing material microstructures and processing paths by managing the different types of uncertainty associated and thereby carry out robust decision-making in the design of complex material, product, and process systems.

The primary requirement for systems-based design architecture for integrated model-based realization of materials, products, and associated manufacturing processes gives rise to the following primary research question in this monograph:

Primary Research Question: *What are the theoretical, mathematical, and computational foundations needed for establishing a comprehensive systems-based design architecture to realize the integrated design of the product, its environment, manufacturing processes, and material as a system?*

The primary hypothesis defined in this monograph to answer the primary research question is:

By establishing the theoretical, mathematical, and computational foundations for,

(1) *forward material, product, and process workflows through systematic identification and integration of models to define the processing–structure–property–performance relationships;*
(2) *a concept exploration framework supporting the systematic formulation of design problems facilitating robust design exploration by bringing together robust design principles and multiobjective decision-making protocols;*
(3) *a generic, goal-oriented, inverse decision-based design method that uses (1) and (2) to facilitate the systems-based inverse design of material microstructures and processing paths to meet multiple product-level performance/property requirements, thereby generating the problem-specific inverse decision workflow; and integrating them with a knowledge-based platform anchored in modeling decision-related knowledge facilitating capture, execution, and reuse of the knowledge associated with (1), (2), and (3), a comprehensive systems-based design architecture to realize the integrated design of the product, its environment, manufacturing processes, and material as a system can be achieved.*

Based on the analysis of the primary research question and hypothesis, the following requirements are specifically laid out:

i. *Integrated design by considering the material, product, and the processes by which the materials/products are made*
ii. *Integration of models (material, process, and product) to establish processing–structure–property–performance relationships—establishing the material information workflow*
iii. *Consideration of end performance requirements for the material, product, and process*
iv. *Support the goal-oriented, inverse decision-based design of material microstructures and processing paths to meet multiple conflicting performance/property requirements*
v. *Facilitation of robust concept exploration and decision-making*

vi. *Accepting the notion that the models are typically incomplete, inaccurate, and not of equal fidelity and managing the uncertainty associated*
vii. *Facilitation of rapid concept exploration when design requirements changes*
viii. *Providing systematic, domain-independent, modular, reconfigurable, reusable, computer interpretable, archivable, and multiobjective decision support in the early stages of design to different users.*

Based on these requirements and the research gaps identified, the primary research question is divided into four secondary research questions: Research Questions 1, 2, 3, and 4. The four secondary research questions, respective research hypotheses, and the associated new contributions in this monograph are summarized in Table 1.5.

The components of the systems-based design architecture for the integrated design of materials, products, and manufacturing processes as proposed in this monograph are shown in Fig. 1.4. Each of these components will be discussed next.

Component 1: Systematic Model Integration and Information Flow
Materials design approaches in its current form do not address the conceptual design phase in a systematic fashion. The conceptual design phase is considered as the most crucial design stage as the decisions taken here affect the entire product's lifecycle and resources. Advanced methodologies for material selection were the focus in the materials design communities before Ashby and Cebon (1993), Ashby et al. (2000), Ashby and Johnson (2013). A paradigm shift has started toward materials design with the objective of tailoring the chemical composition, constituent phases, microstructureestablishing the theoretical, and processing paths to obtain materials with desired properties for particular applications (Rühle et al. 2001; Panchal 2005; Panchal et al. 2005; McDowell et al. 2009; Kalidindi et al. 2011; McDowell and Kalidindi 2016; McDowell 2018). However, even now materials design is mostly exercised in the embodiment phase with a focus on multiscale modeling techniques and simulations (Panchal et al. 2005).

Materials design as an automated search space is a very limited viewpoint as observed by Eberhart and Clougherty (2004), Messer (2008). This is because of the fact that a supercomputer searching for an optimum property using accurate analysis models of an infinite number of materials will still require infinite time to perform the search. Now the question arises whether there is a need for complex multiscale models to carry out materials design? The answer will be "No". Complex multiscale models might not be necessary in many materials design cases because the goal in materials design is to not accurately predict material properties but to satisfy a range of performance requirements. To this argument, we include the second fact that bottom-up modeling is not design but analysis. ***Thus, we believe that the key to materials design is an interplay of multiscale modeling and bottom-up analysis along with top-down, goal-oriented, inverse design and human decision-making***.

A detailed review of materials design is presented in Chap. 2. In order to achieve materials design starting from the conceptual design phase, a function-based systematic approach is needed. This requirement is addressed in Research Question

Table 1.5 Research Gaps, Research Hypothesis, Research Questions discussed in this monograph and the expected contributions

Primary	**Requirement**	Systems-based design ***architecture*** for integrated model-based realization of materials, products, and associated manufacturing processes		
	Research question	What are the theoretical, mathematical, and computational foundations needed for establishing a comprehensive systems-based design architecture to realize the integrated design of the product, its environment, manufacturing processes, and material as a system?		
	Research hypothesis	By establishing the theoretical, mathematical, and computational foundations for, (1) forward material, product, and process workflows through systematic identification and integration of models to define the processing–structure–property–performance relationships; (2) a concept exploration framework supporting the systematic formulation of design problems facilitating robust design exploration by bringing together robust design principles and multiobjective decision-making protocols; (3) a generic, goal-oriented, inverse decision-based design method that uses (1) and (2) to facilitate the systems-based inverse design of material microstructures and processing paths to meet multiple product-level performance/property requirements, thereby generating the problem-specific inverse decision workflow; and integrating them with a knowledge-based platform anchored in modeling decision-related knowledge facilitating capture, execution, and reuse of the knowledge associated with (1), (2), and (3), a comprehensive systems-based design architecture to realize the integrated design of the product, its environment, manufacturing processes, and material as a system can be achieved		
	Research gaps	**Research hypothesis addressing the research gaps**	**Secondary research questions framed from research hypothesis**	**Contributions (New Knowledge)**
I	**G1.** Systematic approaches to defining the forward processing–structure–property–performance relationships and develop material and product workflows **G2.** Systematic approaches to identify and integrate material, process, and product models based on their function structures to frame system-level structure	***H1.1.*** Through a systematic approach from a systems perspective, consisting of concept generation which includes (a) functional decomposition to generate multilevel function structures across the process chain for the end performance requirements, followed by (b) identifying material and process phenomenon associated with function structures and systematically mapping them to solution principles (models identified from literature or developed through experiments), and framing the system structure for problem via, (c) vertical integration of identified/developed material models and horizontal integration of identified/developed process models to systematically map material processing to material microstructure phenomena and next to macrolevel properties and performances, the design of the product, process, and material concepts are integrated, and conceptual materials design is rendered more systematic	***RQ1.*** What are the foundations needed for ***systematically identifying*** and ***integrating*** material models with models of the rest of the system (product, manufacturing processes, and environment), so as to define the ***processing–structure–property–performance*** relationships and associated ***information workflow*** at early stages of design?	Systematic identification and integration of material, process and product models and workflows to define processing–structure–property–performance mapping

(continued)

Table 1.5 (continued)

II	**G3.** Support systematic and rapid concept exploration of materials, products, and processes to generate satisficing design specifications **G4.** Systematic design methods to carry out inverse design exploration of materials, products, and processes meeting end goals	***H2.1.*** Developing a concept exploration framework anchored in decision-based design-construct—the cDSP can support the designer in formulating the design problem systematically and exploring the solution space to generate satisficing design specifications (To address **G3**) ***H2.2.*** Developing a goal-oriented, inverse design method that uses the concept exploration framework to facilitate the systems-based inverse design exploration of material microstructures and processing paths to meet multiple product-level performance/property requirements (To address **G4**)	***RQ2.*** What are the computational foundations needed for performing the ***systematic*** and ***rapid concept exploration*** of complex engineered systems involving the material, product, and manufacturing processes ***satisfying certain end performance requirements***, when simulation models are typically ***incomplete, inaccurate, and not of equal fidelity***?	1. A framework for systematic design and solution space exploration 2. A generic method for inverse design of materials and products across process chains
III	**G5.** Systematic strategies to carry robust concept exploration of material, product, and process system in an inverse manner by managing uncertainty	***H3.1.*** Introduction of specific robust design goals and constraints anchored in the mathematical constructs of error margin indices and design capability indices to determine "satisficing robust design" specifications for given performance requirement ranges using the goal-oriented, inverse design method can bring in robustness for multiple conflicting goals across process chains	***RQ3.*** What are the requirements for an ***inverse, goal-oriented design approach*** for realizing the ***robust design exploration*** of the material, product, and process as a system by managing the associated uncertainties?	1. Metrics, robust design constraints and goals for facilitating robust design across process chains for multiple conflicting goals

(continued)

Table 1.5 (continued)

IV	**G6.** Constructs and tools to capture and reuse the knowledge associated with material and product systems design **G7.** Facilitation of original, adaptive, and variant design decision support **G8.** Facilitation of systematic design exploration through decisions that are robust, adaptive, flexible, and modifiable particularly in the early stages of design	***H4.1.*** Using ontology to represent decision-related knowledge that is modeled as decision support problem (DSP) templates can capture, analyze, archive, and update the decision-based design workflow as per the needs of the individual decision-maker. Separation of declarative (problem-specific) knowledge and procedural (process specific) knowledge in the information flow scheme can help in generalizing the decision models in the design workflow (To address **G6**) ***H4.2.*** Defining three types of users, namely Template Creators, Template Editors, and Template Implementers, and providing customized decision support to these users during the design of engineering systems can help perform original design, adaptive design, and variant design, respectively (To address **G7**) ***H4.3.*** Developing an ontology for design space exploration and a template-based ontological method that supports systematic design space exploration ensuring the determination of the right combination of design information that meets the different goals and requirements set for a process chain (To address **G8**) ***H4.4.*** Using the technologies associated with digital manufacturing and design innovation, such as cloud-based design and manufacturing for seamless, yet controllable information, knowledge, and resource sharing between the stakeholders in decision workflows that are connected in a networked fashion. This facilitates instant communication, feedback, and collaboration that supports adaptive, robust flexible, and modifiable decisions (**G6, G7, and G8**)	***RQ4.*** What are the foundations needed for maintaining ***structural consistency*** of the decision-based design workflow for the manufacturing process chain involving the material and product, ensuring ***robust, adaptive, flexible, and modifiable*** decisions while incorporating newer ***data, information, and knowledge*** associated with the system?	1. Capture knowledge in original design, maintain consistency in adaptive design, and provide a package of documented knowledge in variant design 2. Template-based ontological method for systematic design space exploration 3. Cloud-based decision support platform for instant communication, feedback, and collaboration that supports adaptive, robust, flexible, and modifiable decisions—facilitate co-design

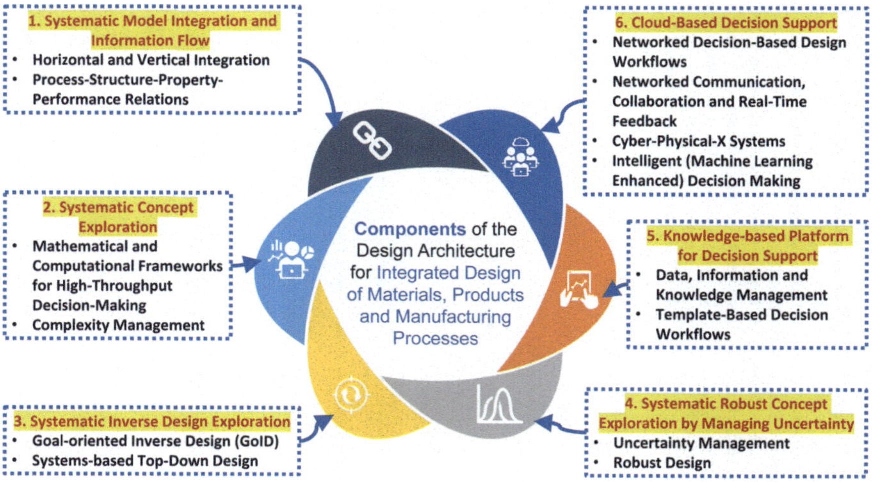

Fig. 1.4 Components of the systems-based design architecture for the integrated design of materials, products, and manufacturing processes as proposed in this monograph

1. The interest therefore by answering this research question is to propose a systematic approach to identify and classify material processing–structure–property–performance relations on multiple length scales to facilitate the design of material concepts to be further explored through systems-based top-down or inverse design exploration methods.

A function-based approach is proposed in this monograph for systematic model integration and information flow based on the foundational work carried out by Messer (2008). The intention in this monograph is to enable designers to identify underlying phenomena and associated solution principles (models) to achieve systematic model integration for materials design exploration. Materials models and simulations evolve over time. Also, interdisciplinary research involving different fields, such as mechanical engineering, materials engineering, systems engineering, and design, forms a part of the emerging materials design research. By focusing on phenomena and associated solution principles embodying identified functional relationships, we are able to overcome such disciplinary boundaries and technological evolution.

Systematic design methods support designers to formulate and solve problems efficiently and effectively (Beitz et al. 1996; Pahl and Beitz 2013). Beitz et al. (1996), describe function-based design, that is, function-based analysis, abstraction, synthesis, and systematic variation as most effective and efficient in mechanical and electrical engineering domains. Achieving integrated material, product, and process design involving phenomena and associated solution principles on the material level to systematically integrate models and establish information flow has not been addressed.

Rendering materials, products, and process design more systematic and domain-independent by enhancing existing function-based systematic design approaches is addressed in Chap. 4 of the monograph.

Component 2: Systematic Concept Exploration

In general, concept exploration is the process of evaluating different design concepts and providing top-level design specifications to meet overall system design requirements (Chen et al. 1995). Determining top-level specifications is strongly influenced by the way in which overall design requirements (goals and constraints) can be used to control solutions. Overall design requirements can be used in two ways (Amarel 1990):

- By constraining a priori, the generation of possible design structures to be consistent with them, and
- By testing a posterior whether a candidate design satisfies them.

This relationship is shown in Fig. 1.5.

The a priori use of design requirements involve *analysis* and *transformation* of the design requirements so as to enable them to control directly the generation of solutions. Relating this to the materials design mapping across processing, structure, property, and performance spaces, this is equivalent to forward mapping using bottom-up modeling (deductive/cause and effect). This is systematically achieved by proposing a concept exploration framework. The a posteriori use of design requirements involves *synthesis* and *evaluation* of a candidate solution or set of candidate solutions, and an assessment to the degree to which the candidate satisfies the design requirements. Relating this to the materials design, this is equivalent to inverse or

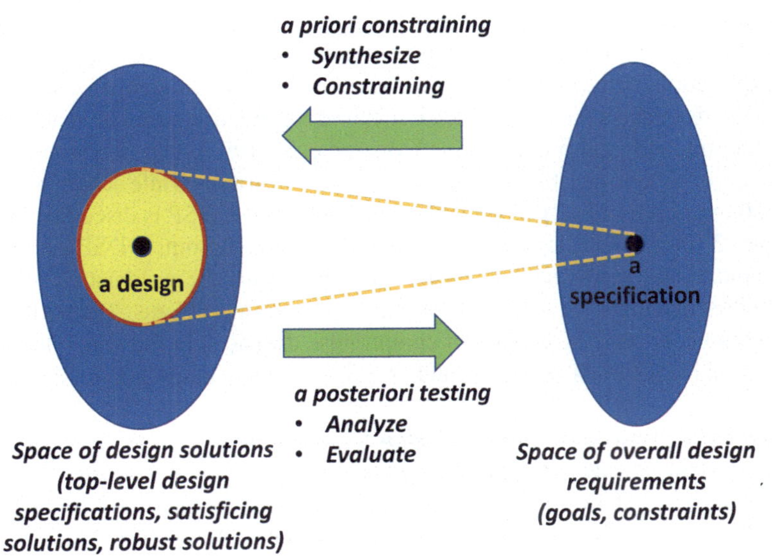

Fig. 1.5 Relationship between space of overall design requirements and space of design solutions

inductive mapping using top-down design methods where the focus is to design the material to satisfy certain performance requirements. To achieve the integrated materials and product design both are required.

In this monograph, a concept exploration framework that is based on a simulation-based design approach for designing complex systems is presented. An ideal concept exploration framework should support both the activities of a priori analysis and a posterior evaluation/synthesis in concept exploration. The focus, however, in this monograph is on using the concept exploration framework to support systems-based top-down design to generate "***satisficing design specifications***". The first proposed concept exploration model was in the ship design field to evaluate design alternatives and generate top-level design specifications (Georgescu and Boonstra 1990). Smith and Mistree (1994) based on their detailed study on modeling and exploration of ship concepts came to an important finding that a single-point design solution/approach as followed in optimization yields limited knowledge of the possibilities of the true solution space. The conclusion that was made, therefore, was that a greater emphasis is hence needed on concept exploration. However, due to the expensive computational burden associated with complex engineering design problems, there should also be a way to manage the complexity associated with concept exploration. Therefore, concept exploration framework for complex, high-dimensional engineering design problems requires efficient ways to manage the complexity—alternative to the extensive factorial grid search and effective data analysis tools, screening experiments. The framework should offer the capability to the designer to visualize the entire design space and further generate information about the *satisficing* regions of the design space.

In this monograph, the compromise decision support problem (cDSP) provides a means for mathematically modeling, formulating, and supporting design decisions that involve seeking *satisficing design solutions* among multiple conflicting goals. The cDSP is the foundational mathematical construct for decision support and concept exploration in this monograph. The compromise DSP is a domain-independent, multiobjective decision model, which is a hybrid formulation based on mathematical programming and goal programming (Mistree et al. 1993). It is used to determine the values of design variables that satisfy a set of constraints while achieving a set of conflicting goals as closely as possible. The compromise DSP is discussed in detail in Chap. 3. In this monograph, it is shown that the compromise DSP can be used as a foundational, mathematical construct for structuring the systematic exploration for families of satisficing solutions for materials design problems. The focus in this monograph is on making consistent compromise decisions in the integrated design exploration of materials, products, and associated manufacturing processes.

Component 3: Systematic Inverse Design Exploration

The need for inverse design exploration in process chains involving complex materials, products, and processes is to identify adjustable ranges of control factor (design variables) that satisfy end performance requirements. As discussed in Sect. 1.1, the inverse methods to identify microstructures and processing paths of a material satisfying a set of specified performance requirements are the focus on materials design communities and there are several works to this credit. Adams et al. (2013) present a framework that utilizes highly efficient spectral representations to arrive at invertible linkages between material structure, its properties, and the processing paths used to alter the material structure. Materials knowledge systems approach by Kalidindi et al. (2010, 2011) showcases advances in rapid inverse design to estimate local responses. However, all these approaches including the strategy proposed by McDowell and Olson (2008) fall to specific classes of materials design problems and demands considerable knowledge and insight in mechanisms, material hierarchy, and information flow. Thus, these classes of inverse design approaches are mostly suited for detailed design and not for "design exploration" (McDowell 2018).

In this monograph, we seek "top-down", goals/means, inductive or inverse methods especially at early stages of design to explore the design space of processing paths and resulting microstructures of a material satisfying a set of specified performance requirements. Approaches to pursue inverse design exploration by employing multiscale modeling and systems-based design especially at the early stages of design are limited and need further evaluations to address hierarchical material design problems with consideration of robustness. Choi et al. (2008a, b) propose the inductive design exploration method (IDEM), a multilevel, robust design method that considers propagation of all three types of uncertainty, such as that arising in hierarchical materials design problems that incorporates process–structure–property relations. The two major design objectives using the IDEM for material and product design are McDowell and Olson (2008): (i) to guide bottom-up modeling so as to conduct top-down, goal-oriented design exploration, (ii) manage the uncertainty in chains of process–structure–property relations. Kern et al. (2017) propose pyDEM, a generalized implementation of the IDEM as an open-source tool in the Python environment. In pyDEM, the authors adopt the general procedure of IDEM as a multilevel robust design tool and expand the algorithm for improved functionalities, like arbitrary feasible boundary representation, improved computational efficiency, multiplatform availability, and so on to suit for practical engineering problems. In this monograph, an inverse decision-based design method to achieve the integrated design exploration of materials, products, and manufacturing processes through the vertical and horizontal integration of models is proposed, see Chaps. 4, 5, 6, and the following articles (Nellippallil et al. 2016, 2017a, b, c, 2018b, 2019b). The inverse method is supported by the concept exploration framework to systematically explore design alternatives and generate "***satisficing***" design solutions across process chains that involve process–structure–property–performance relations (Nellippallil et al. 2018b). The inverse decision-based design method for design exploration is addressed in Chaps. 6 and 7 of the monograph.

Component 4: Systematic Robust Concept Exploration by Managing Uncertainty
Uncertainty could be either *aleatory* (irreducible) or *epistemic* (reducible), depending on their causes. Improving the measurements and/or model formulation and/or increasing the accuracy are ways to diminish epistemic uncertainty. Aleatory uncertainty, however, is inherent in the physical system and can only be quantified in a statistical sense. Extending the classification by Isukapalli et al. (1998), we classify

the types of uncertainty in the simulation-based integrated design of the material, product, and processes as (Choi et al. 2005; McDowell et al. 2009):

- *Natural Uncertainty* (NU): uncertainty due to the inherent randomness or unpredictability of a physical system; aleatory in nature.
- *Model Parameter Uncertainty* (MPU): incomplete knowledge of model parameters/inputs due to insufficient or inaccurate data; reducible by sufficient data or accurate measurements; uncertainty in design variables or control factors.
- *Model Structure Uncertainty* (MSU): uncertain model formulation due to approximations in a model; reducible by improving model formulation; uncertainty in function relationship between control/noise and response.
- *Propagated Uncertainty* (PU): uncertainty compounded by the combination of all the above three types of uncertainty in a chain of models that are connected through input–output relations; interdependent responses; and shared control/noise factors as one model interacts with another, see Allen et al. (2006) for their modes of interaction.

Two approaches are followed in dealing with these sources of uncertainty—mitigating uncertainty and managing uncertainty. In the first approach, the focus is to reduce/mitigate the uncertainty. This is achieved by seeking "perfect" models, collecting more data, developing improved methods to model, calculate and quantify uncertainty through expensive computations. Modern data science methods for materials and microstructure informatics along with multiscale modeling techniques are being developed to provide decision support and address the issue of uncertainty in hierarchical materials design (Panchal et al. 2013; Kalidindi 2015; McDowell and Kalidindi 2016; McDowell and LeSar 2016). Seepersad and co-authors (Shahan and Seepersad 2012; Matthews et al. 2016) consider Bayesian network classifiers to design materials with hierarchy and to treat uncertainty propagation in multilevel material design. Mahadevan and co-authors (Li and Mahadevan 2016; Mullins and Mahadevan 2016) address the issue of uncertainty integration across multilevel and the role of calibration, validation, and relevance in multilevel uncertainty integration for hierarchical material design problems. Even though there are several such recent efforts to address the issue of uncertainty, McDowell (2018) observes that quantifying uncertainty in schemes for linking models at different length and time scale is still an immature field, and formal mathematical approaches for doing this are largely undeveloped. The recommendation, therefore, made for simulation-assisted materials design is to focus on understanding the sensitivity of material properties to material microstructure and to capture dominant mechanisms and transitions that affect material responses or properties instead of focusing on accurately predicting mean properties at higher scales. Sensitivity analysis of responses is important because of several reasons (McDowell 2018):

- It is challenging to isolate response sensitivity experimentally at specific scales in the material hierarchy,
- The identification of key design variables across material structure hierarchy is possible via sensitivity analysis, and

- Core to the concept of robust design is the sensitivity of process–structure and structure–property relations, where the focus is to explore a range of solutions that meet conflicting response requirements and identify satisficing design solutions that are relatively insensitive to uncertainty (McDowell et al. 2009).

This demands the need for the second approach of managing uncertainty in the integrated design of materials, products, and processes with a focus on the sensitivity of key properties or responses to variation in microstructure which in turn connects to variation in processing paths.

In the second approach, the focus is to manage uncertainty by designing the system to be insensitive to the sources without reducing or eliminating them. This is done by exploring the solution space and studying the sensitivity of responses to variations in noise, control factors, and models themselves and understanding the tradeoffs required with various compromises; this is called robust design (Chen et al. 1996; Choi et al. 2005, 2008a; Murphy et al. 2005; Allen 2006; Nellippallil et al. 2017a). There are several practical implications with this approach. Robust solutions are not focused on extensive optimization searches at individual levels and do not necessarily involve a large number of iterations (McDowell and Olson 2008). The practical interest here is for a ranged set of solutions that showcase good performance under variability rather than single-point solutions that are valid for a narrow range of conditions while performing poorly when the conditions are changed slightly. The human designer plays the role of an interpreter of the value of information in this approach. Concepts and mathematical constructs from information economics (Howard 1966; Panchal et al. 2006; Sinha et al. 2013) are used to manage uncertainty by studying the value of information for cost/benefit tradeoff to make design decisions in the presence of uncertainty. In robust design (RD), the quality of products and processes is improved by reducing their sensitivity to variations without eliminating the sources (Taguchi 1986; Taguchi and Clausing 1990; Nair et al. 1992; Tsui 1992). The robust design principles and methods are founded on the philosophy of Taguchi (1986).

Three categories of information interact with the system model in robust design (McDowell et al. 2009): (i) control factors, also known as design variables are parameters that the designer adjusts to move toward the desired product, (ii) noise factors, are exogenous parameters that affect the performance of product/process but cannot be controlled by the designer, and (iii) responses are performance measures for the product or process. Over the years robust designs have been categorized into McDowell et al. (2009):

Type I Robust Design (Taguchi 1993) To identify the control factor (design variable) values that satisfy a set of performance requirements despite variations in noise factors. Though Type I robust design principles as proposed by Taguchi are advocated widely, his statistical techniques that include orthogonal arrays and signal-to-noise ratio are widely criticized. Many researchers (Box 1988; Vining and Myers 1990; Welch et al. 1990; Shoemaker et al. 1991; Tsui 1992; Parkinson et al. 1993; Sundaresan et al. 1995; Chen et al. 1996) have actively worked on improving the statistical techniques in robust design and thus have over the years developed mathematical constructs that bring in robust design into a systematic framework.

Type II Robust Design (Chen et al. 1996, 1999) To identify control factor values that satisfy a set of performance requirements target despite variation in control factors themselves. Chen et al. (1996) propose a procedure for robust design Types I and II by minimizing variations of noise and control factors and formulate the problem as a multiobjective decision problem using mean on targets and variances as separate goals. The foundational mathematical construct for decision support in their work is the compromise decision support problem (cDSP) proposed by Mistree and co-authors (Bras and Mistree 1993; Mistree et al. 1993) for robust design with multiple goals. The robust concept exploration method (RCEM) is further proposed for Type I and II robust designs and includes systematic steps to identify design alternatives and generate robust design solutions (Chen et al. 1997). Mathematical constructs known as design capability indices (DCIs) are further incorporated in RCEM to determine whether a ranged set of design specifications satisfies a ranged set of design requirements (Chen et al. 1999). Nellippallil et al. (2018b) propose the concept exploration framework (CEF); inspired from the RCEM with addition of features (processors) to consider different material and product models and options to explore the solution space for different design scenarios by weighing multiple goals. The CEF is addressed in Chap. 5 of this monograph.

Type III Robust Design (Choi et al. 2005) To obtain design solutions that are insensitive to variability or uncertainty embedded within the model used. Choi et al. (2005) propose the robust concept exploration method with error margin indices (RCEM-EMI) for Type I, II, and III robust designs. Error margin indices are mathematical constructs that indicate the location of the mean response and the spread of the response considering the variability associated with design variables and system models. These are then incorporated as goals in the cDSP formulation to design the system under model structure and model parameter uncertainty. The inverse decision-based design method for robust design exploration is addressed in detail in Chap. 7 of this monograph. Readers are also directed to the following articles for more information (see Nellippallil et al. 2018a, 2020; Wang et al. 2019).

Component 5: Knowledge-based Platform for Decision Support

The design of engineering systems is increasingly recognized as a decision-making process (Daskilewicz and German 2012; Afshari et al. 2016; Berg and Vance 2016; Soria et al. 2017). The fundamental research philosophy in this monograph is that the principal role of a human designer is to make decisions. Providing decision support is of critical importance for augmenting this role, by speeding up the design process and generating quality designs. One of the challenges in providing decision support in the design of engineering systems, especially complex systems that are, by definition, made up of inter-related subsystems (Kuppuraju et al. 1985) arises because of the complexity embodied in the decision workflows that embody multiple coupled decisions networked in various degrees of complexity. The networked decision workflows may include different types of decisions, for example, selection of design alternatives and improvement of an alternative considering multiple goals. The decisions are coupled together due to the dependency existing among systems and subsystems. The different types of decisions and their associated dependencies in the decision workflows make it difficult to provide appropriate decision support.

Decision-making is a knowledge-intensive process; knowledge plays a significant role in speeding up and effecting decisions. Capturing, managing, and reusing of decision-related knowledge such as alternatives, parameters, constraints, goals, dependencies, and the design process in the design of complex systems is an effective way of providing decision support. Many research efforts have been made to develop knowledge-based decision support in designing engineering systems, and these efforts are typically categorized as follows: (1) application of reasoning techniques for dealing with a large number of rules in design (Gero 1990a, b; Tong and Sriram 1997), (2) integrating design knowledge with decision-based design processes (Zha et al. 2003a, b), and (3) representing semantic knowledge to facilitate communication and interoperability of integrated decision support systems (Schoop et al. 2002; Chiu et al. 2005; Rockwell et al. 2010). In addition to the efforts on decision support in design, many others contributed to support organizational decisions by managing enterprise resources, such as Liu et al. (2008, 2009). Despite the fact that many knowledge-based systems have been developed for decision support from different perspectives, the challenge of supporting the decision workflow in the design of complex engineering systems is not yet well addressed, mainly, by the following reasons:

(1) Lack of both reusable and executable decision knowledge representation schemes. Knowledge reusability is critical for adaptive and variant design wherein only a small portion of the original decision workflows needs to change while the rest remains the same and can be reused. Some authors have proposed to represent decision knowledge as ontologies (e.g. Rockwell et al. 2010), but they mainly focus on capturing the semantic information of design decisions while failing to represent the execution process information which is necessary for effecting new decisions, especially in a computational environment whereby some degree of automation is realized.
(2) Lack of classification of users for decision support. The needs of designers for decision support vary according to how much novelty is involved in the design and how much knowledge they have about the design process. For example, an expert has much knowledge about design and can perform the decision-making process independently, thus the support this designer needs from the computer system is very different from a novice designer who only has the basic knowledge about design and needs to get most of the knowledge from the system. Very few of knowledge-based systems recognized this difference and provided appropriate decision support.

To address the aforementioned needs, a knowledge-based platform for decision support in the design of engineering systems (PDSIDES) is proposed (see Ming et al. 2018). The industry-inspired problem for the integrated design of materials, products, and manufacturing processes formulated and studied in this monograph is used as a test example to illustrate the efficacy of the PDSIDES platform; see (Ming et al. 2018; Wang et al. 2018, 2019; Nellippallil et al. 2019a). PDSIDES is discussed in detail in Chap. 8 of the monograph.

Component 6: Cloud-based Decision Support

In order to achieve an intelligent environment for designing complex engineering systems, a good understanding of predicting process behavior is paramount. Achieving this by using decision-based design necessitates a systematic, flexible, dynamic, and adaptive designing of the decision workflows involved. The decision-based design results associated with these workflows should be robust, that is, relatively insensitive to the uncertainties involved. The design results should also be flexible enough to accommodate any risk of errors that may accumulate along with the decision workflows. To address the above demands, an ontology for design space exploration and a template-based ontological method that supports systematic design space exploration in the model-based realization of complex engineered systems is proposed (see Wang et al. 2018). Using this proposed method, a designer is able to determine the right combinations of design information that meets the different goals set, thereby satisfying the end requirements for each stage of the process, and also adjust the design space to achieve solutions that are robust and flexible enough to manage any risk of error propagation in continuous multistage design. The efficacy of this method is illustrated by using the example associated with the design of a multistage hot rod rolling system addressed in this monograph (see Wang et al. 2018).

Along with these tools, technologies associated with the new digital manufacturing and design innovation (DMDI) paradigm, such as cloud-based design and manufacturing, are used for seamless, yet controllable information, knowledge, and resource sharing between the stakeholders in decision workflows that are connected in a networked fashion to facilitate instant communication, feedback, and collaboration. In this monograph, we present the architecture and functionalities of a cloud-based version of the computational platform PDSIDES, known as CB-PDSIDES to facilitate mass collaboration and open innovation thereby supporting integrated material and product realization—that is, facilitate co-design. We illustrate the efficacy of the proposed cloud-based platform using the hot rolling example problem to produce a steel rod. Using this example, we illustrate the utility of the cloud-based platform in carrying out seamless, yet controllable, information, knowledge, and resource sharing, thereby supporting the integrated design of materials, products, and manufacturing processes. Using Table 1.6, we summarize the constituents of the systems-based design architecture proposed in each chapter. The generic form of the table is shown in Table 1.6. The table will be included in each chapter to help readers summarize the key constituents of the architecture discussed in the corresponding chapter.

Table 1.6 Summary of elements discussed and reviewed in each chapter that form part of the architecture

Summary of elements discussed and reviewed in Chapter "x" that form part of the architecture	
Elements	**What?**
Processing elements	Methods, principles, and design constructs that support the components of the architecture and helps in the transformation of design data
Data elements	Elements that contain the data/information that is transformed
Connecting elements	The media/constructs/tools that link the architecture component elements
Form	**How?**
Weights	The relative importance of the elements
Properties	Attributes of the elements used to form the relationship, ex: Type II Robust Design
Relationship	Links within and between the elements in the architecture
Rationale	**Why?**
Motivation	The underlying reason to use this architectural element, driven by user requirements
Characteristics/beliefs	Characteristics, assumptions, and beliefs used to guide the selection of elements and form
Interpretation	Understanding the efficacy and broader impacts

1.8 Fundamental Research Philosophies and Assumptions

The fundamental philosophies on which this monograph is based are summarized in this section. The general philosophies are:

- *Analysis models are incomplete, inaccurate, and of differing fidelity*: The foundational philosophy for this research is accepting the fact that models are but abstractions of reality and are typically incomplete, inaccurate, and not of equal fidelity. Given this, the role of a human designer in this monograph is to provide computational support to make decisions for exploring the solution space.
- *Emergent properties*: A complex system has emergent properties. Complex systems embody systemic features (emergent properties) that cannot be predicted or deduced. Hence, designers need to know how to account for emergent properties associated with the realization of a complex system. The key emergent properties in a complex system are "complexity" and "uncertainty". Therefore, for the model-based realization of complex systems, it is essential to know how to identify and manage complexity and to identify and manage uncertainty.
- *Design is a goal-oriented activity and design requirements are subject to change*: As observed by Gero (1996), design can be conceived of as a purposeful, constrained, decision-making, exploration, and learning activity. Decision-making here refers to the process of deciding the values of a set of design variables.

Exploration here refers to changing the problem or design spaces within which decision-making takes place. Learning here implies the restructuring of the knowledge generated through the design process. The context in which the designer operates partially depends on the designer's perceptions of purposes, constraints, goals, and related requirements. Thus, the designer starts with some goals that he/she wishes to satisfy and thus the design is a goal-oriented activity. These perceptions are bound to change as the designer explores the emerging relationships between possible designs and the context and as learning happens while exploring the design space.

Given the fundamental research philosophies, the question addressed in this monograph from a simulation-based design perspective is,

What is the role of a human designer in simulation-based design?
The question is investigated in this monograph and the answer to which depends on certain fundamental research assumptions, listed as follows:

- A human designer should make decisions using the information provided by computational decision support models that are typically incomplete and inaccurate.
- There are two types of decisions that a human designer can make—selection or compromise.
- Do not eliminate uncertainty—manage it.
- Explore multiple possible solutions and their consequences rather than identifying a single unique solution.

1.9 Contributions in This Monograph

The contributions in the monograph are categorized into new knowledge in four research domains: (a) systematic model integration and information flow (vertical and horizontal) for integrated material and product workflows, (b) goal-oriented, inverse decision support, (c) robust concept exploration of process chains with multiple conflicting goals, and (d) knowledge-based decision support via the cloud for rapid and robust design exploration in simulation-based integrated material, product, and process design.

The creation of new knowledge in this monograph is associated with the development of a systems-based design architecture involving systematic function-based approach of formulating forward material workflows, a concept exploration framework for systematic design exploration, an inverse decision-based design method, and robust design metrics, all integrated with a knowledge-based platform for decision support. The theoretical, mathematical, and computational foundations for the design architecture are proposed in this monograph to facilitate rapid and robust exploration of the design and solution spaces to identify material microstructures and processing paths that satisfy conflicting property and performance for complex materials, products, and processes by managing uncertainty.

Specific contributions in this monograph include:

- Systematic identification and integration of material, process, and product models and workflows to define processing–structure–property–performance mapping and information workflow,
- A reusable, expandable computational framework supporting vertical and horizontal integration of models to identify material structures and processing paths that satisfy ranged set of product and manufacturing process-level property and performance requirements,
- A framework supporting systematic design and solution space exploration,
- A generic method for inverse design of materials and products across process chains,
- Metrics, robust design constraints and goals for facilitating robust design across process chains for multiple conflicting goals,
- Capture knowledge in the original design, maintain consistency in adaptive design, and provide a package of documented knowledge in variant design,
- Template-based ontological method for systematic design space exploration, and
- Cloud-based design decision support.

Based on these contributions a designer now has the following abilities which the baseline designer did not have before:

- Designing materials and products in a systematic fashion during the early stages of design by looking at information flow and mapping across models,
- Designing products, materials, and their processing paths in a function-based, systematic, integrated fashion from a systems perspective by satisfying specific end performance requirements,
- The capability to carry out rapid, integrated design exploration of material and products using simulation models that we accept are typically incomplete and inaccurate,
- The capability to support a human designer under a complex material system's random variability and/or model parameter uncertainty and/or model structure uncertainty in making decisions that satisfy multiple conflicting goals,
- The capability to model decision-related knowledge with templates using ontologies to facilitate execution and reuse,
- The capability to coordinate information and human decision making.

Therefore, crucial to this monograph are:

(i) The requirements-driven, "top-down" design of the system and associated subsystems by taking a goal-oriented, inverse approach which is different from the standard practice of bottom-up modeling and design of material and product systems,

(ii) The management of uncertainty in the system without removing the sources and supporting in identifying robust design solutions across process chains,

(iii) Cloud-based platformization of decision templates to support different types of users to facilitate original, adaptive, and variant designs in materials and product design in a networked collaborative fashion—facilitating co-design.

The systems-based design architecture for integrated model-based realization of materials, products, and associated manufacturing processes is validated using an industry-inspired example problem from the steel manufacturing domain, namely: *the integrated design of steel (material), manufacturing processes (rolling and cooling) and hot rolled rods (product) for automotive gears*. However, potential applications are numerous and compelling, and not limited to the one addressed in this monograph. The framework and method developed in this monograph are generalizable for industries in which mechanical, structural, and thermal systems are essential. The applications include the manufacturing of lightweight, high performance, low cost, and reliable parts and machine components, for example, automobile gearbox, shafts, and so on.

1.10 Organization of the Monograph

The monograph is organized according to the verification and validation strategy followed for the work carried out. The "verification and validation" is based on the validation square framework introduced by Pederson and co-authors (see Pedersen et al. 2000a, b; Seepersad et al. 2006). The "validation square" consists of four quadrants: *Theoretical Structural Validity, Empirical Structural Validity, Empirical Performance Validity,* and *Theoretical Performance Validity*. The corresponding verification involves checking for internal consistency.

The verification and validation square framework is used to verify and validate the chapters in this monograph. This is discussed next and summarized in Fig. 1.6.

Theoretical Structural Verification and Validation (TSV): Accepting the individual constructs constituting the design architecture as well as the internal consistency of the integration of all constructs to form an overall systems-based design architecture. Thus, TSV consists of:

- Establishing requirements for the design architecture,
- Carrying out a literature review,
- Establishing logical soundness of constructs used—individually and integrated.

Empirical Structural Verification and Validation (ESV): Building confidence in the appropriateness of the test example problems chosen for illustrating and verifying the performance of the systems-based design architecture. ESV consists of:

- Checking the appropriateness of the test example problems selected to test the design architecture,
- Accepting the design methods and constructs that form the architecture.

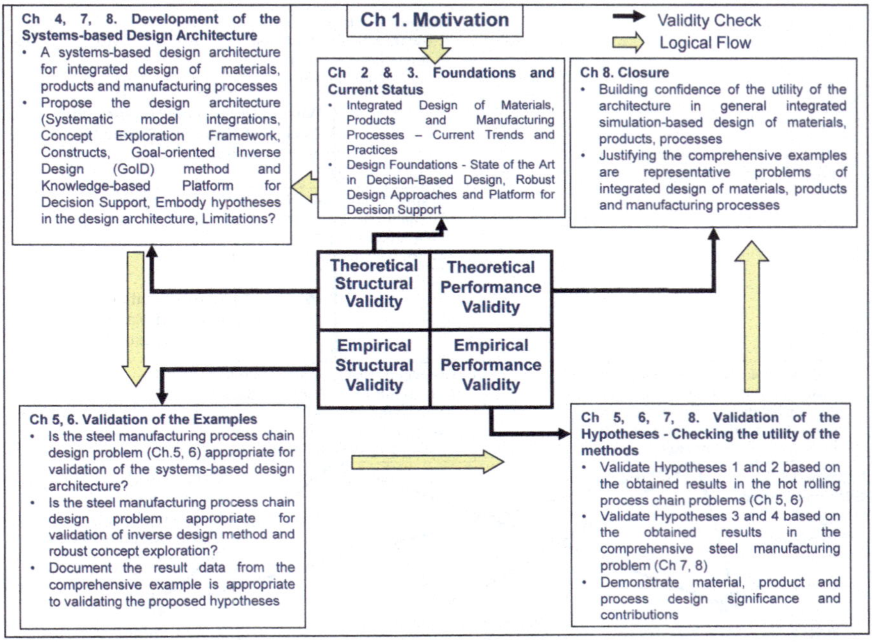

Fig. 1.6 Organization of the monograph chapters according to the verification and validation square

Empirical Performance Verification and Validation (EPV): Building confidence in the usefulness of the design architecture using example problems. EPV consists of:

- Checking the ability of the systems-based design architecture to provide useful results for selected example problems.

Theoretical Performance Verification and Validation (EPV): Building confidence in the generality of the design architecture and accepting that the architecture is useful beyond the example problems considered. EPV consists of:

- Checking the ability to provide useful results beyond example problems,
- Showcasing the generic form of the systems-based design architecture.

In this monograph, we focus on various aspects of systems-based integrated design of materials, products, and associated manufacturing processes. The flow of information between the different chapters is shown in Fig. 1.7. The figure is intended to help navigate through the monograph and develop an overall picture as to what is discussed in each chapter thereby establish context.

In Chap. 2, a review of the existing efforts associated with materials and product design is given. Several critical issues associated with the current capabilities of materials design are discussed in this chapter. Some of the major elements of modern materials design like material informatics, uncertainty management, verification and validation, multiscale modeling, systems design, and so on are discussed in this

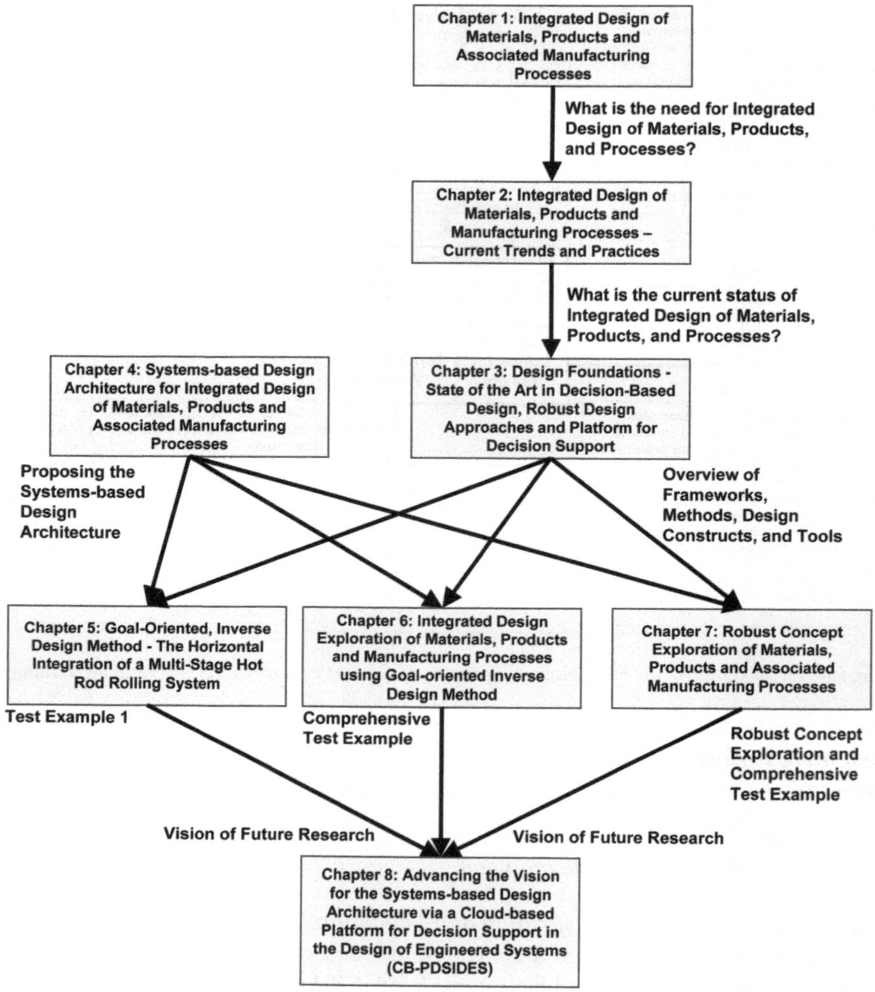

Fig. 1.7 Organization of Monograph

chapter. In Chap. 3, the theoretical foundations for designing simulation-based design processes are discussed. These foundations include existing design-construct related to the decision-based design paradigm, meta-design and decision support problem technique, robust design, compromise decision support problem, robust concept exploration method, robust design metrics, indices for robust design—error margin indices, design capability indices, hyper-dimensional error margin index, inductive design exploration method. Design process modeling is also explored in this chapter. Foundational constructs are reviewed from this area. This includes template-based decision-centric design, design concept flexibility, function-based systematic design, and solution generation approaches. Relevant literatures for each of these areas are referenced, discussed, and critically evaluated to show the appropriateness of use of these constructs for the design architecture presented in the monograph. The literature review in Chap. 3 is used to identify the availability, strengths, and limitations of these constructs in the context of integrated design of materials, products, and design processes, and becomes an essential component of theoretical structural validation. In Chap. 4, the major components of the systems-based design architecture are presented: (1) a systematic function-based approach of integrating models (vertically and horizontally) to formulate forward material workflows establishing process–structure–property–performance relations, and (2) a concept exploration framework for systematic design problem formulation and exploration along with a goal-oriented, inverse design (GoID) method for designing materials for satisfying property and performance goals. In Chap. 5, an example of the horizontal integration of the multistage hot rolling process and product design is presented. The horizontal integration is systematically achieved using well-established theoretical and empirical models and response surface models developed through simulation experiments (finite-element based). The illustration of the efficacy of the constructs proposed is carried out by the decision-based design of a multistage rolling system and the circular rod product. In Chap. 6, the design architecture presented is applied to a multiscale, multistage materials design problem—vertical and horizontal integration and integrated design of hot rod rolling process chain, steel, and rolled rod. In this chapter, the industry inspired problem of focus in this monograph is addressed. The bigger picture of the problem of interest and how integrated materials, product, and process design can be applied at an industrial scale is discussed in the beginning. A discussion on the specific problem (vertical and horizontal integration of hot rolling process chain) is carried out in detail. The requirement of designing the material microstructure and processing paths to satisfy conflicting product and process-related end performances and properties in an inverse manner is addressed. In this chapter, we discuss the validation of the proposed systematic method of model integration, inverse design method, and concept exploration framework. In addition to the validation of design methods, the chapter is also crucial from the standpoint of the major theme addressed in this monograph. In Chap. 7, a variation to the existing goal-oriented, inverse design method (Chap. 6) to bring in robustness for multiple conflicting goals from the standpoint of Type I to III robust design across process chains is introduced. The variation embodies the introduction of specific robust design goals and constraints anchored in the mathematical constructs of error margin indices and design capability indices to determine "satisficing robust design" specifications for given performance requirement ranges using the goal-oriented, inverse design method. The design of a hot rolling process chain for the production of a rod is used as an example to verify and validate the approach proposed. In Chap. 8, we present the architecture and functionalities of a cloud-based computational platform to facilitate mass collaboration and open innovation, thereby supporting the co-design of materials, products, and manufacturing processes. Finally, avenues for future research and broader applications of the fundamental ideas in this monograph are discussed from the context of cloud-based design and manufacturing. The focus in this chapter is on furthering the

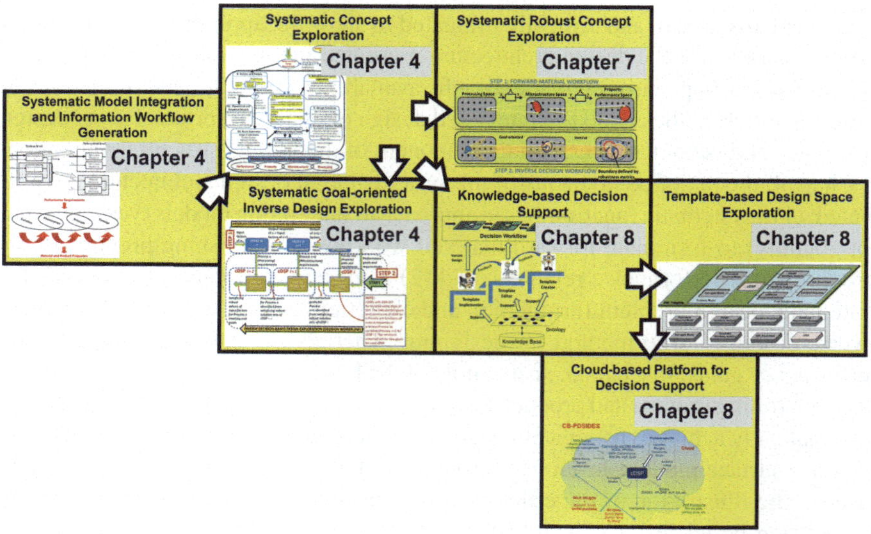

Fig. 1.8 Relationship of research efforts with the constructs of the systems-based design architecture and connection between chapters of the monograph

research vision by exploring the opportunities for the automated realization of decision workflows for materials and product design using the cloud. The relationship of the research efforts with the constructs of the design architecture developed is shown in Fig. 1.8.

References

Adams, B. L., Kalidindi, S., & Fullwood, D. T. (2013). *Microstructure-sensitive design for performance optimization*. Butterworth-Heinemann.

Afshari, H., Peng, Q., & Gu, P. (2016). Design optimization for sustainable products under users' preference changes. *Journal of Computing and Information Science in Engineering, 16*(4), 041001–041007.

Allen, J. K., Seepersad, C., Choi, H., & Mistree, F. (2006). Robust design for multiscale and multidisciplinary applications. *Journal of Mechanical Design, 128*(4), 832–843.

Allen, J. K., Panchal, J., Mistree, F., Singh, A. K., & Gautham, B. (2015). Uncertainty management in the integrated realization of materials and components. In *Proceedings of the 3rd World Congress on Integrated Computational Materials Engineering*. John Wiley & Sons.

Amarel, S. (1990). Artificial intelligence and design. In *Proceedings of the 5th Jerusalem Conference on Information Technology*, IEEE.

Anapagaddi, R., Shukla, R., Mangal, S., & Singh, A. K. (2013). Exploring the design space of a continuous casting tundish for improved inclusion removal and reduced dead volume.

Anapagaddi, R., Shukla, R., Goyal, S., Singh, A. K., Allen, J. K., Panchal, J. H., & Mistree, F. (2014). Exploration of the design space in continuous casting Tundish. *ASME Design Automation Conference*. Paper Number: DETC2014-34254.

Ashby, M. F., & Cebon, D. (1993). Materials selection in mechanical design. *Le Journal de Physique IV, 3*(C7), C7-1–C7-9.

Ashby, M. F., & Johnson, K. (2013). *Materials and design: The art and science of material selection in product design*. Butterworth-Heinemann.

Ashby, M. F., Evans, T., Fleck, N. A., Hutchinson, J., Wadley, H., & Gibson, L. (2000). *Metal foams: A design guide*. Elsevier.

Beitz, W., Pahl, G., & Grote, K. (1996). Engineering design: A systematic approach. *MRS Bulletin, 71.*

Berg, L. P., & Vance, J. M. (2016). An industry case study: Investigating early design decision making in virtual reality. *Journal of Computing and Information Science in Engineering, 17*(1), 011001–011007.

Box, G. E. (1979). *In the strategy of scientific model building. Robustness in statistics*. Elsevier, pp. 201–236.

Box, G. (1988). Signal-to-noise ratios, performance criteria, and transformations. *Technometrics, 30*(1), 1–17.

Bras, B., & Mistree, F. (1993). Robust design using compromise problems. *Engineering Optimization, 21*(3), 213–239.

Chen, W., Tsui, K.-L., Allen, J. K., & Mistree, F. (1995). Integration of the response surface methodology with the compromise problem in developing a general procedure. In *ASME Design Automation Conference,* Boston, Massachusetts, pp. 485–492.

Chen, W., Allen, J. K., Tsui, K.-L., & Mistree, F. (1996). A procedure for: Minimizing variations caused by noise factors and control factors. *Journal of Mechanical Design, 118*(4), 478–485.

Chen, W., Allen, J. K., & Mistree, F. (1997). A robust method for enhancing productivity in concurrent systems design. *Concurrent Engineering, 5*(3), 203–217.

Chen, W., Simpson, T. W., Allen, J. K., & Mistree, F. (1999). Satisfying ranged sets of design requirements using as metrics. *Engineering Optimization, 31*(5), 615–619.

Chiu, D. K. W., Cheung, S. C., Hung, P. C. K., & Leung, H. F. (2005). Facilitating e-negotiation processes with semantic web technologies. In *Proceedings of the 38th Annual Hawaii International Conference on System Sciences (HICSS'05) Track 1*, p. 36a.

Choi, H.-J., Austin, R., Allen, J. K., McDowell, D. L., Mistree, F., & Benson, D. J. (2005). An approach for of reactive power metal mixtures based on non-deterministic micro-scale shock simulation. *Journal of Computer-Aided Materials Design, 12*(1), 57–85.

Choi, H.-J., Mcdowell, D. L., Allen, J. K., & Mistree, F. (2008a). An inductive design exploration method for hierarchical systems design under uncertainty. *Engineering Optimization, 40*(4), 287–307.

Choi, H., McDowell, D. L., Allen, J. K., Rosen, D., & Mistree, F. (2008b). An inductive design exploration method for robust multiscale materials design. *Journal of Mechanical Design, 130*(3), 031402.

Daskilewicz, M. J., & German, B. J. (2012). Rave: A computational framework to facilitate research in design decision support. *Journal of Computing and Information Science in Engineering, 12*(2), 021005–021009.

Eberhart, M. E., & Clougherty, D. P. (2004). Looking for design in materials design. *Nature Materials, 3*(10), 659.

Georgescu, C., & Boonstra, V. F. (1990). Concept exploration models for merchant ships.

Gero, J. S. (1990a). Design prototypes: A knowledge representation schema for design. *AI Magazine, 11*(4), 26.

Gero, J. S. (1990). *Knowledge-based design systems.* Addison-Wesley.

Gero, J. S. (1996). Creativity, emergence and evolution in design. *Knowledge-Based Systems, 9*(7), 435–448.

Horstemeyer, M. F. (2012). *Integrated computational materials engineering () for metals: Using to invigorate engineering design with science*. John Wiley & Sons.

Horstemeyer, M. F. (2018). *Integrated computational materials engineering () for metals: Concepts and case studies*. John Wiley & Sons.

Howard, R. A. (1966). Information value theory. *IEEE Transactions on Systems Science and Cybernetics, 2*(1), 22–26.

Isukapalli, S., Roy, A., & Georgopoulos, P. (1998). Stochastic response surface methods (SRSMs) for uncertainty propagation: Application to environmental and biological systems. *Risk Analysis, 18*(3), 351–363.

Kalidindi, S. R. (2015). *Hierarchical materials informatics: Novel analytics for materials data.* Elsevier.

Kalidindi, S. R., Niezgoda, S. R., Landi, G., Vachhani, S., & Fast, T. (2010). A novel framework for building materials knowledge systems. *Computers, Materials, & Continua, 17*(2), 103–125.

Kalidindi, S. R., Niezgoda, S. R., & Salem, A. A. (2011). Microstructure informatics using higher-order statistics and efficient data-mining protocols. *JOM Journal of the Minerals Metals and Materials Society, 63*(4), 34–41.

Kern, P. C., Priddy, M. W., Ellis, B. D., & McDowell, D. L. (2017). pyDEM: A generalized implementation of the inductive design exploration method. *Materials & Design, 134*, 293–300.

Kuppuraju, N., Ganesan, S., Mistree, F., & Sobieski, J. S. (1985). Hierarchical decision-making in system-design. *Engineering Optimization, 8*(3), 223–252.

Li, C., & Mahadevan, S. (2016). Role of calibration, validation, and relevance in multi-level uncertainty integration. *Reliability Engineering & System Safety, 148*, 32–43.

Liu, S., Duffy, A. H. B., Boyle, I. M., & Whitfield, R. I. (2008). Knowledge re-use for decision support. In *Realising network enabled capability 2008, 2008-10-13–2008-10-14, Oulton Hall Hotel, Leeds.*

Liu, S., Duffy, A. H. B., Whitfield, R. I., Boyle, I. M., & McKenna, I. (2009). Towards the realisation of an integratated decision support environment for organisational decision making. *International Journal of the Decision Support System Technology, 1*(4), 35–58.

Matthews, J., Klatt, T., Morris, C., Seepersad, C. C., Haberman, M., & Shahan, D. (2016). Hierarchical design of negative stiffness metamaterials using a Bayesian network classifier. *Journal of Mechanical Design, 138*(4), 041404.

McDowell, D. L. (2018). *Microstructure-sensitive computational structure-property relations in materials design* (pp. 1–25). Computational Materials System Design: Springer.

McDowell, D. L., & Kalidindi, S. R. (2016). The materials innovation ecosystem: A key enabler for the materials genome initiative. *MRS Bulletin, 41*(4), 326–337.

McDowell, D. L., & LeSar, R. A. (2016). The need for microstructure informatics in process–structure–property relations. *MRS Bulletin, 41*(8), 587–593.

McDowell, D. L., & Olson, G. (2008). *Concurrent design of hierarchical materials and structures* (pp. 207–240). Scientific Modeling and Simulations: Springer.

McDowell, D., & Story, T. (1998). New directions in materials design science and engineering. In *Report of a Workshop Sponsored by the US National Science Foundation, Georgia Institute of Technology and Morehouse College, Atlanta, GA.*

McDowell, D. L., Choi, H. J., Panchal, J., Austin, R., Allen, J., & Mistree, F. (2007). *Plasticity-related microstructure-property relations for materials design.* Key Engineering Materials: Trans Tech Publ.

McDowell, D. L., Panchal, J., Choi, H.-J., Seepersad, C., Allen, J., & Mistree, F. (2009). *Integrated design of multiscale, multifunctional materials and products.* Butterworth-Heinemann.

Messer, M. (2008). *A systematic approach for integrated product, materials, and design-process design.* Ph.D. Dissertation, The George W. Woodruff School of Mechanical Engineering, Georgia Institute of Technology.

Ming, Z., Nellippallil, A. B., Yan, Y., Wang, G., Goh, C. H., Allen, J. K., et al. (2018). PDSIDES—A knowledge-based platform for decision support in the design of engineering systems. *Journal of Computing and Information Science in Engineering, 18*(4), 041001.

Mistree, F., Hughes, O. F., & Bras, B. (1993). Compromise problem and the adaptive linear programming algorithm. *Progress in Astronautics and Aeronautics, 150*, pp. 251–251.

Mullins, J., & Mahadevan, S. (2016). Bayesian uncertainty integration for model calibration, validation, and prediction. *Journal of Verification, Validation and Uncertainty Quantification, 1*(1), 011006.

Murphy, T. E., Tsui, K.-L., & Allen, J. K. (2005). A review of methods for multiple responses. *Research in Engineering Design, 15*(4), 201–215.

Nair, V. N., Abraham, B., MacKay, J., Box, G., Kacker, R. N., Lorenzen, T. J., et al. (1992). Taguchi's parameter design: A panel discussion. *Technometrics, 34*(2), 127–161.

Nellippallil, A., Song, K., Goh, C.-H., Zagade, P., Gautham, B., Allen, J., & Mistree, F. (2016). A goal oriented, sequential process design of a multi-stage hot rod rolling system. *ASME Design Automation Conference*. Paper Number: DETC2016-59402.

Nellippallil, A. B., Allen, J. K., Mistree, F., Vignesh, R., Gautham, B. P., & Singh, A. K. (2017a). A goal-oriented, inverse decision-based design method to achieve the vertical and horizontal integration of models in a hot-rod rolling process chain. *ASME Design Automation Conference*, Paper Number: DETC2017-67570.

Nellippallil, A. B., Song, K. N., Goh, C.-H., Zagade, P., Gautham, B., Allen, J. K., & Mistree, F. (2017b). A goal-oriented, sequential, inverse design method for the horizontal integration of a multistage hot rod rolling system. *Journal of Mechanical Design, 139*(3), 031403.

Nellippallil, A. B., Vignesh, R., Allen, J. K., Mistree, F., Gautham, B. P., & Singh, A. K. (2017c). A decision-based design method to explore the solution space for microstructure after cooling stage to realize the end mechanical properties of hot rolled product. In *4th World Congress on Integrated Computational Materials Engineering,* Ypsilanti, Michigan, USA.

Nellippallil, A. B., Mohan, P., Allen, J. K., & Mistree, F. (2018a). Robust of materials, products and associated manufacturing processes. *ASME Design Automation Conference*, Paper Number: DETC2018-85913.

Nellippallil, A. B., Rangaraj, V., Gautham, B. P., Singh, A. K., Allen, J. K. & Mistree, F. (2018b). An inverse, decision-based design method for integrated design exploration of materials, products and manufacturing processes. *ASME Journal of Mechanical Design, 140*(11), 111403-111403-17

Nellippallil, A. B., Ming, Z., Allen, J. K., & Mistree, F., (2019a). Cloud-based materials and product realization—Fostering ICME via Industry 4.0, *Integrating Materials and Manufacturing Innovation, 8*(2), 107–121.

Nellippallil, A. B., Mohan, P., Allen, J. K., & Mistree, F. (2019b). Inverse Thermo-Mechanical Processing (ITMP) design of a steel rod during hot rolling process. *ASME Design Automation Conference*, Paper Number: IDETC2019-97390.

Nellippallil, A. B., Mohan, P., Allen, J. K., & Mistree, F. (2020). An inverse, decision-based design method for robust concept exploration. *Journal of Mechanical Design*, https://doi.org/10.1115/1.4045877.

Norton, R. L. (1996) *Machine design: An integrated approach*. Prentice-Hall Inc: NJ, USA.

Olson, G. B. (1997). Computational design of hierarchically structured materials. *Science, 277*(5330), 1237–1242.

Olson, G. B. (2000). Designing a new material world. *Science, 288*(5468), 993–998.

Pahl, G., & Beitz, W. (2013). *Engineering design: A systematic approach*. Springer Science & Business Media.

Panchal, J. H. (2005). A framework for simulation-based integrated design of multiscale products and design processes, Ph.D. Dissertation, The George W. Woodruff School of Mechanical Engineering, Georgia Institute of Technology.

Panchal, J. H., Choi, H.-J., Shephard, J., Allen, J. K., McDowell, D. L., & Mistree, F. (2005). A strategy for simulation-based multiscale, multi-functional products and associated design processes. *ASME Design Automation Conference*, Paper Number: DETC2005-85316.

Panchal, J. H., Paredis, C. J., Allen, J. K., & Mistree, F. (2006). Simulation model refinement for decision making via a value-of-information based metric. *ASME Design Automation Conference*, Paper Number: DETC2006-99433.

Panchal, J. H., Kalidindi, S. R., & McDowell, D. L. (2013). Key computational modeling issues in integrated computational materials engineering. *Computer-Aided Design, 45*(1), 4–25.

Parkinson, A., Sorensen, C., & Pourhassan, N. (1993). A general approach for robust optimal design. *Journal of Mechanical Design, 115*(1), 74–80.
Pedersen, K., Emblemsvag, J., Bailey, R., Allen, J., & Mistree, F. (2000a). The 'Validation Square'–Validating Design Methods. In *ASME Design Theory and Methodology Conference*, New York: ASME.
Pedersen, K., Emblemsvag, J., Bailey, R., Allen, J. K., & Mistree, F. (2000b). Validating design methods and research: The validation square. *ASME Design Engineering Technical Conferences*, DETC2000/DTM-14579.
Pollock, T.M., Allison J. E., et al. (2008). Integrated computational materials engineering: a transformational discipline for improved competitiveness and national security. National Academies Press.
Rockwell, J. A., Grosse, I. R., Krishnamurty, S., & Wileden, J. C. (2010). A semantic information model for capturing and communicating design decisions. *Journal of Computing and Information Science in Engineering, 10*(3), 031008.
Rühle, M., Dosch, H., Mittemeijer, E., & Van de Voorde, M. (2001). European White book on fundamental research in materials science.
Schoop, M., Becks, A., Quix, C., Burwick, T., Engels, C., & Jarke, M., (2002) Enhancing decision and negotiation support in enterprise networks through semantic web technologies, www.sewasie.org/documents/xsw2002.pdf.
Seepersad, C. C., Pedersen, K., Emblemsvåg, J., Bailey, R., Allen, J. K., & Mistree, F. (2006). *The validation square: How does one verify and validate a design method?* (pp. 303–314). New York: ASME Press.
Shahan, D. W., & Seepersad, C. C. (2012). Bayesian network classifiers for set-based collaborative design. *Journal of Mechanical Design, 134*(7), 071001.
Shigley, J. E. (1972). Mechanical engineering design. McGraw-Hill, New York, NY.
Shoemaker, A. C., Tsui, K.-L., & Wu, C. J. (1991). Economical experimentation methods for. *Technometrics, 33*(4), 415–427.
Shukla, R., Anapagaddi, R., Singh, A. K., Panchal, J. H., Allen, J. K., & Mistree, F. (2015). Exploring the design set points of refining operation in ladle for cost effective desulfurization and inclusion removal. *ASME Design Automation Conference*, Paper Number: DETC2015-46265.
Sinha, A., Bera, N., Allen, J. K., Panchal, J. H., & Mistree, F. (2013). In the design of multiscale systems. *Journal of Mechanical Design*, 135(1), 011008.
Smith, W., Mistree, F. (1994). The development of top level ship specifications: A decision-based approach.
Soria, N., Colby, M. K., Tumer, I. Y., Hoyle, C., & Tumer, K. (2017). Design of complex engineered system using multiagent coordination. *Journal of Computing and Information Science in Engineering, 18*(1), 011003-1-13.
Sundaresan, S., Ishii, K., & Houser, D. R. (1995). A robust optimization procedure with variations on design variables and constraints. *Engineering Optimization + A35, 24*(2), 101–117.
Taguchi, G. (1986). Introduction to quality engineering: Designing quality into products and processes. Quality Resources.
Taguchi, G. (1993). Robust technology development. *Mechanical Engineering-CIME, 115*(3), 60–63.
Taguchi, G., & Clausing, D. (1990). Robust quality. *Harvard Business Review, 68*(1), 65–75.
Tong, C., & Sriram, R. D. (1997). *Artificial intelligence in engineering design*. Academic Press.
Tsui, K.-L. (1992). An overview of Taguchi method and newly developed statistical methods for. *IIE Transactions, 24*(5), 44–57.
Vining, G. G., & Myers, R. H. (1990). Combining Taguchi and response surface philosophies: A dual response approach. *Journal of Quality Technology, 22*(1), 38–45.
Wang, R., Nellippallil, A. B., Wang, G., Yan, Y., Allen, J. K., & Mistree F. (2018) Systematic design space exploration using a template-based ontological method. *Advanced Engineering Informatics, 36*,163–177.

Wang, R., Nellippallil, A. B., Wang, G., Yan, Y., Allen, J. K., & Mistree, F. (2019). Ontology-based uncertainty management approach in designing of robust decision workflows. *Journal of Engineering Design, 30*(10–12), 726–757.

Welch, W., Yu, T., Kang, S. M., & Sacks, J. (1990). Computer experiments for quality control by parameter design. *Journal of Quality Technology, 22*(1), 15–22.

Zha, X. F., Lim, Y. E., & Lu, W. F. (2003a). A knowledge intensive multi-agent system for cooperative/ collaborative assembly modeling and process planning. *Journal of Integrated Design and Process Science, 7*(1), 99–122.

Zha, X. F., Sriram, R. D., & Lu, W. F. (2003b). Knowledge-Intensive Collaborative Decision Support For Design Processes. *ASME Design Automation Conference*, Paper Number: DETC2003/DAC-48747.

Chapter 2
Integrated Design of Materials, Products, and Manufacturing Processes—Current Trends and Practices

In this chapter, we review the current trends and practices in the integrated design of materials, products, and manufacturing processes. In Sect. 2.1, we discuss the field of integrated materials and product design. A detailed discussion on the current capabilities, the associated limitations, and the research opportunities that are worthy of investigation from the standpoint of material models, simulations, and databases; multiscale materials models and information linking; and materials design under uncertainty is carried out. In Sect. 2.2, the need for the research addressed in this monograph is established. In Sect. 2.3, a review of vertical and horizontal integration of models as defined by different authors is carried out and the definition of these terms for the monograph is established. In Sect. 2.4, verification and validation from the perspectives materials design and model-based realization of multiscale systems are discussed. In Sect. 2.5, remarks on the current status of materials design are presented based on the review carried out in this chapter.

2.1 The Emerging Field of Integrated Materials and Products Design

There has been a rapid reduction in the time needed to develop a product due to the advancements in design and manufacturing procedures further enabled by advances in computing and digital infrastructures in the last two to three decades. The concept of *integrated computational materials Engineering* (ICME) (Pollock et al. 2008; Horstemeyer 2012) is being widely discussed in the materials, manufacturing, and design communities as a tool to reduce the lead time in the development of a new product or component. The schematic structure of ICME is shown in Fig. 2.1; adopted from, Pollock et al. (2008). A major area of focus where ICME is intended to play a crucial role is in the process scale-up of mill products at lower costs (Cotton et al. 2012). Even though there is a fall in the time required to develop a product, the time required to develop and deploy a new material in a product, however, has remained

A. B. Nellippallil et al., *Architecting Robust Co-Design of Materials, Products, and Manufacturing Processes*, https://doi.org/10.1007/978-3-030-45324-4_2

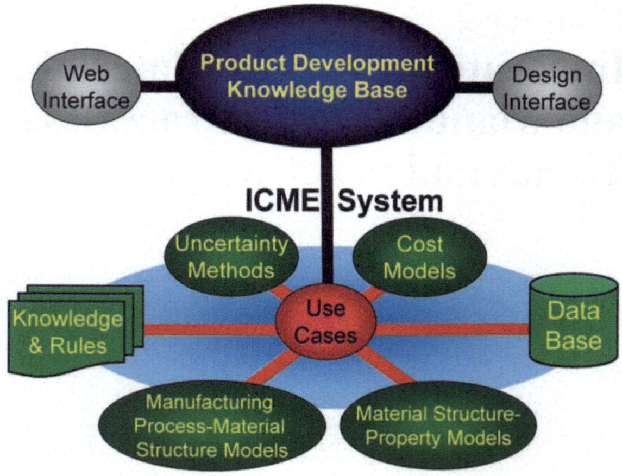

Fig. 2.1 Schematic structure of an ICME system that unifies materials information into a holistic system that is linked by means of a software integration tool to a designer knowledge base containing tools and models from other engineering disciplines (Pollock et al. 2008)

relatively constant. This has necessitated the need to ask this question—*Have we fully realized the potential of ICME at an industrial scale?*

The answer to this question can vary when asked to different communities as the definition and scope of ICME and the boundary that it holds itself has not been set properly as of now. However, there is a strong realization that to develop new materials for use in new products, there is a need to develop the ***capability of developing the product and material concurrently and in an integrated fashion taking into consideration the associated manufacturing processes***. This requires us to address the challenges involved in realizing ICME fully on an industrial scale. If we look at any major manufacturing route for a product, we see that there are a host of unit operations involved and the end properties and performances of the product are influenced and linked by the processing steps at the final as well as intermediate stages. An example of this is the manufacture of a steel product mix (rod, bar, sheet) that involves a series of unit operations like secondary steel making, continuous casting, reheating, rolling, annealing, and so on, which would further influence the products made using these mill products. Often, lack of understanding and traceability for the evolution of properties across these steps in a quantifiable manner is a major concern for the industry. In recent years there has been a tremendous increase in computational power available along with physics-based models that are "good" enough to define the processing–structure–property–performance relationships of a material. Also, there have been advances in comprehensive robust, multidisciplinary, system exploration methods that facilitate the development of multiscale materials that can achieve robust, multifunctional performances in varying product development environments. Enhancing these available models across multiscales and integration at various length scales (vertical integration) ensures the flow of information from one unit of operation to another, thereby ensuring the integration of these individual processes (horizontal integration) and thereby address the issues highlighted earlier. Accordingly, the designs of products and materials are not mutually exclusive and independent events but synergistic components of an integrated product, process, and materials design endeavor, as noted by McDowell et al. (2009). This necessitates a philosophical and cultural shift toward **inductive (inverse), goal-oriented** synthesis of products, their constituent materials, and their processing paths from a systems perspective. Several critical issues that need to be addressed to achieve this are identified in Chap. 1. In this chapter, we discuss some of the current capabilities, the associated limitations, and the research opportunities that are worthy of investigation:

- Models and simulations on different length and time scales for a different set of functions that a material and product system must satisfy;
- Tools, techniques, and systematic approaches for integrating and linking models and simulation tools across length and time scales, functional domains, and material hierarchies;
- Systems design tools, constructs, methods, and frameworks that support the integrated design of materials, products, and manufacturing processes;
- Techniques, constructs, and methods for characterizing and managing uncertainty in material and product variables, models, and their propagation across workflows.

In Chap. 1, we discussed the components of the systems-based design architecture as proposed in this monograph. In Table 2.1, we summarize the elements discussed in this chapter that form part of the architecture proposed. The interpretation of this table is shown in Chap. 1, see Table 1.6.

2.1.1 Material Models, Simulations, and Databases

In integrated materials and product design, greater emphasis is placed on developing and linking models, simulations, and databases for processing–structure–property relations at multiple length scales to satisfy specific performance requirements of products. To meet application-specific performance requirements, the hierarchical scheme proposed by Olson is essential with top-down, goal-oriented approaches to carry out the microstructure-mediated design of materials. An early vision for this was laid out at the 1998 National Science Foundation (NSF) workshop (McDowell and Story 1998). In 2008, the National Academy of Engineering (NAE) National Materials Advisory Study Group (Pollock et al. 2008) recognized ICME as a way to the integrated, concurrent design of materials and products.

The essential building blocks of simulation-supported materials design are: (i) thermodynamics, (ii) kinetics, and (iii) kinematics (McDowell and Olson 2008). Thermodynamics provides information on stable and meta-stable phases, characterization of structures and energies of interfaces, and driving forces (transition states) for rearrangement of structure due to thermally activated processes; and therefore it is considered as the fundamental building block of simulation-supported materials design (Olson 1997; McDowell et al. 2007; McDowell and Olson 2008). Preliminary design exploration of solutions to concurrent materials and product design problems is facilitated by it. Along with this, several other areas and research efforts play

Table 2.1 Summary of elements discussed and reviewed in this chapter that form part of the architecture

Elements	**What?**
Processing Elements	Elements of modern materials design like material selection, multiscale modeling, microstructure design using invertible linkages, optimal materials design, systems-based materials design, material informatics, uncertainty management, verification, and validation in materials design
Data Elements	Applications to materials design problems are discussed
Connecting Elements	Material selection approaches, multiscale modeling frameworks, material informatics and data-driven approaches, decision-based design constructs
Form	**How?**
Weights	Review of the foundations and current status of materials design approaches
Properties	Focus on current capabilities in materials design approaches, the associated limitations, and the research opportunities that are worthy of investigation
Relationship	Concepts discussed are used as theoretical foundations and as research philosophies to address the research gaps in the monograph
Rationale	**Why?**
Motivation	To identify critical issues associated with the current capabilities of materials design and address the challenges and gaps
Characteristics/ Beliefs	The design of products and materials are not mutually exclusive and independent events but synergistic components of an integrated product, process, and materials design
Interpretation	Understanding the efficacy and broader impacts

a key role in ICME. First-principles calculations support the exploration of multi-component systems for which empirical models are not yet established (Asta et al. 2001; Liu and Chen 2007). Data mining tools are also established and integrated with material modeling software to explore several candidate solutions (Rajan 2005; Liu et al. 2006). Multiscale modeling methods are developed to model multiple levels of materials hierarchy (Cuitino and Ortiz 1993; Liu et al. 2004; McDowell 2008; Horstemeyer 2012, 2018). Current research efforts in each of these modeling areas are rather extensive and highly domain-dependent (see McDowell et al. 2009).

Improving the fidelity and accuracy of models is the focus of material scientists in this domain. However, it is not practical to allocate time to improve all models and simulations. It is also difficult to prioritize which all models need to be selected for investing in improving fidelity and accuracy. In this regard, a very important player in the integrated model-based realization of materials and products is uncertainty. McDowell and co-authors observe that uncertainty is an area that is usually neglected and not given much attention (see McDowell et al. 2009). We review uncertainty from a materials design perspective in this chapter and discuss the management of uncertainty from a design perspective in Chap. 7 of the monograph.

2.1.2 *Multiscale Models and Information Linking in Materials Design*

For integrated materials and product design, it is essential to integrate and link models, databases, and simulation tools across length and time scales, functional domains, and material hierarchies. Materials design at multiple levels of hierarchy is a much broader activity than multiscale modeling (McDowell et al. 2009).

Using Fig. 2.2, Olson (2000) depicts the process–structure–property–performance hierarchy for the design of high strength steels for multiple objectives. ***Now is this a representation of the hierarchy of length scales in materials design?*** McDowell (2018) clarifies that the processing–structure–property–performance mapping shown in Fig. 2.2 should not be confused with the hierarchy of length scales in materials design. The hierarchy of length scales is shown in Fig. 2.3; adopted from Horstemeyer (2012).

To establish processing–structure–property–performance mapping at a single scale or across two scales will probably require models or sets of simulation experiments that span the entire hierarchy of length scales. The collection of all such models that serve the purpose of bridging information to higher-scale response is called hierarchical multiscale models. Several research labs and government organizations have created their own computational frameworks focused on multiscale

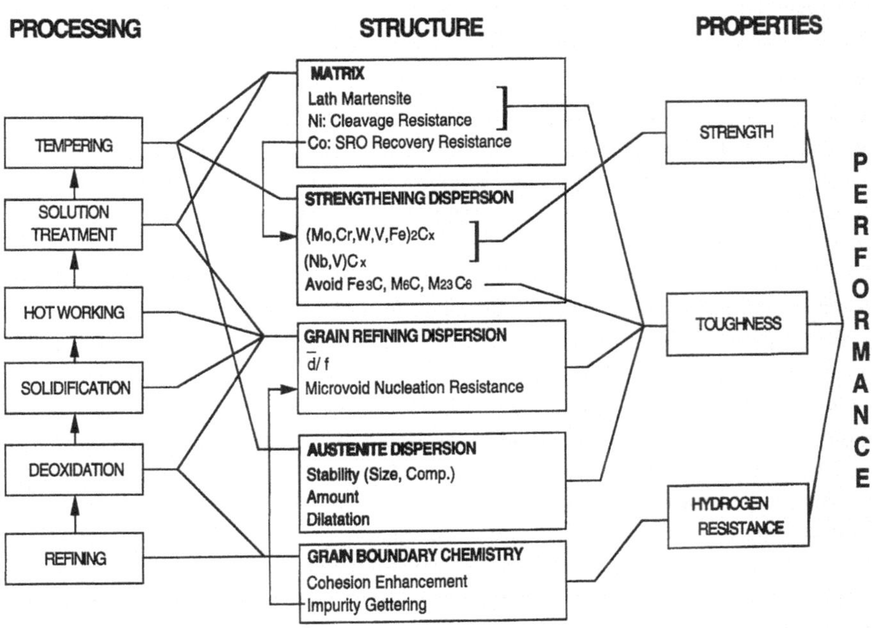

Fig. 2.2 Process–structure–property–performance hierarchy for design of high strength steels for multiple objectives of strength, toughness, and hydrogen resistance. From the SRG at Northwestern University (Olson 1997, 2000)

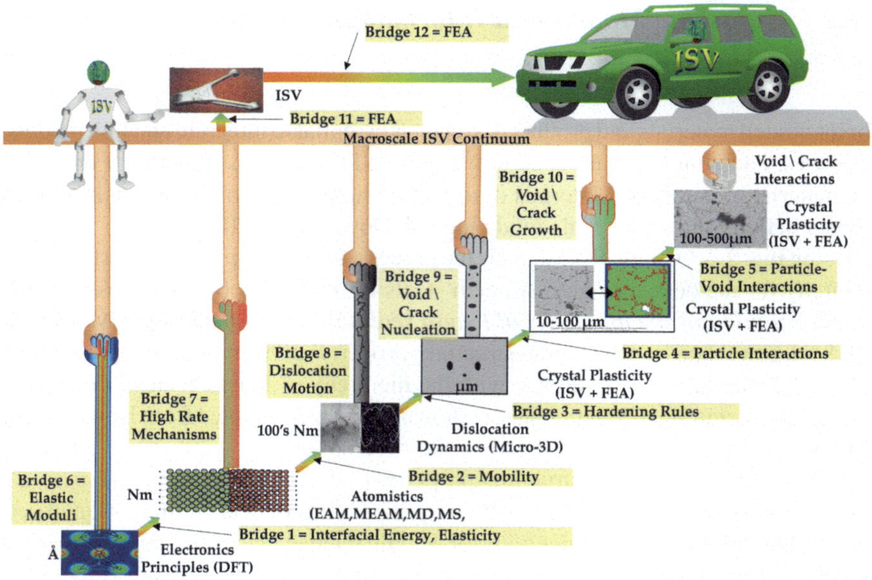

Fig. 2.3 Multiscale modeling example of a metal alloy used for design in an automotive component. The hierarchical methodology illustrates the different length scale analyses used and various bridges needed. (ISV = Internal State Variable, FEA = Finite Element Analysis, EAM = Embedded Atom Method, MEAM = Modified Embedded Atom Method, MD = Molecular Dynamics, MS = Molecular Statics, and DFT = Density Functional Theory) (Horstemeyer 2012)

materials design. Some examples of these include: The *Materials by Design*™ initiative by Olson (1997) for advanced steels using the THERMOCALC software system to integrate models from the quantum level to continuum level. The MatCASE (Material Computation and Simulation Environment) developed by Liu et al. (2004, 2006) integrates software from the atomic scale to continuum scale and evaluates finite element mechanical responses of designed microstructures. The MSU DMG ISV model developed by Horstemeyer and co-authors (Bammann et al. 1993; Horstemeyer et al. 2000) is another framework to capture the history effects of stresses, strains, and microstructures to model the sequential processing of material during manufacturing processes. The model is based on the ISV plasticity formulation of Bammann et al. (1993). The viscoplasticity model by Bammann (1990) coupled with the damage model by Horstemeyer (2012) has the capability to capture the nonlinear response of material during processing. Horstemeyer (2012) presents the hierarchical multiscale modeling carried out to relate the history effects of a material through its processing to the performance lifecycle. The framework addresses the issue of handling bridging between scales by using thermodynamically constrained internal state variables (ISVs) that are physically based on microstructure–property relations. The ISV modeling framework takes a top-down approach as the ISVs exist at macroscale but reach down to various subscales to receive pertinent information. Using Fig. 2.3, Horstemeyer depicts the hierarchical multiscale modeling methodology illustrating the different bridges and analyses required to capture the pertinent plasticity, damage, and failure aspects of metal alloys for use in the design of an automotive component.

The frameworks reviewed till now are aimed at "hard" computing by the use of physics-based models that are linked at multiple scales. There is also an increasing focus on "soft" computing aspects of materials design. Statistical or heuristic-based relationships between materials design parameters for different applications are the major focus in the soft computing research domain. The *Material Informatics* approach by Rajan and co-authors (Rajan 2005, 2013; Liu et al. 2006) has paved the way for objective (data-driven) formulation of surrogate PSP linkages. Kalidindi et al. (2019) propose Python-based materials knowledge systems (PyMKS)—a Python-based open-source materials data analytics framework to create high-fidelity, reduced-order, process–structure–property linkages for a broad range of material systems. PyMKS framework has a rich hierarchy of internal material structures that span multiple length scales. These linkages exhibit a remarkable combination of high accuracy and low computational cost through advances in material data sciences. The application of the same for the design of new alloys and catalysts is demonstrated (see Kalidindi et al. 2019). These frameworks can be categorized as novel data science approaches for solving multiscale materials science problems. Definitely, increasing a materials designer's insight and perspective can happen through these soft computing techniques, but they still have to be paired with hard computing techniques for the true realization of new materials and products.

What are the difficulties associated with these hard and soft computing techniques for new material and product realization from a design perspective?

- They are customized for specific applications and are not transferrable;
- They are highly domain and platform-dependent;
- They facilitate the detail design of materials and products and is not suitable for the early stages of design;
- There are issues of uncertainty that need to be addressed;
- Developing such models/frameworks and integration of models consumes a lot of time and resources; and
- These techniques demand problem/platform-specific knowledge and skills which are not generally the focus/expertise of design engineers.

2.1.3 Materials Design Under Uncertainty

Uncertainty can be classified as stochastic (aleatory) and epistemic in materials design. Improving the measurements and/or model formulation and/or increasing the accuracy are the ways to diminish epistemic uncertainty. Aleatory uncertainty, however, is inherent in the physical system and can only be quantified in a statistical sense. In order to realize the integrated model-based realization of materials, products, and processes, it is very essential to characterize and manage both these types of uncertainty.

Engineering design communities have focused over the years on simulation-based design in the presence of uncertainty, and a lot of research works are there to this credit. However, this body of knowledge has not found many applications in materials design under uncertainty due to certain significant challenges. We observe that the major reason for this is the lack of expertise that exists for researchers in engineering design domains on materials design. In order to characterize and manage the uncertainty in materials design problems, there should be close collaboration and transfer of knowledge between materials engineers/scientists and design engineers so that the nuances and sources of uncertainty related to designing materials can be clearly understood. Several challenges associated with materials design under uncertainty are highlighted by Allen et al. (2006). Extending the classification by Isukapalli et al. (1998), the types of uncertainty in simulation-based integrated design of material, product, and processes can be classified as Choi et al. (2005), McDowell et al. (2009): (i) *Natural Uncertainty*, ii) *Model Parameter Uncertainty* (MPU), iii) *Model Structure Uncertainty* (MSU), and iv) *Propagated Uncertainty* (PU). Techniques for addressing model uncertainty are addressed by Du and Chen (2000, 2002). Metrics known as design capability indices (DCIs) for managing parameter uncertainty were developed by Chen et al. (1999). These metrics were later expanded to manage model uncertainty and its propagation by Choi et al. (2005, 2008a), and named as error margin indices (EMIs). Gu et al. (2000, 2006) studied the propagation of uncertainty through a series of models spanning across scales and/or disciplines. Studies on collaborative decision-making under uncertainty by coupling robust design methods and game theory protocols were carried out by Chen and Lewis (1999). This was later extended by Xiao (2003) by using design capability indices and game theory protocols to facilitate flexible, robust, and interactive decision-making among multiple, distributed engineers. Kalsi et al. (2001) carry out collaborative, multidisciplinary systems design by treating shared variables as noise factors in their study. Chang et al. (1994), Chang and Ward (1995) in their work support designers to carry out robust collaborative decisions by considering the coupled parameters as noise and using robust design Types I and II to manage the effects of coupling.

Even though there is research focused on addressing the issue of uncertainty, there are several challenges associated with the successful application of this knowledge in the simulation-based design of materials under uncertainty. One major challenge is the propagation of uncertainty as information is shared from one model to another. This is amplified by the fact that the models in materials design are already sensitive to uncertainty due to the underlying physics-based assumptions, the lack of proper input data, and the approximations made during model development because of computational cost and time. Materials design due to all these reasons should be viewed as human decision-making and collaborative activity with the assistance of computers. We define the capability to carry out this collaborative activity between different stakeholders in an integrated fashion as co-design (see Chaps. 1 and 8). There are some recent efforts in this direction as briefly discussed in Chap. 1. Modern data science methods for materials and microstructure informatics along with multiscale modeling techniques are being developed to provide decision support and address the issue of uncertainty in hierarchical materials design (Panchal et al. 2013; Kalidindi 2015; McDowell and Kalidindi 2016; McDowell and LeSar 2016). Panchal et al. (2013) address some of the key computational modeling issues in integrated computational materials engineering and address the research efforts in mitigating and managing uncertainty. One important message delivered in their work is that ***materials design in its current form with the advent of ICME is not just an assemblage of tools as such tools do not have natural interfaces to material structure nor are they framed in a way that quantifies sources of uncertainty and manages uncertainty in representing physical phenomena to support decision-based design*** (Panchal et al. 2013). Seepersad and co-authors (Shahan and Seepersad 2012; Matthews et al. 2016) consider Bayesian network classifiers to design materials with hierarchy and to treat uncertainty propagation in multilevel material design. The approach incorporates Bayesian network classifiers (BNC) for mapping design spaces at each level and flexibility metrics for intelligently narrowing the design space as the design process progresses. The approach supports top-down design exploration of material hierarchy under uncertainty to make design decisions. Mahadevan and co-authors (Li and Mahadevan 2016; Mullins and Mahadevan 2016) address the issue of uncertainty integration across multiple levels and the role of calibration, validation, and relevance in multilevel uncertainty integration for hierarchical material design problems. Systematic approaches supporting optimal materials design exploration in an efficient manner along with the consideration of the uncertainty quantification and management are limited. One approach that has gained wide attention in the last several years is the Bayesian optimization (BO) approach. We describe the approach in the context of materials design next.

Bayesian Optimization for Materials Design Under Uncertainty

Bayesian optimization (BO) is an iterative global optimization approach that is used when an exact functional form for a function y to be optimized (explored and exploited) is not available. BO is performed by maintaining a probabilistic belief about y and formulating an acquisition function to determine where the function needs to be evaluated next. BO approach involves developing: (i) the surrogate model $\hat{y}$, for an actual function, say y, and (ii) an acquisition function computed from the surrogate $\hat{y}$ that will be used for guiding the selection of the next evaluation point for exploration and exploitation of the surrogate model. Even though BO is addressed from the context of global optimization of expensive surrogate models, we can relate this to materials design as materials design exploration typically involves optimizing (exploring and exploiting) such expensive black-box functions to meet certain material property/performance requirements. One of the most widely used surrogate models for BO is the Gaussian process (GP) meta-models (see Jones et al. 1998; Simpson et al. 1998; Hollingsworth and Mavris 2003; Rasmussen 2003; Rasmussen and Williams 2006; Forrester et al. 2008; Kleijnen 2017). GP models have recently gained popularity as a supervised machine learning model. They are known to capture highly nonlinear responses over a design space in a computationally efficient manner while also providing information on the uncertainty measure of the prediction thereby supporting analysis, design exploration, and uncertainty quantification (see Yan et al. 2018). GP models require fewer data compared to other machine learning methods, especially when GP models are developed by pairing with optimal design of experiments, such as the optimal Latin hypercube sampling (OLHS) (Jin et al. 2005). Optimal experimental designs are known for their good "space-filling" and projective properties. The OLHS uses a fast global search algorithm, known as enhanced stochastic evolutionary (ESE) algorithm, to efficiently maximize the sample uniformity over the input design space, while at the same time, minimizing the number of samples necessary (see Jin et al. 2005). Gaussian process models thus developed are characterized by its mean and the covariance matrix as $\mathrm{y}(\mathbf{X}) \sim \mathrm{GP}(m(\mathbf{X}), \mathbf{\Sigma})$. They are developed as the sum of a linear regression model and a stochastic process (see Forrester et al. 2008):

$$\mathrm{y}(\mathbf{X}) = \sum_{j=1}^{p} \beta_j f_j(\mathbf{X}) + Z(\mathbf{X}) = \mathbf{f}^T(\mathbf{X})\boldsymbol{\beta} + Z(\mathbf{X}), \tag{2.1}$$

where $f_j(\mathbf{X})$ are the basis functions and β_j are the unknown coefficients (Rasmussen and Williams 2006). The error in the model due to lack of data is captured using the stochastic part, $Z(\mathbf{X})$. It follows a Gaussian distribution with zero-mean and a covariance function $c(\mathbf{x}, \mathbf{x}') = \sigma^2(R(\mathbf{x}, \mathbf{x}'))$, where σ^2 is the unknown variance and $R(\mathbf{x}, \mathbf{x}')$ is a spatial correlation function (SCF). The Gaussian spatial correlation function is defined as Forrester et al. (2008), Yan et al. (2018), Kleijnen (2020):

$$R_{ij} = R(\mathrm{x}_i, \mathrm{x}_j) = \exp\left\{-\sum_{k=1}^{d} \theta_k (x_{ik} - x_{jk})^2\right\}, i, j \in \{1, 2, \ldots, n\}, \tag{2.2}$$

where $\boldsymbol{\theta} = [\theta_1, \theta_2, \ldots, \theta_d]$ are "roughness" hyperparameters. This gives $m(\mathbf{X}) = \mathbf{f}^T(\mathbf{X})\boldsymbol{\beta}$ and $\sum_{ij} = c_{ij} = \sigma^2 R_{ij}$. To fit the GP surrogate model, maximum likelihood estimation (MLE) is used to find the optimal values of the hyperparameters $\boldsymbol{\beta}, \boldsymbol{\theta}, \sigma^2$. Using the closed-form equation for the best linear unbiased predictor (BLUP) of Eq. (2.1), the mean prediction at a new point, $\mathbf{x}_0$, is made as (see Yan et al. 2018):

$$\hat{y}(\mathbf{x}_0) = 1 \cdot \hat{\beta} + \mathbf{r}^T(\mathbf{x}_0)\mathbf{R}^{-1}\left(\mathbf{y} - 1_{n\times 1}\hat{\beta}\right) \tag{2.3}$$

where $\mathbf{r}(\mathbf{x}_0)$ is a vector of size $n \times 1$ such that $r_i(\mathbf{x}_0) = R(\mathrm{x}_0, \mathrm{x}_i), i = 1, \ldots, n$. The uncertainty of each prediction point is quantified using the mean squared error (MSE) (Yan et al. 2018):

$$\mathrm{MSE}[\hat{y}(\mathbf{x}_0)] = \hat{s}^2(\mathbf{x}_0) = \hat{\sigma}^2\left[r(\mathbf{x}_0, \mathbf{x}_0) - \mathbf{r}^T(\mathbf{x}_0)\mathbf{R}^{-1}\mathbf{r}(\mathbf{x}_0) + \mathbf{W}^T\left(1_{n\times 1}^T \mathbf{R}^{-1} 1_{n\times 1}\right)^{-1}\mathbf{W}\right] \tag{2.4}$$

where $\mathbf{W} = 1 - 1_{n\times 1}\mathbf{R}^{-1}\mathbf{r}(\mathbf{x}_0)$. The prediction interval of mean prediction will be obtained as:

$$\hat{y}(\mathbf{x}_0) \pm z\sqrt{\mathrm{MSE}\big[\hat{y}(\mathbf{x}_0)\big]} \tag{2.5}$$

Once the surrogate model is developed, an acquisition function computed from the surrogate $\hat{y}$ is used in BO for guiding the selection of the next evaluation point for exploration and exploitation of the surrogate model. Some major acquisition functions used in BO are statistical lower bound, probability of improvement, and expected improvement (see Forrester et al. 2008). Among these, expected improvement has gained much popularity in BO for materials and product design applications (see Xue et al. 2016; Kristensen et al. 2019). Expected improvement (EI) defines the improvement of a new design point $\mathbf{x}$ as (Forrester et al. 2008):

$$I(\mathbf{x}) = \max\big\{y_{\min} - \hat{y}(\mathbf{x}), 0\big\} \tag{2.6}$$

where $y_{\min}$ is the current best point (CBP). The expected improvement is calculated as:

$$E[I(\mathbf{x})] = \begin{cases} \left(y_{\min} - \hat{y}(\mathbf{x})\right)\phi\left(\frac{y_{\min}-\hat{y}(\mathbf{x})}{\hat{s}(\mathbf{x})}\right) + s\emptyset\left(\frac{y_{\min}-\hat{y}(\mathbf{x})}{\hat{s}(\mathbf{x})}\right) & if s > 0 \\ 0 & if s = 0 \end{cases} \tag{2.7}$$

where the mean prediction $\hat{y}(\mathbf{x})$ is calculated based on Eq. (2.3) and the square root of the predicted variance $\hat{s}(\mathbf{x})$ is calculated as the square root of Eq. 2.4. $\phi(\cdot)$ and $\emptyset(\cdot)$ are the standard normal cumulative distribution function and standard normal probability density function, respectively. In Fig. 2.4, we capture a BO approach applicable to materials design problems. The point with maximum EI is the point that needs to be further evaluated. The value of EI will increase with either the increase in $y_{\min} - \hat{y}(\mathbf{x})$ or $\hat{s}(\mathbf{x})$. Thus, maximizing EI achieves a tradeoff between exploitation (evaluating at points with low mean) and exploration (evaluating at points with high uncertainty). Based on the calculation of EI, the point with maximum EI is identified and added to the design. The new design point thus identified is added to the original design and simulations are carried out for this point. The obtained simulation response is added to the expanded design. Once the expanded design training dataset with the new optimal points is generated, the Gaussian process surrogate model is rebuild. The newly built Gaussian process model is further evaluated, and the expected improvement is again calculated. The new points with maximum EI are identified and the process is repeated. The iteration is typically continued until a stopping criterion is met, such as checking if the expected improvement is less than 1% of the current best point (see Jones et al. 1998).

Arróyave and McDowell (2019) report that the key advantage of BO-based approach to materials design discovery compared to other traditional approaches is the optimal allocation of resources. However, a key observation made by them

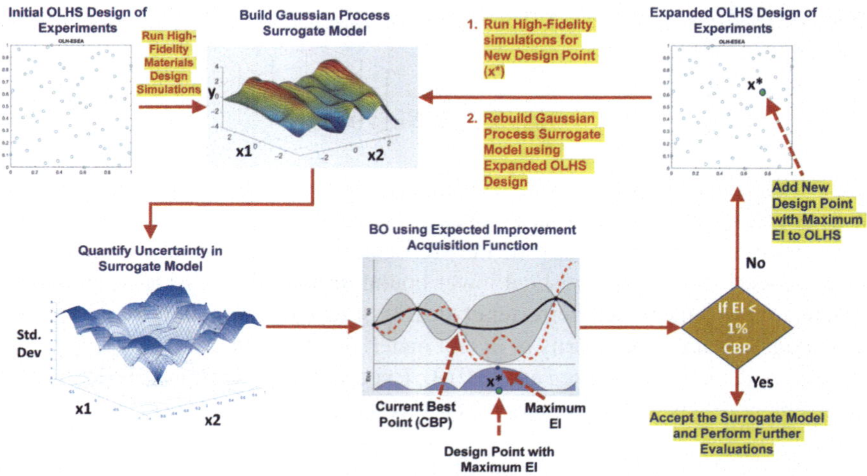

Fig. 2.4 Bayesian optimization approach for design exploration and uncertainty management. The process starts with an initial DOE, such as OLHS. The high-fidelity materials simulations are run for the OLHS design, and GP models are built. The uncertainty in the GP model is quantified. BO with EI acquisition function is used to optimize the objective—maximize EI. The points with maximum EI are identified to be run and then the GP model is re-build. This completes one iteration. The process is repeated until the EI is less than 1% of the CBP or any other stopping criterion is met

here is the fact that most of the Bayesian optimization applications in materials design are focused on single-objective problems rather than multiobjective. We also observe that the whole iterative approach of design space exploration and exploitation will be tedious when dealing with complex multiscale, multiobjective materials design problems. Given these drawbacks of BO in materials design applications, we recognize three major capabilities needed in computational approaches/frameworks that address materials design exploration as required by ICME. They are: (i) the capability to address multiple conflicting objectives, (ii) the capability to systematically and efficiently explore the multiobjective solution space and identify robust satisficing solutions that are relatively insensitive to the uncertainty present, and (iii) the capability to partition the multiscale, multistage, and multiobjective materials design problems into subproblems and represent them via an integrated information workflow. Some recent efforts demonstrating the successful application of GP and BO-based approaches for materials and product design problems include: Solomou et al. (2018), Yan et al. (2018), Kristensen et al. (2019).

Even though there are several such recent efforts to address the issue of uncertainty, McDowell (2018) observes that quantifying uncertainty in schemes for linking models at different length and time scale is still an immature field and formal mathematical approaches for doing this are largely undeveloped. The recommendation, therefore, made for simulation-assisted materials design is to focus on understanding the sensitivity of material properties to material microstructure and to capture dominant mechanisms and transitions that affect material responses or properties instead of focusing on accurately predict mean properties at higher scales. Even though techniques are available for sensitivity analysis and uncertainty quantification, there still need to be research efforts to address materials design problems and infuse these knowledges into computational mechanics and materials science domains. In this monograph, fundamentals of robust design and uncertainty management are discussed in Chaps. 3 and 7 with robust concept exploration techniques for materials applications in Chap. 7.

2.2 The Need for the Research Addressed in This Monograph

Materials design is viewed as fundamentally ***a goal-oriented*** synthesis activity, where requirements or goals for a product (or material or component or part or assembly or system) in terms of performance or properties are translated into suitable material microstructures and corresponding processing paths. A systems perspective is therefore needed for linking the information associated with multiscale and multilevel models, tools, and databases that are developed. It is equally important to develop design methods that allow systematic and strategic coordination of information and human decision-making for the design of the material structure, processing paths, and parent products. Design methods thus developed should support:

- Co-ordination of information and human decision-making across material and product hierarchies/levels;
- Bridging the gap between materials design and product (system) design;
- Goal-oriented design exploration, where the designer starts with the end goals that need to be achieved for product performance;
- Systematic solution space exploration and design tradeoffs from a large set of possible solutions and visualization of solution regions of interest;
- Management of uncertainty associated with materials and products;
- Management of complexity via reduced-order material, process, and product models and simulations;
- Propagation of robust solution regions of interest across the material process chain;
- Support decision-making and distributed solution space exploration for distributed decision-makers;
- Support the inverse design exploration thereby achieving the integrated microstructure-mediated design of materials and products; and
- Domain-independent paradigms/frameworks for the design of any complex material-product system.

As discussed in previous sections, a lot of researchers over the years have looked at materials design from different perspectives. Most of these are focused on the detailed design of material behavior and not on early-stage design exploration. There are only very few who have addressed the need for developing domain-independent design methods, tools, and frameworks that support the systematic and comprehensive early-stage design of broad classes of materials and products, as noted by McDowell et al. (2009). A systems approach for materials design by taking into account the processing–structure–property–performance relationship was advocated by Olson in his seminal work on materials-by-design (Olson 1997, 2000). This was later reiterated by many others in their work and forms one of the underlying philosophies for the emerging ICME domain. However, strategies to explore the design space and coordinate distributed human decision-making remains an open question in Olson's work. Olson's lead was followed by Subbarayan and Raj (1999) to design system and subsystems sharing variables as links between the levels of hierarchy. The design of a tungsten filament is used as an example to illustrate the approach. To formalize the relationships between the subsystems in the tungsten filament example, Lu and Deng (2004) use variable dependency graphs and thereby adding to the systems-based approach. Gall and Horstemeyer (2000) picks up Olson's systems philosophy and illustrate the design of a cast component using a multiscale modeling approach by establishing relationships at the component level and the microstructural level.

Adams et al. (2013) present a framework that utilizes highly efficient spectral representations to arrive at invertible linkages between material structure, its properties, and the processing paths used to alter the material structure. The authors address how the material microstructure can be tailored to meet stringent properties and performance requirements of complex components and systems. Materials knowledge systems approach by Kalidindi et al. (2010, 2011) showcases advances in rapid inverse design to estimate local responses. However, these approaches are focused on problems where the structure–property relations can be inverted in some manner. For problems involving nonlinear relationships and path dependencies across length and time scales and nonequilibrium microstructural evolution, the application of these ideas remains an open question. Other efforts in systems-based microstructure-mediated design include the design of aluminum alloy microstructures for targeted properties using genetic algorithms, by Kulkarni et al. (2004) and topology and microstructure design of materials and products for specified properties by Sigmund (1994, 1995), Sigmund and Torquato (1997), Hyun and Torquato (2002). Although these systems-based approaches are the focus in materials design communities, they are still in the infancy stage.

Thus, we recognize in this monograph that,

Systematic methods for materials design are needed that support a human designer in exploiting the power of computational materials and product models to carry out simulation-supported, integrated, decision-based, inverse design exploration of material microstructures and processing paths to satisfy specific product- and process-level properties and performance requirements.

Systems-based design architecture (complete with frameworks, mathematical constructs, design methods, and computational tools) is presented in Chap. 4 of this monograph addressing these requirements.

2.3 Vertical and Horizontal Integrations and Information Flow in Materials Design

The development of computational tools has led to systems-based design methods that are simulation-based to realize complex products with less lesser development cycle times and improved quality. The simulation models available with the revolution in computational techniques are often "good" enough to predict the behavior of systems at multiple scales and the requirement at this stage is to systematically use these simulation models to design the behavior of the overall system by integrating the information that is obtained from each of these individual simulation models. Complex systems are mostly characterized by the hierarchical coupling between systems, subsystems, and components. The hierarchical systems design is meant to simulate the performance of the system over multiple levels of the hierarchically partitioned system to reduce the time needed to achieve that system-level performance. The hierarchical system is differentiated from the multiscale system in this study. By multiscale system we mean the systems simulated at multiple lengths and time scale (see Fig. 2.5). In Fig. 2.5, we see that the end performance of the automobile depends on the functionality of its individual components, like the gearbox assembly and the gear parts in it. The functionality of these parts, in turn, depends on the corresponding material properties. These material properties depend on the behavior of the material at different scales ranging from the quantum scale to the microstructure of the material. The interactions at the quantum scale affect the microstructure of the material, which in turn affects the continuum behavior, thereby affecting the mechanical properties of the material and the product. The variation of time scale may be in the order of few femtoseconds in the quantum scale and can go up to months or years at the system level. The need for simulation models at each of these levels is very essential due to this mapping and since these scales are linked, there needs to be a coupling between the models. The coupling of these models to allow information flow has been categorized as "vertical integration" and "horizontal integration". These integrations will allow us to carry out a goal-oriented, inverse design exploration starting from the systems level (see Fig. 2.5). This requires the bridging of the two domains, namely multiscale modeling and systems-based multilevel design, as illustrated in Fig. 2.5.

Different researchers have defined these two types of integrations from different perspectives. In this section, we review the definitions by Panchal (2005), Shukla et al. (2015), and Horstemeyer and Wang (2003).

Panchal (2005) describes the coupling in hierarchical system as primarily between physical components and the multiscale coupling as that which exists within physical

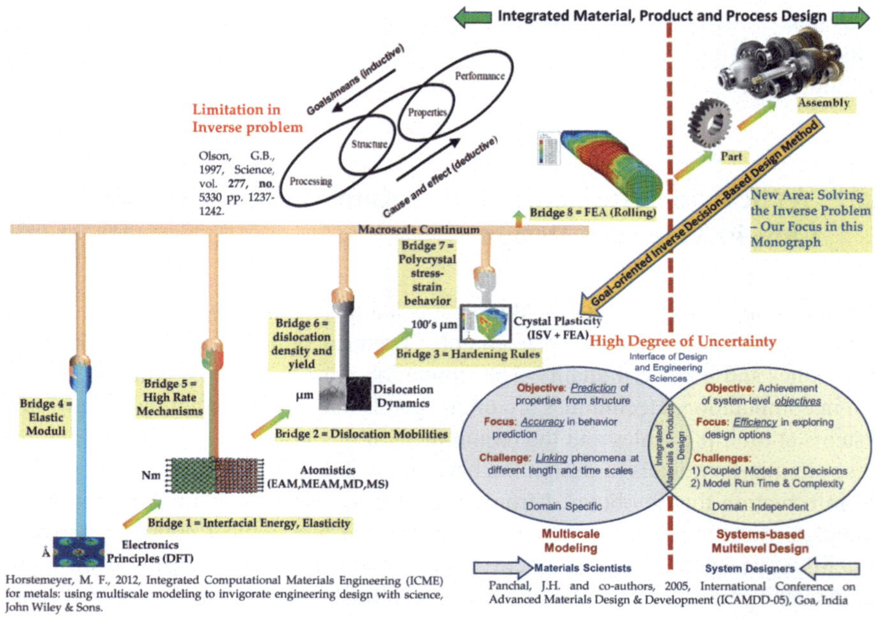

Fig. 2.5 A multiscale automobile system depiction with information flow across scales. The new area of solving the inverse problem leading to integrated materials and product design requires bridging of the two domains, namely multiscale modeling and systems-based multilevel design

phenomena at different scales for the same component. In hierarchical systems, coupling exits between systems, subsystems, and components in the same scale known as horizontal coupling, while in multiscale systems an additional element of coupling exists over the length and time scale to designing the individual components known as vertical coupling (see Fig. 2.6; Panchal 2005). Hence the requirement here

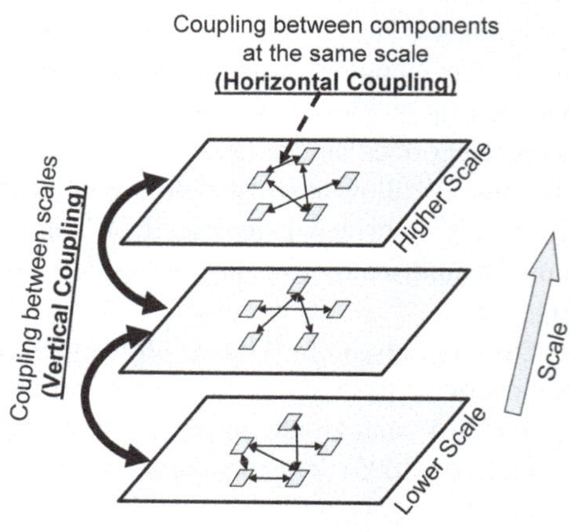

Fig. 2.6 Horizontal and vertical couplings in multiscale systems (Panchal 2005)

is to establish vertical coupling along with horizontal coupling in multiscale systems to effectively establish the integrated information flow from the physical phenomena at different scales in order to make decisions.

An example of this can be visualized from the hot rolling problem perspective. The main aim in hot rolling is to breakdown the columnar grain structure to equiaxed grains by rolling the hot semi-product (slab, billet, or bloom) through a set of rollers. The formation of new grains depends on the rolling load applied, the percentage reduction that is targeted, the rolling speed, temperature, and so on at the macro-level. At the micro-level, recrystallization phenomena (dynamic, meta-dynamic, and static) will take place resulting in new grain formation. A finite element model at the macro-level can give significant information regarding all this. The finite element model thus needs to integrate all the rolling information at the macro-level to simulate the process accurately. The analysis results from the finite element model, say the rolling load, the temperature profile along the workpiece or the strain developed, and so on are used by a microstructural evolution model at the micro-level to predict the recrystallization and the evolution of grain size, which in turn may impact macroscopic behavior. This requires integration at the micro-level between recrystallization models for dynamic, static recrystallizations and the grain growth model. There needs to be a constant back and forth information flow between models at these two scales while carrying out the simulation. Thus, vertical integration at the two different scales and horizontal integration within models at the same scale help in simulating the process effectively. Some of the challenges associated with multiscale modeling are (leveraged from Panchal's dissertation 2005 and Sinha's thesis 2011):

- balancing the prediction accuracy with the computational cost;
- modeling physical phenomena and interaction between scales;
- achieving collaborated decision-making; achieving collaborative computational infrastructure;
- managing uncertainty and its propagation; and
- framing and solving the inverse problem (Sinha 2011).

Shukla et al. (2015) have defined vertical and horizontal integration based on their study of the steelmaking process chain. The manufacture of a steel product mix (rod, bar, or sheet) involves a series of unit operations, like continuous casting, reheating, rolling, annealing and so on. The flow of information from one unit of operation to another is very essential to carry out an integrated study of each of these unit operations (Shukla et al. 2015; Tennyson et al. 2015). This has been termed as horizontal integration—the integration of different unit operations. To carry out horizontal integration, there needs to be information in far greater detail from each of these unit operations. Thus there should be the modeling of the important phenomenon that occurs during a particular unit operation which is at different length scales with a very deep insight (Tennyson et al. 2015). This has been termed as vertical integration (see Fig. 2.7). Thus, vertical integration of a unit operation helps in adding information to the unit operation by integrating the different phenomenon and also results in information that can be passed to other unit operations, thus helping in achieving horizontal integration. An example of this type of integration

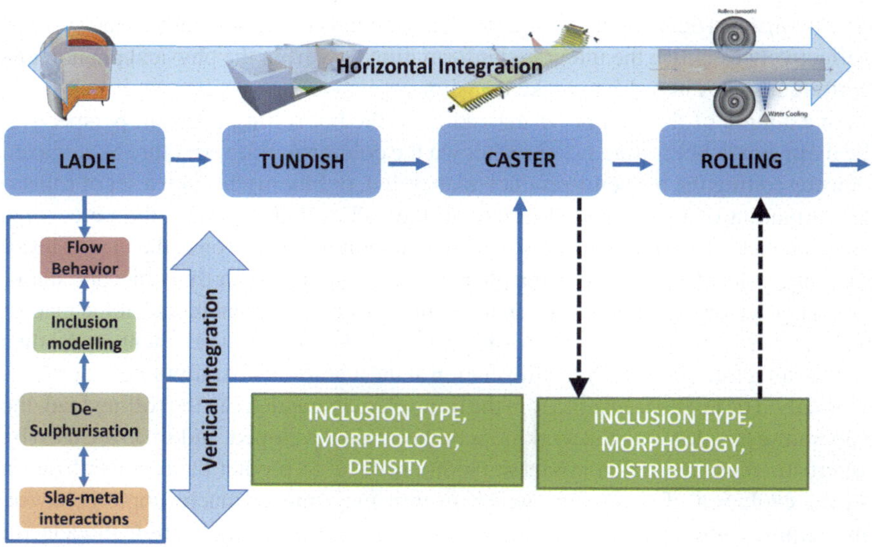

Fig. 2.7 Vertical and horizontal integration as defined by Shukla et al. (2015), Tennyson et al. (2015)

is illustrated in Fig. 2.7. Inclusions are generally non-metallic compounds that are formed during the refining and pouring stages in the upstream process (ladle-tundish) stage of steelmaking. They are usually formed due to chemical reactions or due to the presence of impurities in the melt. The total oxygen content present is a measure of the inclusions present. A simple chemistry model is used to control the ladle refining and predict the operating setpoints for the ladle that meets the required oxygen content. However, during an integrated study, the size, shape, and morphology of the inclusions need to be known and are to be modeled and this information has to pass to the subsequent stages like rolling where it is important to know the size and morphology of the inclusions if present.

Similarly, desulfurization is another important phenomenon during ladle refining. Argon purging at a high rate is essential for sulfur removal. However, argon purging at high rate results in inclusions through oxygen pick up. Hence it is essential to model the desulfurization phenomena and integrating it with the inclusion model before information flow to rolling operation. Thus, a detailed vertical integration of unit operation is essential for information flow to subsequent operations and also to understand and model the corresponding unit operation in the best way.

Horstemeyer and Wang (2003) define ICME as the bridging of information from two or more experimentally validated models or simulation codes in which structure–property information passes from one code to another. They describe "horizontal ICME" as the integration in which simulation codes connect the sequential material processes with their structure which is at multiscale to their mechanical properties that are used in the performance lifecycle evaluation (Horstemeyer and Wang 2003). The

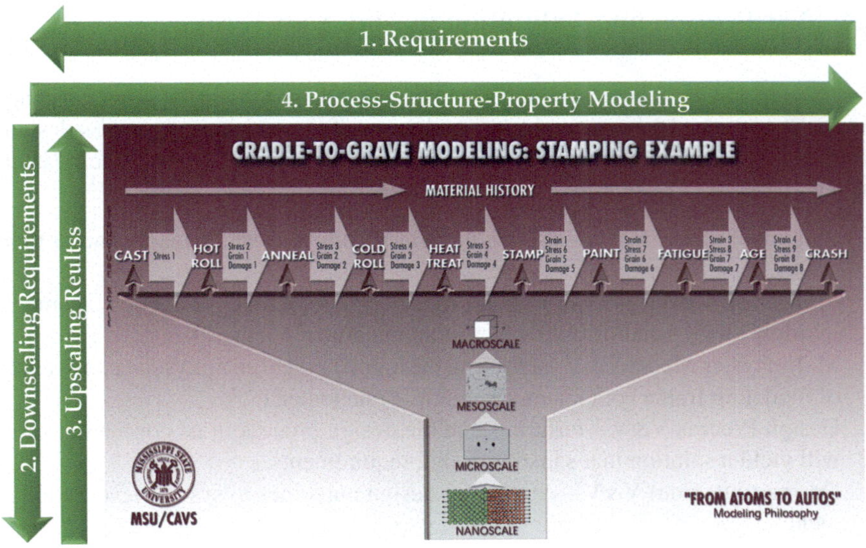

Fig. 2.8 Vertical and horizontal integration as depicted by Horstemeyer (2012)

"vertical ICME" is described as the integration in which simulation codes connect the multiple length scales (see Fig. 2.8). The integration of both "horizontal ICME" and "vertical ICME" is described as the "hybrid ICME".

On analyzing the three definitions of vertical and horizontal integration, we understand that vertical integration has been defined as the integration of models at multiple length and time scales by all the authors. There is a slight variation in the definition of horizontal integration by these authors. In this monograph, we accept the definition of vertical and horizontal integration by Tennyson et al. (2015) as it more closely relates to the problem at hand and the modeling approach followed in this monograph—where we focus on the flow of information across models and systematically establishing linkages via workflows and provide design decision support. The focus in this monograph is not on multiscale modeling. Thus, to carry out design space exploration across the material processing–structure–property–performance spaces, there should be the flow of information via simulation models integrated across multiple scales and across multiple manufacturing processes—defined as the vertical and horizontal integration of models. We define *vertical integration* as the integration of models and simulations of the different phenomena that occur at multiple length scales for a specific manufacturing process to generate information that can be passed to other manufacturing processes that follow. We define *horizontal integration* as the integration of different such manufacturing processes using simulation models ensuring proper flow of the information generated through vertical integration at each manufacturing process, thereby establishing the processing–structure–property–performance route to realize an end product. This will be addressed in detail in Chap. 6 of this monograph using the hot rolling process chain problem.

2.4 Verification and Validation in Materials Design

Verification and validation were addressed from the perspective of monograph chapters in Chap. 1. In the context of integrated materials, products, and process design, verification and validation (V&V) consist of the following activities (Panchal et al. 2013):

1. Individual Model V&V—a single model focusing on a single length and/or time scales.
2. Multiscale Model V&V—a single model or coupled set of models spanning multiple lengths and/or time scales in an integrated manner.
3. Multiphysics Model V&V—ensuring the mathematical and physical consistency of modeling framework spanning multiple phenomena.
4. Design Process V&V—ensuring that the design process in its configured form will yield a solution that satisfies design requirements.
5. Design (outcome) V&V—comparing design outcomes to system-level requirements.

2.4.1 Individual Model Verification and Validation

Model verification and validation have received significant attention in the past years due to the advent of simulation-based design technologies. The following tasks are associated with model verification and validation (Sargent 2009; Panchal et al. 2013); the process of V&V is illustrated in Fig. 2.9:

- *Conceptual model verification and validation*: The process of validating whether the theories and assumptions underlying a model and its submodels are correct

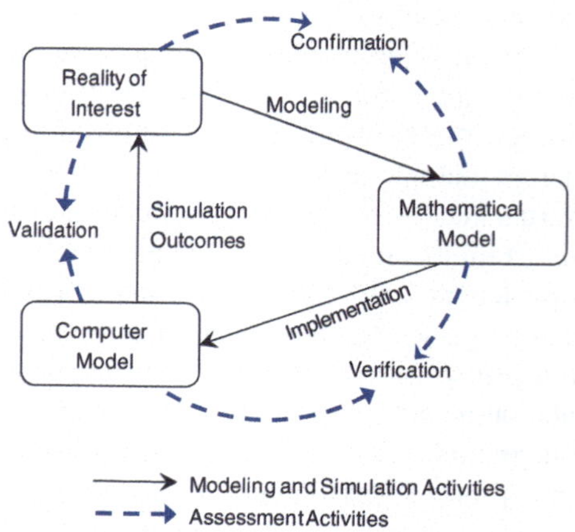

Fig. 2.9 Model verification and validation process (Sargent 2009; Panchal et al. 2013). Republished with permission from Computer Aided Design, 45(1):25

and that the representation of the system including models and submodels are correct and reasonable for the intended study.

- *Model verification*: The process of assuring that the computer model is "good enough" in terms of accuracy of the representation of a conceptual model.
- *Operational validation*: The process of determining whether the computerized model is sufficiently accurate for the needs of the simulation study.
- *Data validation*: The process of checking the accuracy and consistency of the numerical data used to support the models in the simulation study.

2.4.2 Multiscale Model Verification and Validation

Multiscale model V&V is very important because valid individual models for specific length and time scales won't necessarily result in valid multiscale models across scales. The following tasks are involved with multiscale model V&V:

- *Compatibility validity*: Compatibility validity is the process of determining whether the input ranges of an upper-level model is consistent with the domain of outputs of a lower-level model. This ensures whether the output domain of the lower-level model is a subset of the valid input domain of the upper-level model.
- *Uncertainty propagation check*: The goal here is to check that the effects of uncertainty at lower length scales do not amplify beyond the desired uncertainty bounds or limits set for which the design decisions are to be made. This can be viewed both from bottom-up and top-down perspectives. From a top-down perspective, the uncertainty limits allowable for a system are used to determine the allowable uncertainty limits for lower scales and thereby manage the propagation across a chain of models.

2.4.3 Design Process Verification and Validation

The goal in design process V&V is to ensure that the design process will yield design solutions worthy of investigation satisfying the design requirements. In the simulation-based design of complex systems, design processes represent the manner in which design decision networks and simulation models are configured to achieve the design task. One approach to verify and validate a design process is with the help of the verification and validation square framework introduced in Chap. 1 of the monograph. The V&V square consists of four quadrants, as shown in Fig. 2.10:

1. *Theoretical Structural Validity*: *Is the design method internally consistent*? Internal consistency of the design method is checked—this includes, checking the logical soundness of the constructs used in the design method both individually and integrated.

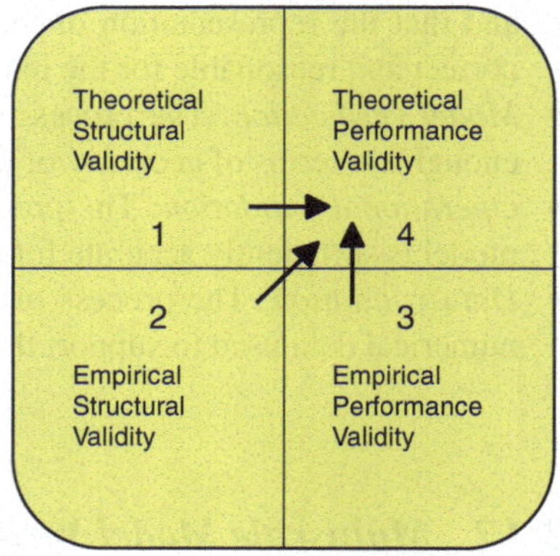

Fig. 2.10 The verification and validation square framework (Pedersen et al. 2000; Seepersad et al. 2006)

2. *Empirical Structural Validity*: *Are the example problems appropriately chosen for testing the design method*? The appropriateness of the chosen example problems to test the efficacy of the design method is checked.
3. *Empirical Performance Validity*: *Does the application of the method to the sample problems produce practical results*? Checking the ability of the design method to produce useful results worthy of investigation for the chosen example problems.
4. *Theoretical Performance Validity*: *Is the design method applicable to other problems*? Here the ability of the design method to produce useful results beyond the chosen example problems is established. This requires the designer to take a "leap of faith" which is supported by the confidence gained by carrying out V&V processes 1–3 in establishing the generic nature of the design method.

2.4.4 Design Outcome Verification and Validation

The goal here is to ensure the validity of the design outcome rather than the simulation models used for the design. The process involves gaining confidence in the resulting design of the material when compared with the system-level design requirements. Experiments are generally carried out to test the design outcomes. Li and co-authors propose an approach for design outcome validation. The approach is illustrated by them for a simple cantilever beam design subject to vibration.

2.4.5 Verification and Validation in This Monograph

In this monograph, the different verification and validation approaches described are used to verify and validate the design methods, simulation models, and design results. The verification and validation square framework is used to verify and validate the systems-based design architecture proposed in Chap. 4 of this monograph. The systems-based design architecture is tested for the hot rod rolling process chain problem introduced in Chap. 1. Example problems are used for achieving this. The first example involves the macrostructural design of a hot rolling process chain involving the product. The horizontal integration of manufacturing processes is illustrated via this example (see Chap. 5). The comprehensive example problem discussed in Chap. 6 involves the integrated design exploration of rod—product, steel—material, hot rolling and cooling processes—manufacturing processes. This comprehensive example is used to test the inverse design method developed in this monograph and serves the theoretical and empirical structural and performance validations of the design method. Individual, multiscale, and multiphysics models used in the example problems are tested in terms of concept, accuracy, operation, and data and will be discussed in detail in Chaps. 5 and 6. Interested readers are also directed to the following articles for more details on the testing carried out using example problems (see Nellippallil et al. 2016, 2017a, b, c, d, 2018a, b, 2019, 2020).

2.5 Remarks on the Current Status of Materials Design

A paradigm shift is happening from the classical material selection (Ashby and Johnson 2013), with a focus on designing the material by tailoring the chemical composition, constituent phases, microstructure, and processing paths of a material to obtain desired properties and performances at the product/system level, subject to the dynamic changes in customer requirements and market. Most efforts in this domain are focused on multiscale modeling techniques that allow designers to come up with the material processing–microstructure–property–performance relationships. However, materials design is still addressed at the embodiment and detail design phases and does not support early-stage conceptual design exploration.

The integration of design engineering and materials science is still limited to the selection of appropriate materials from a set of material databases. Product development cycles for new products and materials are still consuming huge time and resources and thus are highly expensive. Even now the performances of many systems and products are limited to the properties offered by the available set of materials from which a selection can be made. New materials development has primarily occurred via empirical, trial-and-error experiments that are usually time-consuming and costly. The real performance of a material in a product would also depend on the manufacturing route the product is subjected to. Here material remains a black-box that is subjected to repeated experiments resulting in the population of material databases for material selection. Methods to select materials by analysis, synthesis, similarity, or inspiration are proposed by Ashby. These selection methods are useful for selecting material properties and mapping to the performance that is possible based on composition and assumptions on the manufacturing route. However, these methods do not support tailoring the material microstructure or processing paths to satisfy certain end performance goals. The necessary combination of material properties needed to satisfy a performance requirement may not be available in the material database. There is enormous potential here if material selection approaches are integrated with materials design techniques and methods that allow tailoring of material processing paths and microstructures to meet performance goals. This is an important research gap that is worthy of further investigation and this is addressed in **Research Gaps 1, 2, 3,** and **4** identified in Chap. 1 of this monograph.

Multiscale modeling efforts are the next class of research efforts focused on materials design. However, it can be said that multiscale modeling efforts are simply a tool that may be used in materials design and does not comprise the whole of materials design and its goals. The objective of multiscale modeling is to accurately predict the response of material microstructures with a focus on specific applications. This usually corresponds to the detail design of different scales associated with material and generates information that can be passed to subsequent upper scales. Thus, as mentioned in Chap. 1, complex multiscale models might not be necessary in many materials design cases because the goal in materials design is not necessarily to accurately predict material properties but to satisfy a range of performance requirements. To this argument, we include the second fact that bottom-up modeling is not design but analysis. Thus, we believe that the key to materials design is an interplay of multiscale modeling and bottom-up analysis along with top-down, goal-oriented, inverse design and human decision-making. All these are investigated as **Research Gaps 3** and **4** in this monograph. Research Gaps 1, 2, 3, and 4 identified in Chap. 1 are summarized below.

- G1. Systematic approaches to identify and integrate material, process, and product models based on their function structures to frame system-level structure
- G2. Systematic approaches to define the forward processing–structure–property–performance relationships and develop material and product workflows
- G3. Support systematic and rapid concept exploration of materials, products, and processes to generate satisficing design specifications
- G4. Systematic design methods to carry out inverse design exploration of materials, products, and processes meeting end goals.

To address these research gaps both deductive mapping and inductive design exploration are necessary to support materials design. In this monograph, the challenge of incorporating the design of the material is addressed as part of a larger overall systems design process embodying the hierarchy of process–structure–property–performance set forth by Olson (1997) with consideration on supporting the coordination of information and human decision-making, as illustrated in Fig. 2.11.

Olson's hierarchy includes as shown in Fig. 2.11:

- Processing-structure mapping: This includes relationships that can map the way a material is processed from the manufacturing process side to the corresponding microstructures, composition, and phases that can be generated;

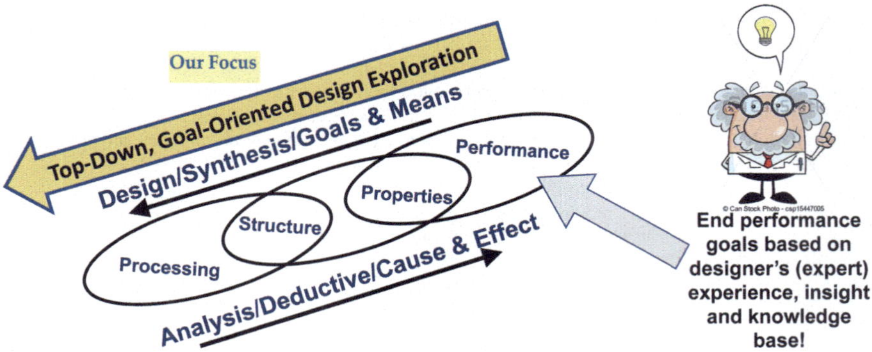

Fig. 2.11 The focus in this monograph founded on Olson's materials-by-design

- Structure-property mapping: This includes relationships that can map the composition, phases, microstructures of the material to the properties of relevance to achieve desired performance attributes; and
- Property-performance mapping: This includes relationships between properties and the specific performance requirements that are desired for the end product/system.

However, our major focus in this monograph is to use this foundational philosophy to support systematic goal-oriented materials design for a designer (who is considered an expert in the domain) based on his/her experience, insight, and knowledge base. Foundational to the work in this monograph is the decision-based design philosophy of finding "satisficing" and robust solutions based on work of Mistree et al. (1993), Chen et al. (1997), Seepersad et al. (2005), and Choi et al. (2005). However, the work on decision-based design carried out till now does not address the need for goal-oriented, inverse design exploration of materials, products and processes and identification robust "satisficing" solutions for multiple conflicting goals. These research gaps are addressed in this monograph and will be explained in detail in Chaps. 4–7. An overview of the research efforts discussed in this chapter is provided in Table 2.2.

2.6 Role of This Chapter in This Monograph

In this chapter, a review of the existing efforts associated with materials and product design is carried out. Several critical issues associated with the current capabilities of materials design are discussed in this chapter. Some of the major elements of modern materials design like material selection, multiscale modeling, microstructure design using invertible linkages, systems-based materials design, material informatics, uncertainty management, verification, and validation in materials design are discussed in this chapter. Based on the review carried out, the limitations of the existing

Table 2.2 Limitations and use of existing materials design research efforts

Research effort	Limitations	Use in monograph
Material selection (Ashby and Cebon 1993; Ashby and Johnson 2013)	• Lacks the ability to tailor material microstructures and processing paths for defined property or performance requirements • Choice of only selecting from a set of material databases to identify the achievable performances for the material	Used as a theoretical foundation to develop the systems-based design architecture for integrated materials, product, and process design
Multiscale modeling (Khaleel 2004; Horstemeyer 2012, 2018)	• Focus on bottom-up modeling spanning different material scales to predict responses of structure and properties • Highly focused on the detailed design of materials and in the accurate prediction of their behavior • The focus here is not on achieving system-level design objectives and inverse design exploration	Used as theoretical foundations for establishing processing–structure–property–performance relationships that are further explored using goal-oriented, inverse design exploration
Microstructure design using invertible linkages (Kalidindi et al. 2010, 2011; Adams et al. 2013)	• Focused only on the design of those microstructures for which invertible linkages between property and structure can be established; • Focus is on detail design and not on the early stages of design	Used as theoretical foundations for establishing inverse mapping
Material informatics (Rajan 2005, 2013)	• Focused only on the data-driven formulation of surrogate PSP linkages ("soft" computing) • Not possible if sufficient data to establish linkages are not present • Need to be configured with "hard" computing practices to address materials design challenges	Used as a theoretical foundation—the need for databases and models

(continued)

Table 2.2 (continued)

Research effort	Limitations	Use in monograph
Systems-based materials design (Olson 1997, 2000; Panchal et al. 2005)	• Require systematic methods to carry out inverse, goal-oriented design to tailor microstructure and processing paths for specified performances • Require approaches to manage uncertainty across material process chains • Require efficient ways to merge product design and materials design in an integrated fashion	Philosophy used to address the research gaps 1, 2, 3, and 4 identified in this monograph
Robust materials design under uncertainty (Chen et al. 1999; Choi et al. 2005, 2008; Seepersad et al. 2005)	• The current focus is on embodiment and detail design stages—require methods for early stages of design • Focus on mitigating uncertainty rather than managing it • Limited methods to address the propagation of uncertainty and management of uncertainty across process chains • Current practices do not or rarely address situations where multiple conflicting goals are present	Philosophy used to address the research gaps 5 identified in this monograph (Chap. 7)

capabilities are identified and reported in Table 2.2. The research gaps that are worthy of investigation are identified based on this review for further exploration. In the next chapter, the design foundations for addressing some of the research gaps identified in this chapter are reviewed and limitations that exist in the current capabilities are identified.

References

Adams, B. L., Kalidindi, S., & Fullwood, D. T. (2013). *Microstructure-sensitive design for performance optimization*. Butterworth-Heinemann.

Allen, J. K., Seepersad, C., Choi, H., & Mistree, F. (2006). Robust design for multiscale and multidisciplinary applications. *Journal of Mechanical Design, 128*(4), 832–843.

Arróyave, R., & McDowell, D. L. (2019). Systems approaches to materials design: past, present, and future. *Annual Review of Materials Research, 49*, 103–126.

Ashby, M. F., & Cebon, D. (1993). Materials selection in mechanical design. *Le Journal de Physique IV, 3*, C7, C7-1–C7-9.

Ashby, M. F., & Johnson, K. (2013). *Materials and design: the art and science of material selection in product design*. Butterworth-Heinemann.

Asta, M., Ozolins, V., & Woodward, C. (2001). A first-principles approach to modeling alloy phase equilibria. *JOM Journal of the Minerals Metals and Materials Society, 53*(9), 16–19.

Bammann, D., Chiesa, M., Horstemeyer, M., & Weingarten, L. (1993). Failure in ductile materials using finite element methods. *Structural Crashworthiness and Failure*, 1–54.

Bammann, D. J. (1990). Modeling temperature and strain rate dependent large deformations of metals. *Applied Mechanics Reviews, 43*(5), S312–S319.

Chang, T.-S., & Ward, A. C. (1995). Conceptual robustness in simultaneous engineering: a formulation in continuous spaces. *Research in Engineering Design, 7*(2), 67–85.

Chang, T.-S., Ward, A. C., Lee, J., & Jacox, E. H. (1994). Conceptual robustness in simultaneous engineering: an extension of Taguchi's parameter design. *Research in Engineering Design, 6*(4), 211–222.

Chen, W., Allen, J. K., & Mistree, F. (1997). A robust concept exploration method for enhancing productivity in concurrent systems design. *Concurrent Engineering, 5*(3), 203–217.

Chen, W., & Lewis, K. (1999). Robust design approach for achieving flexibility in multidisciplinary design. *AIAA Journal, 37*(8), 982–989.

Chen, W., Simpson, T. W., Allen, J. K., & Mistree, F. (1999). Satisfying ranged sets of design requirements using design capability indices as metrics. *Engineering Optimization, 31*(5), 615–619.

Choi, H.-J., Austin, R., Allen, J. K., McDowell, D. L., Mistree, F., & Benson, D. J. (2005). An approach for robust design of reactive power metal mixtures based on non-deterministic micro-scale shock simulation. *Journal of Computer-Aided Materials Design, 12*(1), 57–85.

Choi, H.-J., Mcdowell, D. L., Allen, J. K., & Mistree, F. (2008a). An inductive design exploration method for hierarchical systems design under uncertainty. *Engineering Optimization, 40*(4), 287–307.

Choi, H., McDowell, D. L., Allen, J. K., Rosen, D., & Mistree, F. (2008b). An inductive design exploration method for robust multiscale materials design. *Journal of Mechanical Design, 130*(3), 031402.

Cotton, J., Frohlich, C., & Glamm, R. (2012). What boeing wants from integrated computational materials engineering for metallic materials. In *53rd AIAA/ASME/ASCE/AHS/ASC Structures, Structural Dynamics and Materials Conference 20th AIAA/ASME/AHS Adaptive Structures Conference 14th AIAA*.

Cuitino, A. M., & Ortiz, M. (1993). Computational modelling of single crystals. *Modelling and Simulation in Materials Science and Engineering, 1*(3), 225.

Du, X., & Chen, W. (2000). Methodology for managing the effect of uncertainty in simulation-based design. *AIAA Journal, 38*(8), 1471–1478.

Du, X., & Chen, W. (2002). Efficient uncertainty analysis methods for multidisciplinary robust design. *AIAA Journal, 40*(3), 545–552.

Forrester, A., Sobester, A., & Keane, A. (2008). *Engineering Design via Surrogate Modelling: A Practical Guide*. John Wiley & Sons.

Gall, K., & Horstemeyer, M. F. (2000). Integration of basic materials research into the design of cast components by a multi-scale methodology. *Journal of Engineering Materials and Technology, 122*(3), 355–362.

Gu, X., Renaud, J. E., Batill, S. M., Brach, R. M., & Budhiraja, A. S. (2000). Worst case propagated uncertainty of multidisciplinary systems in robust design optimization. *Structural and Multidisciplinary Optimization, 20*(3), 190–213.

Gu, X. S., Renaud, J. E., & Penninger, C. L. (2006). Implicit uncertainty propagation for robust collaborative optimization. *Journal of Mechanical Design, 128*(4), 1001–1013.

Hollingsworth, P., & Mavris, D. (2003). Gaussian process meta-modeling: Comparison of gaussian process training methods. In *AIAA's 3rd Annual Aviation Technology, Integration, and Operations (ATIO) Forum*.

Horstemeyer, M., Lathrop, J., Gokhale, A., & Dighe, M. (2000). Modeling stress state dependent damage evolution in a cast Al–Si–Mg aluminum alloy. *Theoretical and Applied Fracture Mechanics, 33*(1), 31–47.

Horstemeyer, M., & Wang, P. (2003). Cradle-to-grave simulation-based design incorporating multiscale microstructure-property modeling: Reinvigorating design with science. *Journal of Computer-Aided Materials Design, 10*(1), 13–34.

Horstemeyer, M. F. (2012). *Integrated computational materials engineering (ICME) for metals: Using multiscale modeling to invigorate engineering design with science*. John Wiley & Sons.

Horstemeyer, M. F. (2018). *Integrated computational materials engineering (ICME) for metals: Concepts and case studies*. Wiley.

Hyun, S., & Torquato, S. (2002). Optimal and manufacturable two-dimensional, Kagome-like cellular solids. *Journal of Materials Research, 17*(1), 137–144.

Isukapalli, S., Roy, A., & Georgopoulos, P. (1998). Stochastic response surface methods (SRSMs) for uncertainty propagation: application to environmental and biological systems. *Risk Analysis, 18*(3), 351–363.

Jin, R., Chen, W., & Sudjianto, A. (2005). An efficient algorithm for constructing optimal design of computer experiments. *Journal of Statistical Planning and Inference, 134*(1), 268–287.

Jones, D. R., Schonlau, M., & Welch, W. J. (1998). Efficient global optimization of expensive black-box functions. *Journal of Global Optimization, 13*(4), 455–492.

Kalidindi, S. R. (2015). *Hierarchical materials informatics: Novel analytics for materials data*. Elsevier.

Kalidindi, S. R., Khosravani, A., Yucel, B., Shanker, A., & Blekh, A. L. (2019). Data infrastructure elements in support of accelerated materials innovation: ELA, PyMKS, and MATIN. *Integrating Materials and Manufacturing Innovation*, 1–14.

Kalidindi, S. R., Niezgoda, S. R., Landi, G., Vachhani, S., & Fast, T. (2010). A novel framework for building materials knowledge systems. *Computers, Materials, & Continua, 17*(2), 103–125.

Kalidindi, S. R., Niezgoda, S. R., & Salem, A. A. (2011). Microstructure informatics using higher-order statistics and efficient data-mining protocols. *JOM Journal of the Minerals Metals and Materials Society, 63*(4), 34–41.

Kalsi, M., Hacker, K., & Lewis, K. (2001). A comprehensive robust design approach for decision trade-offs in complex systems design. *Journal of Mechanical Design, 123*(1), 1–10.

Khaleel, M. (2004). *Multiscale mathematics initiative: A roadmap*.

Kleijnen, J. P. (2017). Design and analysis of simulation experiments: Tutorial. In *Advances in modeling and simulation* (pp. 135–158), Springer.

Kleijnen, J. P. (2020). Simulation optimization through regression or kriging metamodels. In *High-performance simulation-based optimization* (pp. 115–135). Springer

Kristensen, J., Subber, W., Zhang, Y., Ghosh, S., Kumar, N. C., Khan, G., et al. (2019). *Industrial applications of intelligent adaptive sampling methods for multi-objective optimization*. IntechOpen: Design Engineering and Manufacturing.

Kulkarni, A. J., Krishnamurthy, K., Deshmukh, S. P., & Mishra, R. S. (2004). Microstructural optimization of alloys using a genetic algorithm. *Materials Science and Engineering A, 372*(1–2), 213–220.

Li, C., & Mahadevan, S. (2016). Role of calibration, validation, and relevance in multi-level uncertainty integration. *Reliability Engineering & System Safety, 148*, 32–43.

Liu, W. K., Karpov, E., Zhang, S., & Park, H. (2004a). An introduction to computational nanomechanics and materials. *Computer Methods in Applied Mechanics and Engineering, 193*(17–20), 1529–1578.

Liu, Z.-K., & Chen, L.-Q. (2007). Integration of first-principles calculations, CALPHAD modeling, and phase-field simulations. In *Applied computational materials modeling* (pp. 171–213). Springer.

Liu, Z.-K., Chen, L.-Q., Raghavan, P., Du, Q., Sofo, J. O., Langer, S. A., et al. (2004b). An integrated framework for multi-scale materials simulation and design. *Journal of Computer-Aided Materials Design, 11*(2–3), 183–199.

Liu, Z.-K., Chen, L.-Q., & Rajan, K. (2006). Linking length scales via materials informatics. *JOM Journal of the Minerals Metals and Materials Society, 58*(11), 42–50.

Lu, W., & Deng, Y.-M. (2004). A system modelling methodology for materials and engineering systems design integration. *Materials and Design, 25*(6), 459–469.

Matthews, J., Klatt, T., Morris, C., Seepersad, C. C., Haberman, M., & Shahan, D. (2016). Hierarchical design of negative stiffness metamaterials using a bayesian network classifier. *Journal of Mechanical Design, 138*(4), 041404.

McDowell, D., & Story, T. (1998). New directions in materials design science and engineering. *Report of a Workshop Sponsored by the US National Science Foundation, Georgia Institute of Technology and Morehouse College, Atlanta, GA*.

McDowell, D. L. (2008). Viscoplasticity of heterogeneous metallic materials. *Materials Science and Engineering: R: Reports, 62*(3), 67–123.

McDowell, D. L. (2018). Microstructure-sensitive computational structure-property relations in materials design. In *Computational aterials system design* (pp. 1–25). Springer.

McDowell, D. L., Choi, H. J., Panchal, J., Austin, R., Allen, J., & Mistree, F. (2007). Plasticity-related microstructure-property relations for materials design. In *Key engineering materials*, Trans Tech Publ.

McDowell, D. L., & Kalidindi, S. R. (2016). The materials innovation ecosystem: a key enabler for the materials genome initiative. *MRS Bulletin, 41*(4), 326–337.

McDowell, D. L., & LeSar, R. A. (2016). The need for microstructure informatics in process–structure–property relations. *MRS Bulletin, 41*(8), 587–593.

McDowell, D. L., & Olson, G. (2008). Concurrent design of hierarchical materials and structures. In *Scientific modeling and simulations* (pp. 207–240). Springer.

McDowell, D. L., Panchal, J., Choi, H.-J., Seepersad, C., Allen, J., & Mistree, F. (2009). *Integrated design of multiscale, multifunctional materials and products*. Butterworth-Heinemann.

Mistree, F., Hughes, O. F., & Bras, B. (1993). Compromise decision support problem and the adaptive linear programming algorithm. *Progress in Astronautics and Aeronautics, 150*, 251–251.

Mullins, J., & Mahadevan, S. (2016). Bayesian uncertainty integration for model calibration, validation, and prediction. *Journal of Verification, Validation and Uncertainty Quantification, 1*(1), 011006.

Nellippallil, A. B., Song, K. N., Goh, C.-H., Zagade, P., Gautham, B., Allen, J. K., & Mistree, F. (2016). A goal oriented, sequential process design of a multi-stage hot rod rolling system. *ASME Design Automation Conference*, Paper Number: DETC2016-59402.

Nellippallil, A. B., Rangaraj, V., Allen, J. K., Mistree, F., Gautham, B., & Singh, A. K. (2017a). A decision-based design method to explore the solution space for microstructure after cooling stage to realize the end mechanical properties of hot rolled product. In *Proceedings of the 4th World Congress on Integrated Computational Materials Engineering (ICME 2017)*. Springer.

Nellippallil, A. B., Rangaraj, V., Gautham, B., Singh, A. K., Allen, J. K., & Mistree, F. (2017b). A goal-oriented, inverse decision-based design method to achieve the vertical and horizontal integration of models in a hot rod rolling process chain. *ASME Design Automation Conference*, Paper Number: DETC2017-67570.

Nellippallil, A. B., Shukla, R., Ardham, S., Goh, C.-H., Allen, J. K., & Mistree, F. (2017c). Exploration of solution space to study thermo-mechanical behavior of AA5083 Al-alloy during hot rolling process. *ASME Design Automation Conference*, Paper Number: DETC2017-68173.

Nellippallil, A. B., Song, K. N., Goh, C.-H., Zagade, P., Gautham, B., Allen, J. K., Mistree, F. (2017d). A goal-oriented, sequential, inverse design method for the horizontal integration of a multistage hot rod rolling system. *Journal of Mechanical Design, 139*(3), 031403.

Nellippallil, A. B., Mohan, P., Allen, J. K., & Mistree, F. (2018a). Robust concept exploration of materials, products and associated manufacturing processes. *ASME Design Automation Conference*, Paper Number: DETC2018-85913.

Nellippallil, A. B., Rangaraj, V., Gautham, B., Singh, A. K., Allen, J. K., & Mistree, F. (2018b). An inverse, decision-based design method for integrated design exploration of materials, products, and manufacturing processes. *Journal of Mechanical Design, 140*(11), 111403-1-111403-17.

Nellippallil, A. B., Mohan, P., Allen, J. K., & Mistree, F. (2019). Inverse thermo-mechanical processing (ITMP) design of a steel rod during hot rolling process. *ASME Design Automation Conference*, Paper Number: IDETC2019-97390.

Nellippallil, A. B., Mohan, P., Allen, J. K., & Mistree, F. (2020). An inverse, decision-based design method for robust concept exploration. *Journal of Mechanical Design*, https://doi.org/10.1115/1.4045877.

Olson, G. B. (1997). Computational design of hierarchically structured materials. *Science, 277*(5330), 1237–1242.

Olson, G. B. (2000). Designing a new material world. *Science, 288*(5468), 993–998.

Panchal, J. H. (2005). A framework for simulation-based integrated design of multiscale products and design processes, Ph.D. Dissertation, The George W. Woodruff School of Mechanical Engineering, Georgia Institute of Technology.

Panchal, J. H., Choi, H.-J., Shephard, J., Allen, J. K., McDowell, D. L., & Mistree, F. (2005). A strategy for simulation-based multiscale, multi-functional products and associated design processes. *ASME Design Automation Conference*, Paper Number: DETC2005-85316.

Panchal, J. H., Kalidindi, S. R., & McDowell, D. L. (2013). Key computational modeling issues in integrated computational materials engineering. *Computer-Aided Design, 45*(1), 4–25.

Pedersen, K., Emblemsvag, J., Bailey, R., Allen, J., & Mistree, F. (2000). The 'Validation Square'–Validating design methods. In *ASME Design Theory and Methodology Conference*, New York: ASME.

Pollock, T. M., Allison J. E., et al. (2008). Integrated computational materials engineering: A transformational discipline for improved competitiveness and national security. National Academies Press.

Rajan, K. (2005). Materials informatics. *Materials Today, 8*(10), 38–45.

Rajan, K. (2013). *Informatics for materials science and engineering: Data-driven discovery for accelerated experimentation and application*. Butterworth-Heinemann.

Rasmussen, C. E. (2003) Gaussian processes in machine learning. In *Summer school on machine learning*. Springer.

Rasmussen, C. E., & Williams, C. (2006). *Gaussian processes for machine learning*, (vol. 1, 39, pp. 40–43). MIT Press.

Sargent, R. G. (2009). Verification and validation of simulation models. In *Proceedings of the 2009 Winter Simulation Conference (WSC)*. IEEE.

Seepersad, C. C., Allen, J. K., McDowell, D. L., & Mistree, F. (2005). Robust design of cellular materials with topological and dimensional imperfections. *ASME Design Automation Conference*, Paper Number: DETC2005-85061.

Seepersad, C. C., Pedersen, K., Emblemsvåg, J., Bailey, R., Allen, J. K., & Mistree, F. (2006). *"The validation square: How does one verify and validate a design method?* (pp. 303–314). New York: ASME Press.

Shahan, D. W., & Seepersad, C. C. (2012). Bayesian network classifiers for set-based collaborative design. *Journal of Mechanical Design, 134*(7), 071001.

Shukla, R., Kulkarni, N. H., Gautham, B., Singh, A. K., Mistree, F., Allen, J. K., et al. (2015). Design Exploration of Engineered Materials, Products, and Associated Manufacturing Processes. *JOM Journal of the Minerals Metals and Materials Society, 67*(1), 94–107.

Sigmund, O. (1994). Materials with prescribed constitutive parameters: an inverse homogenization problem. *International Journal of Solids and Structures, 31*(17), 2313–2329.

Sigmund, O. (1995). Tailoring materials with prescribed elastic properties. *Mechanics of Materials, 20*(4), 351–368.

Sigmund, O., & Torquato, S. (1997). Design of materials with extreme thermal expansion using a three-phase topology optimization method. *Journal of the Mechanics and Physics of Solids, 45*(6), 1037–1067.

Simpson, T., Mistree, F., Korte, J., & Mauery, T. (1998). Comparison of response surface and kriging models for multidisciplinary design optimization. In *7th AIAA/USAF/NASA/ISSMO Symposium on Multidisciplinary Analysis and Optimization*.

Sinha, A. (2011). *Uncertainty Management in Design of Multiscale Systems*. The George W. Woodruff School of Mechanical Engineering. Atlanta, GA, Georgia Institute of Technology. M. S. Thesis.

Solomou, A., Zhao, G., Boluki, S., Joy, J. K., Qian, X., Karaman, I., Arróyave, R., & Lagoudas, D. C. (2018). Multi-objective bayesian materials discovery: Application on the discovery of precipitation strengthened niti shape memory alloys through micromechanical modeling. *Materials & Design, 160*, pp. 810–827.

Subbarayan, G., & Raj, R. (1999). A methodology for integrating materials science with system engineering. *Materials and Design, 20*(1), 1–12.

Tennyson, G., Shukla, R., Mangal, S., Sachi, S., & Singh, A. K., (2015). ICME for process scale-up: Importance of vertical and horizontal integration of models. In *Proceedings of the 3rd World Congress on Integrated Computational Materials Engineering (ICME)*. Wiley.

Xiao, A. (2003). Collaborative multidisciplinary decision making in distributed environment, Ph.D. Dissertation, The George W. Woodruff School of Mechanical Engineering, Georgia Institute of Technology.

Xue, D., Balachandran, P. V., Hogden, J., Theiler, J., Xue, D., & Lookman, T. (2016). Accelerated search for materials with targeted properties by adaptive design. *Nature Communications, 7*(1), 1–9.

Yan, F., Chan, Y.-C., Saboo, A., Shah, J., Olson, G. B., & Chen, W. (2018). Data-driven prediction of mechanical properties in support of rapid certification of additively manufactured alloys. *Computer Modeling in Engineering & Sciences, 117*(3), 343–366.

Chapter 3
Design Foundations—State-of-the-Art in Decision-Based Design, Robust Design Approaches, and Platform for Decision Support

The objective in this chapter is to introduce and review the design foundations based on which the systems-based design architecture for integrated design of materials, products, and manufacturing processes design is developed. Besides the underlying decision-based design, systems design, and robust design approaches, methods and tools reviewed are classified in terms of concept, application to design process, and value in design. The relationship of these research efforts reviewed in this chapter with the constructs of the systematic approach developed in this monograph is highlighted in Fig. 3.1.

In Sect. 3.1, the need for systems-based design architecture is emphasized from the perspective of materials and product design. In Sect. 3.2, the perspective with which design is viewed in this monograph is defined with a review of the different design paradigms that exist. In Sect. 3.3, the decision-based design paradigm adopted in this monograph is reviewed. This is followed by a review of robust design and design under uncertainty in Sect. 3.4. In Sect. 3.5, the foundations for a platform for decision support are reviewed and discussed. In Table 3.1, we summarize the architecture component constituents discussed and reviewed in this chapter.

3.1 Designing Design Processes in Conjunction with Materials, Products, and Manufacturing Processes

In Chap. 2, we emphasized the point that "*Materials Design goes beyond Materials Selection*". The goal in the integrated design of materials, products, and processes is to tailor material microstructures and processing paths to satisfy specific system-level (material, product, or assembly) properties and performances. In accordance with the above point, the fundamentals of engineering design are introduced in this chapter. The concepts, tools, and constructs reviewed in this chapter are used in the

A. B. Nellippallil et al., *Architecting Robust Co-Design of Materials, Products, and Manufacturing Processes*, https://doi.org/10.1007/978-3-030-45324-4_3

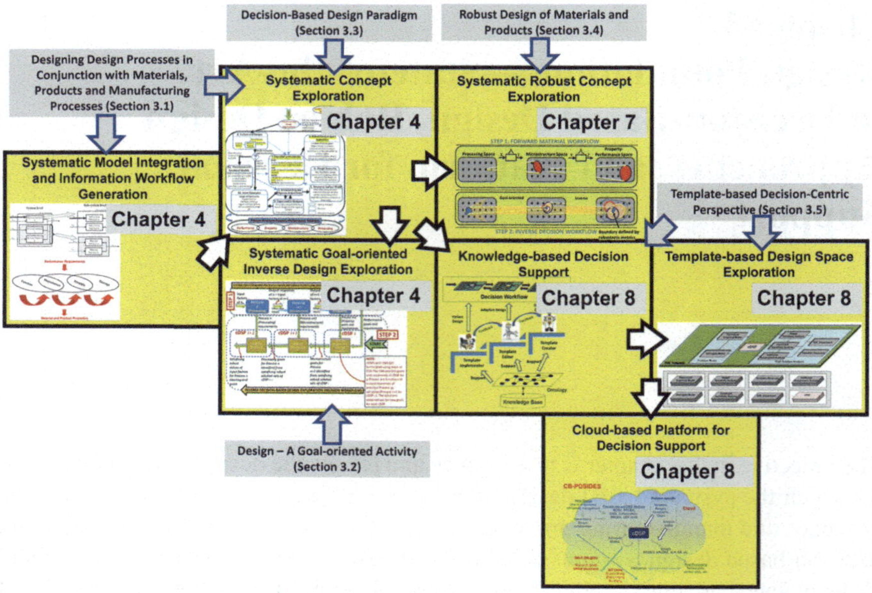

Fig. 3.1 Relationship of research efforts discussed in this chapter with the constructs of the systems-based design architecture developed in this monograph

Table 3.1 Summary of elements discussed and reviewed in this chapter that form part of the architecture

Elements	*What?*
Processing elements	Elements of the design foundations based on which the systems-based design architecture for integrated design of materials, products, and manufacturing processes design is developed: (i) Design as a goal-oriented activity, (ii) Pahl and Beitz design method, (iii) Suh's axiomatic design, (iii) Gero's model of design as a process, (iii) decision-based design paradigm and DSP technique, (iv) robust design, and (v) template-based decision-centric view of design
Data elements	Decision-based design
Connecting elements	Besides the underlying decision-based design, systems design, and robust design approaches, methods and tools reviewed are classified and connected in terms of concept, application to design process, and value in design
Form	*How?*
Weights	Review of the foundations of decision-based design, systems design, and robust design approaches
Properties	Focus on current capabilities in design approaches, the associated limitations and the research opportunities that are worthy of investigation

(continued)

Table 3.1 (continued)

Relationship	Concepts discussed are used as theoretical foundations and as research philosophies to address the research gaps identified in the monograph
Rationale	*Why?*
Motivation	To identify critical issues associated with the current capabilities of design and address the challenges and gaps
Characteristics/beliefs	Design is a goal-oriented synthesis activity and essentially a decision-making process
Interpretation	Understanding the efficacy of design constructs and broader impacts

rest of the chapters of the monograph to develop the systems-based design architecture for integrated material, product, and process design. We begin this section with the goal-oriented nature of design processes. There are several schools of thought for engineering design. One school of thought is decision-based design (DBD). Two different perspectives are popularly identified within DBD: one articulated by Hazelrigg (1996, 1998) and the other by Mistree et al. (1990, 1991, 1993). In this monograph, we follow the perspective of Mistree et al.; the details of which will be discussed in later sections. In the next section, we discuss design as a goal-oriented activity and its implications in materials and product design.

3.2 Design—A Goal-Oriented Synthesis Activity

See Fig. 3.2.

The following conversation takes place between Alice and the Cheshire Cat in Lewis Carroll's "Alice's Adventures in Wonderland" (Carroll and Tenniel 1865):
"Cheshire Puss, would you tell me, please, which way I ought to go from here?"
'That depends a good deal on where you want to get to,' said the Cat.
"I don't much care where–" said Alice.
'Then it doesn't matter which way you go,' said the Cat.
"–so long as I get SOMEWHERE," Alice added as an explanation.
'Oh, you're sure to do that,' said the Cat, 'if you only walk long enough'

Charles Lutwidge Dodgson (better known by his pen name Lewis Carroll), an English writer and mathematician, presents a fundamental aspect of design decision-making in *Alice's Adventures in Wonderland* (Carroll and Tenniel 1865). Alice comes to a fork after walking through the wonderland forest. She stands to ponder about the

Fig. 3.2 Image depicting a conversation between Alice and Cheshire Cat in Lewis Carroll's "Alice's Adventures in Wonderland" (Image source: Internet)

decision she should make, not knowing which way to go. This is when the Cheshire Cat appears on the tree above her.

Alice asks the Cheshire Cat to make a decision for her, *"Cheshire Puss, would you tell me, please, which way I ought to go from here?"*. The Cat instead of offering a decision responds ***"That depends a good deal on where you want to get to"***. The Cat here recognizes the fact that it cannot make the choice for Alice. It is completely upon the person actually taking the action resulting directly from the decision can make the decision (Hazelrigg 1996). The Cat instead offers a piece of advice in general which is foundational to design decision-making: "***That depends a good deal on where you want to get to***". The choice taken by the decision-maker should be based on namely the decision-maker's preference over the outcome of the decision (Hazelrigg 1996). This is a very powerful statement and has wide implications across design and decision-making.

A very important implication that can be derived here is that the decision-maker should have a preference on the outcomes (or a goal that he/she wants to achieve/satisfy) and only the decision-maker's preference matters here. Here the Cat makes it clear to Alice that the preferences are on the outcome and not on the choices of decisions as to which path to take.

Alice then responds, "*I don't much care where–*". Alice suggests that she doesn't have a preference over the outcomes. To this, the Cat responds, "***Then it doesn't matter which way you go***". The Cat basically is saying: "*If you don't know where you are going … any road will take you there*". The implication that can be derived from this statement is that unless the decision-maker has a preference over the outcomes, any choice can be accepted equally. The Cat here reminds us that one choice is better than the other only because we have a preference over the outcomes (Hazelrigg 1996); this is the perspective of modern normative decision theory.

From the standpoint of this monograph, an important implication that we derive out of this conversation is that every designer (or decision-maker) should have a goal that he/she wants to satisfy and that the designer should start the path that he/she wants to traverse with this goal in mind. If there is no such goal that the designer wants to satisfy, then the designer is free to choose any path that he/she wishes to traverse. This idea is very relevant to decision-based design and forms the foundational research philosophy for this monograph from the context of integrated materials and product design.

"*While natural sciences are concerned with how things are, an engineer, and more generally a designer is concerned with* ***how things ought to be in order to attain goals and to function***" (Simon 1996). The distinction between natural science and engineering science is pointed out by Braha and Maimon (1997). Natural science is theory-oriented and focuses on analysis. Engineering science is result-oriented and focuses on synthesis. This difference reflects in the definitions of "design" by many researchers.

3.2.1 Some Background on Models for Design

Descriptive and Prescriptive Models of Design

Research in engineering design can be categorized as design philosophies, models, and methods. **We believe that the term "Design Theory" is an oxymoron**. This is because there cannot be a unique theory in design that proposes how design must be carried out. However, we review design theory in the way it has been defined in the past. Design theory as per popular belief is a collection of principles that are useful for explaining a design process and provide a foundation for basic understanding required to propose useful methodologies (Panchal 2005). Design theory is used to explain what design is. Design methodology, on the other hand, is a collection of procedures, tools, and techniques for designers to use when designing. **Again, we also believe that the term "Design Methodology" is an oxymoron.** "Methodology" is the systematic, theoretical analysis of the methods applied to a field of study, and the use of the term in the context of design as a collection of procedures, tools, and techniques is an oxymoron. This also holds for the term "**Design Research**" as discussed by Krippendorff (2007). Krippendorff (2007) makes the point that design research is an oxymoron whose contradictions, because they are not obvious to everyone, can lead its naïve users into thinking of it as a kind of research similar to what reputable scientists do. Ignoring the issue of the usage of terms being an oxymoron, design methodology is prescriptive, while design theory is descriptive (Finger and Dixon 1989a; Sivaloganathan et al. 1996).

Design methods are developed from different viewpoints emphasizing various facets of the overall design process. An ideal design method supports all the following viewpoints of design as (Sivaloganathan et al. 1996):

- a top-down and bottom-up process;
- an incremental activity (evolutionary);
- a knowledge-based exploratory activity;
- an investigative process (research);
- a creative process (art);
- a rational process;
- a decision-making process;
- an iterative process; and
- an interactive process.

Design methods are usually developed with some of these viewpoints in mind. An ideal design, however, should support all of these.

Pahl and Beitz Design Process

Pahl and Beitz (1996) and Beitz et al. (1996) define four phases of design. These four phases are common to any prescriptive model of design. These phases include planning and clarification of tasks, conceptual design, embodiment design, and detail design. In planning and clarification, the designer identifies requirements that the outcome of design should fulfill. From these requirements, a problem statement is formulated. Conceptual design includes the generation of solution principles to satisfy the problem statement. In embodiment design, these solution principles are refined until the final solution remains. In detail design, all the details of the final design are specified, and manufacturing drawings and documentation are produced. Steps of the Pahl and Beitz approach are shown in Fig. 3.3.

Pahl and Beitz design process is a systematic design process and is based on ***discursive thinking*** and ***intuitive thinking***. We believe that for systematic design to happen both discursive thinking and intuitive thinking are needed. Discursive thinking is a conscious process in which scientific knowledge and relationships are consciously analyzed, varied, combined in new ways, logically checked, rejected, and considered further to come up with reasonings or conclusions (Pahl and Beitz 1996). In this systematic process, **information is transformed successively** via successive steps—which helps in making problem-solving systematic. Intuitive thinking

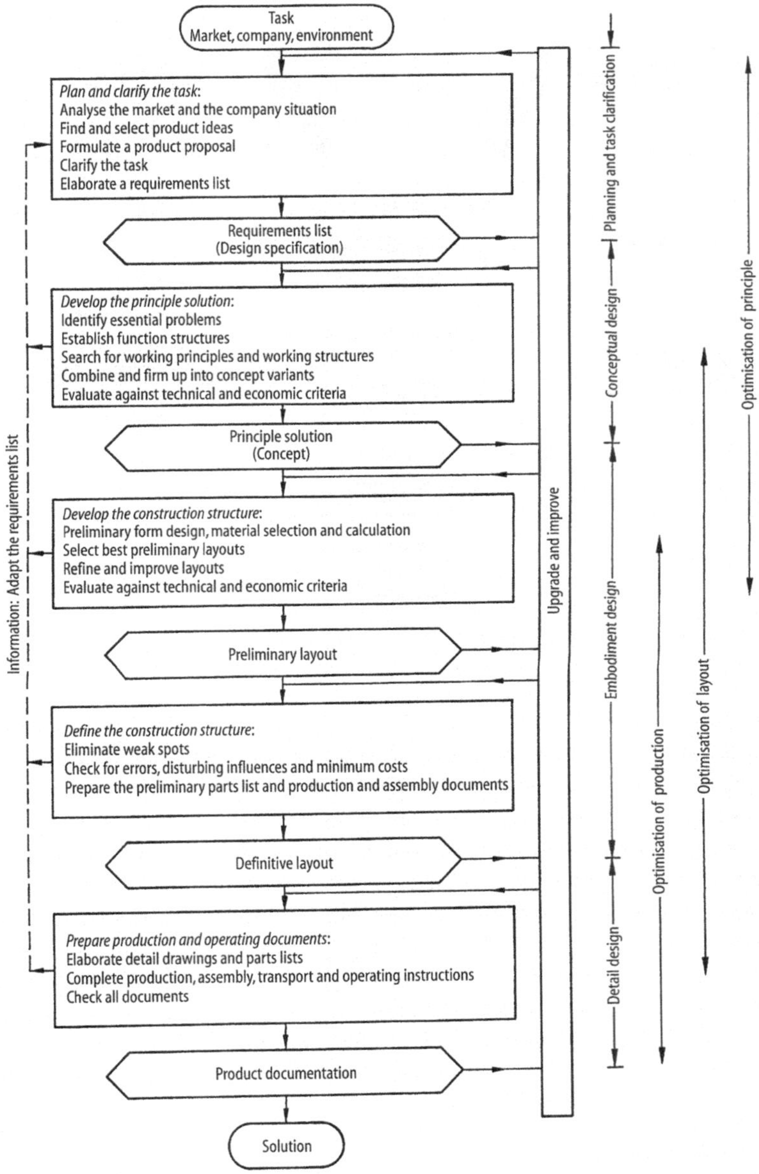

Fig. 3.3 Pahl and Beitz design process (Pahl and Beitz 1996)

is strongly associated with flashes of inspiration required to fulfill various information transformation in systematic design; see Fig. 3.4 (adopted from Messer 2008) for illustrations on discursive thinking and intuitive thinking for systematic design.

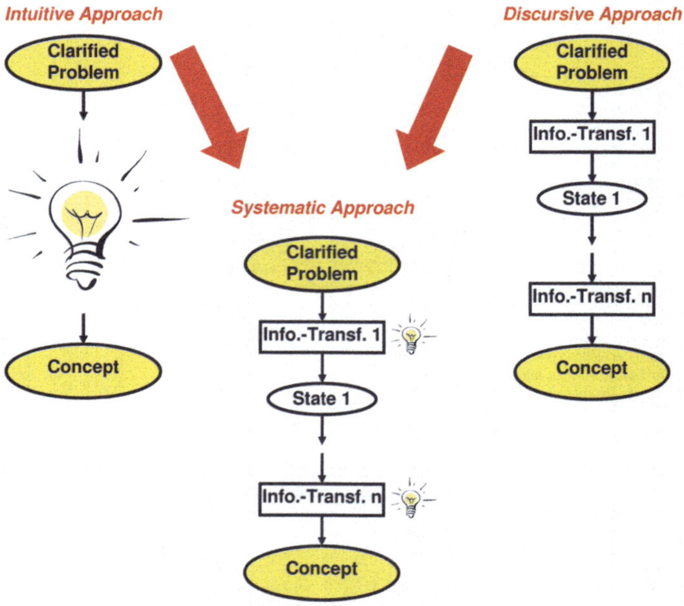

Fig. 3.4 Systematic design (Messer 2008)

After the planning and clarification of the task phase of the Pahl and Beitz design process, the designer comes up with a requirements list and starts the conceptual design phase. In the conceptual design phase, the designer determines a principal solution or a concept. This is achieved by abstracting the essence of the problem, establishing ***function structures***, searching suitable ***working inter-relationships*** (working principles, phenomenon, etc.), and then combining those (integration of information) to develop a **system-level function structure** or working structure or a **workflow**—from the context of this monograph. The working structure is then transformed into a more comprehensive representation so as to evaluate the essentials of the principal solution and review constraints, goals, and other design objectives. Once the specification of concepts is done at the conceptual design stage, the design process re-starts at a much more concrete level in embodiment and detail design phases.

Gero's Model of Design as a Process

The prescriptive models of design are mostly based on the assumption that a design activity consists of three core activities—**Analysis**, **Synthesis**, and **Evaluation** (**ASE**). "Analysis" is defined as the resolution of anything complex into its elements and the study of these elements and of their relationships. Synthesis is putting together of parts or elements to produce new effects and to demonstrate that these parts create an order (Pahl and Beitz 1996). Design can be visualized as an iterative feedback loop of synthesis, analysis, and evaluation. Gero (1990) describes the

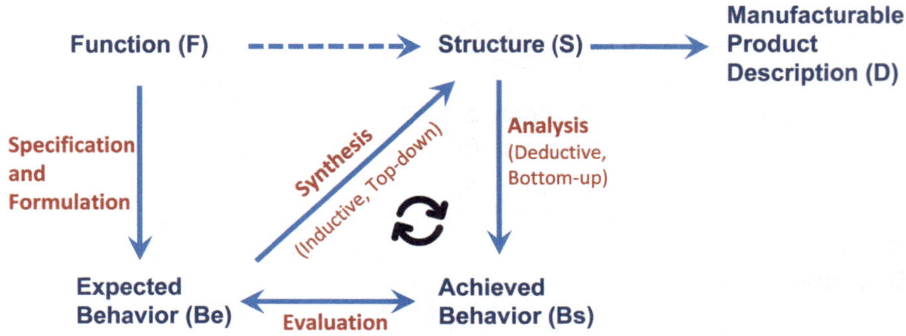

Fig. 3.5 Gero's model of design as a process (Gero 1990)

ideas of analysis, synthesis, and evaluation as a **series of information transformation** starting with requirements and ending with descriptions of design that satisfy those requirements. Gero's view of design is foundational for many research efforts focused on product design.

Gero defines the key aspects of product information as function (F), structure (S), expected behavior (Be), achieved behavior (Bs), and product descriptions (D) (see Fig. 3.5).

- **Function (F)**: Function F is the relation between the goal of a human designer and the behavior of the system. It specifies a relation between input and output in terms of material (matter), energy, and signal (information).
- **Expected Behavior (Be)**: Expected behavior (Be) represents the physical behavior that the artifact being designed should have in order to satisfy the functional requirements (F).
- **Structure (S)**: Structure (S) represents the artifact's elements and their relationships. It is also called the form of the artifact. It represents the proposed design solutions—information about geometry, material, configurations, and so on.
- **Achieved Behavior (Bs)**: Behavior of the structure derived directly using engineering principles.
- **The manufacturable product descriptions** are derived from the product structure.

Connecting Gero's Model of Design with Olson's Materials-By-Design

The process of designing a material starts with the customer needs and requirements for the end product. These customer needs are captured as functional requirements or functions. Using engineering parameters, these functional requirements are mapped into the expected behavior. Based on the expected behavior formulated, the structure of the material is synthesized (synthesis transformation). The synthesis here may be using expert's knowledge, previous designs, or through computations, and is a top-down inductive approach. Simulation models may be used to predict/analyze the achieved behavior from the structure synthesized from expected behavior. This refers to analysis and is a bottom-up deductive approach. The expected behavior is

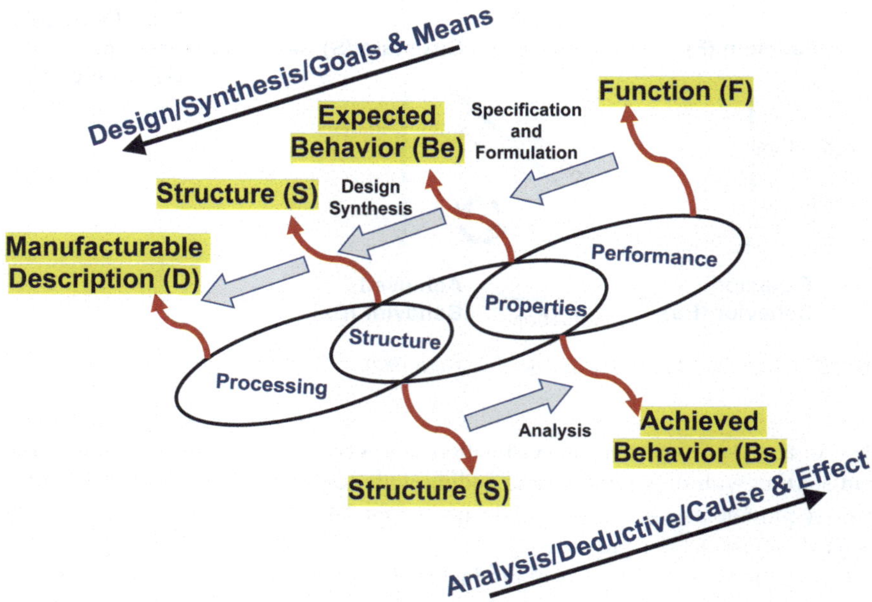

Fig. 3.6 Connecting Gero's model (1990) of design with Olson's (1997) diagram

then evaluated with the achieved behavior. This is referred to as evaluation. This whole cycle of synthesis, analysis, and evaluation is repeated, and the structure is refined until the required performance is satisfied. From the final structure identified the manufacturable product descriptions are derived. This model of design usually applies to product design. However, it has strong ties to the systems-based materials design proposed by Olson. In terms of Olson's processing–structure–property–performance relationships, the function (F) corresponds to the performance of the system/product/material. Behavior (B) corresponds to the properties of the material system. The specification and formulation step in the process is equivalent to performance and to property mapping. The design synthesis step involves mapping from property to microstructure. The structure (S) in Gero's model corresponds to the structure domain in Olson. The design synthesis is a top-down, inductive process. The microstructure obtained is then analyzed by mapping structure to properties. This is the bottom-up, deductive process of predicting properties from the structure. Finally, the product descriptions correspond to the processing information of the product/material. This whole scheme is depicted in Fig. 3.6.

The design model by Gero is foundational for function-based design and will be discussed in detail in later sections. In the last chapter, Gero's model will be addressed from the perspective of robust product design.

Suh's Axiomatic Design

Suh defines design as an **interplay** between "**what we want to achieve**" and "**how we want to achieve it**" (Suh 1990). Here what we want to achieve is the goal, and how we

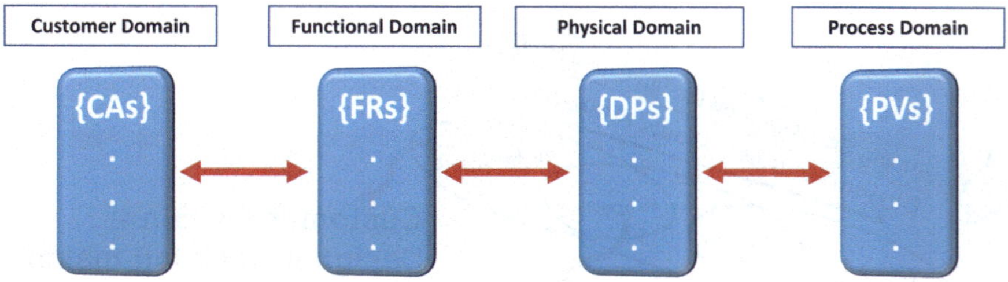

Fig. 3.7 Relationship of domains, mapping, and design spaces in Suh's axiomatic design (Suh 1990)

want to achieve is the path that needs to be taken to reach/satisfy that goal. Suh proposes the engineering sequence in which this happens via four design domains—the four domains of the design world. This includes customer attributes (CAs)—customer domain, functional requirements (FRs)—functional domain, design parameters (DPs)—physical domain, process variables (PVs)—process domain, as shown in Fig. 3.7. CAs are customer needs. CAs are transformed into functional requirements FRs. This is equivalent to "what we want to achieve" or the goals identified by the designer. FRs are satisfied by identifying appropriate design parameters (DPs) in the physical domain. In a similar manner process variables (PRs) are identified from the DPs. All this process is carried out by effectively mapping from one domain to another. A good design process here is defined by means of the efficiency of the mapping process.

Suh proposes two design axioms that form the basic principles that govern design decision-making. The two axioms are (Suh 1990):

- "The Independence Axiom" (maintain the independence of functional requirements); and
- "The Information Axiom" (minimize the information necessary to meet the functional requirements).

The two axioms can be interpreted as follows (Park 2007):

- A good design according to Suh's Axiom 1 always maintains the independence of functional requirements (FRs). This means that in an acceptable design, DPs and FRs are related in such a way that a specific DP can be adjusted to satisfy its corresponding FR without affecting other functional requirements.
- A good design according to Suh's Axiom 2 is a functionally uncoupled design (satisfying Axiom 1) that has minimum information content.

The axioms help designers to structure and understand design problems, thereby facilitating the synthesis and analysis (interplay) of suitable design requirements, solutions, and processes. However, the axiomatic character of these two design axioms is found to be flawed, as discussed by Olewnik and Lewis (2005). From a validation perspective, the issue with Suh's axiomatic design is that it forces the designer to conform to a particular preference structure, thereby biasing the designer (Olewnik

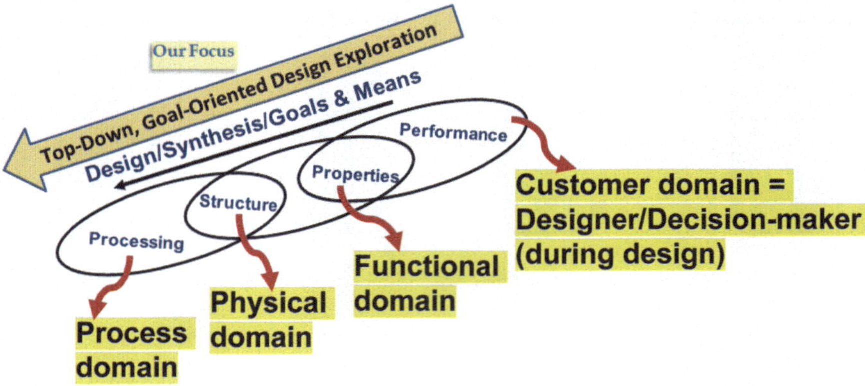

Fig. 3.8 Translation of Suh's axiomatic design domain mappings to Olson's Materials-by-Design

and Lewis 2005). Even though Suh's axioms are flawed, the four domains and the way mappings are carried out across these domains to satisfy customer attributes find application in Olson's process–structure–property–performance mapping from a top-down design perspective. Suh's axiomatic design domain mappings can be translated to Olson's Materials-by-Design as shown in Fig. 3.8. This relationship will be explored further in Chap. 8 in terms of robust product design.

Suh (1990) presents a design equation to mathematically represent the design process in terms of design equations. The design equation is as follows:

$$\{FR\} = [A]\{DP\} \tag{3.1}$$

Using the equation, Suh represents the mapping between the functional domain and the physical domain. Here functional domain refers to "what we want" and the physical domain refers to the means for satisfying what we want. Functional requirements in Suh's design equation refer to the "minimum" (Axiom 2) set of "independent" (Axiom 1) requirements that completely characterize the functional needs of the product in the functional domain that are translated from the customer attributes. Design parameters (DP) are the "key" (Axiom 2) design variables that characterize the physical domain that can satisfy the FRs. The matrix [A] is called the design matrix. A similar vector equation can be written for the mapping from the physical domain to process domain.

FR–DP relationships could be uncoupled (diagonal A matrix), decoupled (triangular A matrix), or coupled (neither diagonal nor triangular) designs. Independence Axiom is satisfied only by uncoupled or a decoupled design, while the coupled design does not satisfy Axiom 1. Thus, Suh's design equation has the following drawbacks:

- Inability to capture complex relationships: For complex problems and relationships, there does not exist an information model associated with Suh's design equation to represent FRs and DPs. Thus, relationships between functional requirements and design parameters are not captured following the axioms.

- Inability to capture design activities other than mapping: In Suh's design equation the only transformation of information captured is the mapping from one domain to another. Other activities associated with design such as decomposition, abstraction, and evaluation are unable to be modeled using the design equation.
- Computational implementation of a design process is not easy using Suh's design equation.
- The design equation serves as a guideline for what a good design is but does not provide guidelines for designing design processes or meta-design.
- The evolution of product information along the design process cannot be understood via Suh's design equation.
- The reusability of design process-related information is also not addressed in Suh's design equation.
- Suh's design equation forces the designer to conform to a particular preference structure, thereby biasing the designer.

3.3 The Decision-Based Design Paradigm—Our Frame of Reference

Mistree et al. (1990) define design as the conversion of information that characterizes the needs and requirements for a product into knowledge about the product. **The underlying philosophy in the definition of design by Mistree et al.** (1990) **is that the designer starting with the functional requirements that are desired (the goal that designer wishes to achieve) should be able to work backward to explore effective design solutions.** This philosophy is adopted in this monograph for design—as a goal-oriented activity. As noted by Gero (1990), the goal of design is to transform requirements—generally termed *functions*—into design descriptions.

The work on decision-based design by Mistree et al. (1990) is anchored in the works of Herbert Simon and James Miller. Simon (1969), in his book, suggests that design is decision-based and one of the sciences of the artificial. The development of any science is anchored on a body of beliefs, hypothesis, and knowledge. In the case of Mistree et al. this is anchored in the exposition of the Living Systems Theory by Miller (1978).

Decision-based design (Shupe 1988; Mistree et al. 1990) is a term coined to emphasize a different perspective to develop methods for design. The principal role of a human designer in decision-based design (DBD) is to make decisions given the information available. Now how do we define a decision here? From an engineering perspective, decisions exclusively deal with the allocation of resources in some form, usually as capital expenditures. Thus, the definition of a decision here is "an irrevocable allocation of resources" (Hazelrigg 1996).

"*There are two important characteristics of a decision:*

- *A decision is made at an instant in time.*
- *A decision must be made based on the information available at the time it is made*" (Hazelrigg 1996).

In this definition of design and decision, the term product is used in its most general sense; it may include processes as well (Mistree and Allen 1997). Through a process of decision-making, there occurs the conversion of information into knowledge. Several characteristics associated with design decisions are identified and are summarized as descriptive sentences (Mistree et al. 1990):

- Decisions in design are invariably multileveled and multidimensional in nature.
- Decisions involve information that comes from different sources and disciplines.
- Decisions are governed by multiple measures of merit and performance.
- All the information required to make a decision may not be available.
- Some of the information used in making a decision may be hard, that is, based on scientific principles; and some information may be soft, that is, based on the designer's judgment and experience.
- The problem for which a decision is being made is invariably loosely defined and open and is characterized by the lack of a singular, unique solution. The decisions are less than optimal and represent ***satisfying solutions***.

3.3.1 The Design Equation (The Decision-Based Design Equation)

Bras (1993) developed a generalization of Suh's design equation. Design is viewed as a process of converting information that characterizes the needs and requirements of products into knowledge about a product. Knowledge is derived by human beings from information by reasoning, discussion, and other mind-involving processes. Thus, knowledge here is specific information. Data is the simple form here and is characterized by a sense of hardness. Information is data, but not all data is information. In the design equation by Bras, a single transformation in the design process is represented as an algebraic design equation, as shown in Eq. 3.2 and Fig. 3.9.

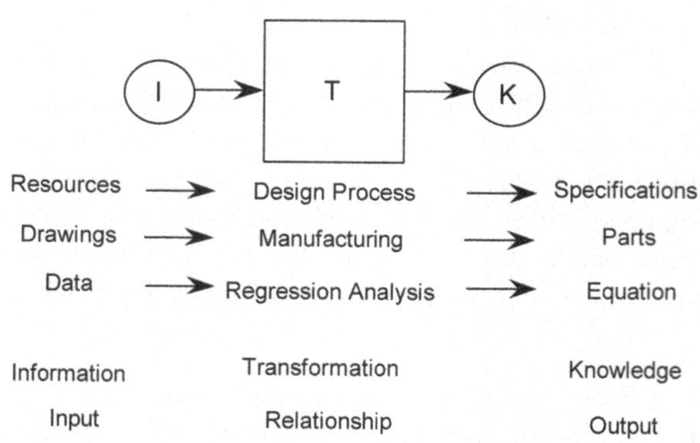

Fig. 3.9 The design equation (Bras 1993)

$$K = T(I) \tag{3.2}$$

where

I is a vector with n components representing the information,
K is a vector with m components representing the knowledge,
T() is a function to transform the vector **I** into vector **K**; the transformation function **T()** comprises a set of m functions, that is, $\mathbf{T}() = (T_1(), T_2(), \ldots, T_m())$.

The design equation by Suh can be viewed as a special case of the design equation developed by Bras and Mistree (1993). This is because of the ability of the design equation by Bras and Mistree to capture nonlinear transformations, unlike Suh's design equation which can capture only linear transformations.

A meta-design equation is also developed as an approximation of the design equation. The meta-design equation is represented as:

$$\Delta K = [T]\Delta I \tag{3.3}$$

where ΔI represents the difference in information and ΔK represents the difference in knowledge, respectively. The [T] is a $m \times n$ transformation matrix. The conversion of information to knowledge is embodied in the transformation matrix. The meta-design transformation matrix can be interpreted as an equivalent of Suh's axiomatic design matrix, [A], which converts design parameters (DPs) into functional requirements (FRs). However, the [T] matrix is not limited to functioning as an approximation to Suh's axiomatic design equation but has the capability to provide an approximate relationship between ΔK and ΔI for any design. The function T() in Eq. 3.2 is satisfied by multiple decision-support problems (DSPs). Hence, DSPs are the implementation of the design equations within DBD. Note there is a difference between matrix [T] and function T(). The DSP technique uses function, whereas Suh uses matrix.

3.3.2 The DSP Technique

In this section, we discuss the DSP technique which is one of the implementation approaches for DBD. According to Muster and Mistree (1988), the decision support problem technique supports a human designer in making rational design decisions using human judgment. There are several methods/approaches to plan, establish goals, and plan systems. However, independent of these approaches, designers are and will continue to be involved in two primary activities, namely *processing symbols* and *making decisions* (Mistree et al. 1993). The assertion here is that the process of design is basically a network of decisions. Designers and engineers need an approach for negotiating ***satisficing solutions*** for their problems rather than optimal solutions due to increasing complexity and interactions of system with its environment resulting in more and more uncertainty within the system. The DSP technique supports a human designer in portioning and formulating the problems in simple terms so that it is

Table 3.2 The phases of DSP technique (Mistree and Muster 1990)

<table>
<tr><th>Phase I: Meta-Design</th><th>Phase II: Design</th></tr>
<tr><td rowspan="2">STEP 1: IDENTIFY/CLARIFY PROBLEM
(Characteristics and design type)
Problem story
↓
Technical brief
↓
Abstracts</td><td>STEP 3 & 4: STRUCTURE
• Organize domain-dependent information and formulate DSP templates
• Develop DSP word formulations
• Develop DSP mathematical formulations</td></tr>
<tr><td>STEP 5: SOLVE
• Obtain solutions
• Solve the DSPs using appropriate means</td></tr>
<tr><td>STEP 2: PARTITION AND PLAN
• Partition each abstract into problem statements and identify decisions associated with each problem statement
• Devise plan for solution in terms of DSPs corresponding to decisions</td><td>STEP 5: POST-SOLUTION ANALYSIS
• Verify and validate solutions
• Sensitivity analysis
• Check for consistency
• Check for need for iteration
• Make design decisions</td></tr>
</table>

possible to find satisficing solutions for it, while being close to the actual system without removing its sources of uncertainty.

Meta-Design and Design Phases in the DSP Technique
The DSP technique requires designers to implement two phases: (a) *meta-design* and (b) *design phase*. The meta-design phase consists of planning and structuring of decision support problems. The meta-design phase is accomplished via problem partitioning into its elemental DSPs and devising a plan of action. In this phase, there is no attempt to make or pursue product-specific decisions. The goal in this phase is to ***design the design process*** (meta-design) to be implemented. The base entities used to specify a design process are phases, events, decisions, tasks, and so on. The information generation comes from input–output relationships. The base entities are used to model the design process as a network that can be managed. In the design phase, the decision support problem is actually solved, and post-solution analysis is carried out. In this phase, the solutions to the design process are sought and these solutions are further verified and validated. Decision support problems provide a means for modeling decisions encountered in design, and the domain-specific mathematical models so built are called ***templates*** or decision support problem templates (Mistree and Allen 1997).

The two phases of the DSP technique can be represented as shown in Table 3.2.

Design Process Modeling Using DSP Technique Palette
In DSP technique, the entities for carrying out meta-design are contained in a DSP palette. These entities are domain-independent and support hierarchical modeling of design processes. There are three classes of entities—***potential support problem entities***, ***base entities***, and ***transmission entities***. The *potential support problem entities* include phases, events, tasks, decisions, and system, as shown in Fig. 3.10.

Phases, denoted by the icon "P", are used to represent elements of the partitioned process. Some examples of phases are the different phases of design in Pahl and Beitz

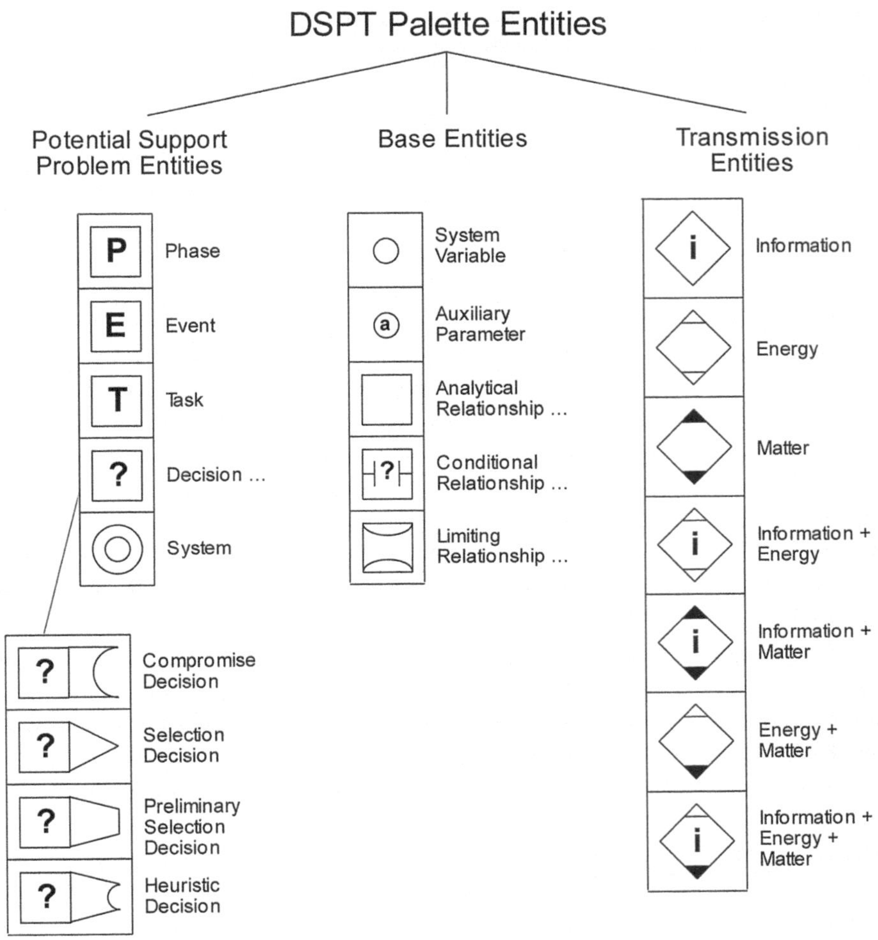

Fig. 3.10 Decision support problem technique palette entities (Mistree et al. 1993)

design process like conceptual design, embodiment design, detail design, and so on. Events, denoted by "E", occur within a phase. An example of an event is—"check for system feasibility". Accomplishments of phases and events occur by tasks and decisions for which a human designer is required. A task here refers to any activity that needs to be accomplished. According to Mistree et al. (1991), Kamal et al. (1992), all decisions identified in the DSP technique are categorized as ***selection, compromise***, or a ***combination*** of these. The selection and compromise decisions are considered as ***primary decisions*** and others as **derived decisions**.

Base entities are the most elementary entities in the DSP technique. They can be implemented on a computer and are used to describe constraints, bounds, relationships between design variables, and so on. Base entities are shown in Fig. 3.10.

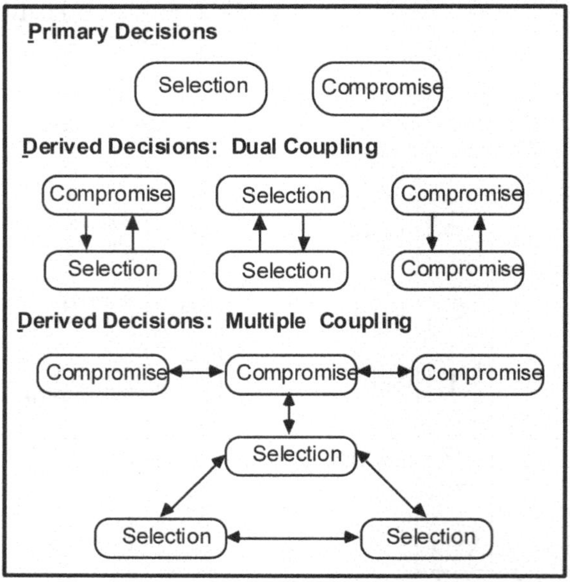

Fig. 3.11 Primary and derived decisions (Mistree and Allen 1997)

Transmission entities are used to define the connections between various other entities used to model the design process and include three types of basic transmissions—mass, energy, information, and their combinations. These entities are based on Miller's Living Systems Theory. Transmissions entities are shown in Fig. 3.10.

3.3.3 Decision-Making in Decision-Based Design

According to the types of decisions, there are the following types of DSPs, see Fig. 3.11.

- ***Selection DSP***—a primary DSP—making a choice between a number of possibilities taking into account and a number of measures of merit or attributes (Kuppuraju et al. 1985; Mistree et al. 1988; Vadde et al. 1995). The emphasis in selection is on the acceptance of certain alternatives through the rejection of others based on different measures of merit, called attributes, which represent the functional requirements.
- ***Compromise DSP***—a primary DSP—the determination of the "right" values (or a combination) of design variables to describe the ***best satisficing*** system design with respect to constraints and multiple goals (Hughes et al. 1993). The cDSP is discussed in greater detail in Sect. 3.3.4 and is the foundational mathematical construct used in this monograph.

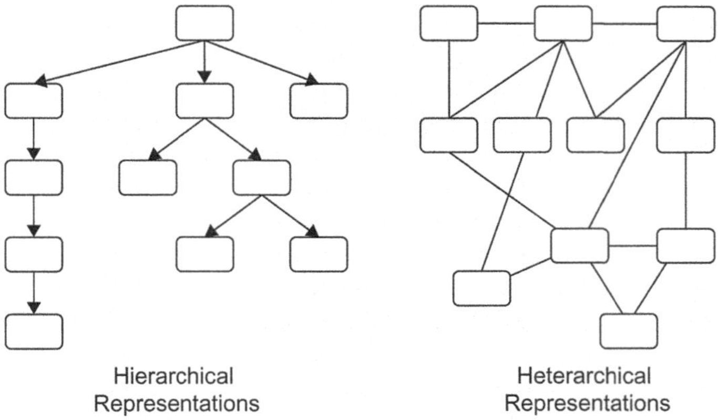

Fig. 3.12 Heterarchical and hierarchical representations (Mistree et al. 1990) [Originally presented at FAST 99, Seattle, August 99. Reprinted with the permission of the Society of Naval Architects and Marine Engineers (SNAME)]

- ***Derived DSPs***—combination of primary DSPs to model a complex decision, for example, selection/selection, compromise/compromise, and selection/compromise decisions (Bascaran et al. 1989; Karandikar and Mistree 1993; Mistree et al. 1993; Vadde et al. 1994).

In the DSP technique, the selection and compromise DSPs are used to address independent decisions. Coupled DSP s are used to model hierarchies of decisions, thus forming a network of decisions (Bascaran 1991). Given a network and its information interactions, two types of modeling of decisions relationship are possible—hierarchical and heterarchical (Mistree et al. 1990; see Fig. 3.12). Heterarchical decision relationships are unordered and it is difficult for designers to define precedence in such decision networks. Hierarchical decisions, on the other hand, are clear in terms of information flow and the sequence of decisions is well defined. Coupled DSPs can be used to implement such hierarchical decisions.

Coupled decisions occur during collaborative design scenarios. One example will be the design of a product with coupling between design engineer and materials engineer. Such coupled decision support problems are applied in various problems like the design of composite material structures (Karandikar and Mistree 1993).

3.3.4 The Compromise Decision Support Problem Construct

In the model-based realization of complex systems, we have to deal with models that are typically incomplete, inaccurate, and not of equal fidelity. This brings into the design process different types of uncertainties associated with the system, the parameters considered, the models considered, and the uncertainties due to their interactions (McDowell et al. 2009). From the decision-based design perspective, the fundamental role of a human designer is to make decisions given the uncertainties associated. In this regard, we define robust design as a design that is relatively insensitive to changes. This involves achieving a desired performance for the system while the sensitivity of the performance objectives with respect to the system variables are minimized (Ebro and Howard 2016). Thus, the designer's objective here is to find "*satisficing*" solutions that showcase good performance given the presence of uncertainties and not optimum solutions that are valid for a narrow range of conditions while performing poorly when the conditions are changed slightly. The cDSP is proposed by Mistree et al. for robust design with multiple goals (Bras and Mistree 1993; Mistree et al. 1993). The fundamental assumption here is that the models are not complete and accurate; opposed to the fundamental assumption in optimization where the models are complete and accurate, and the objective function can be modeled accurately so that solution obtained is implementable. Hence the cDSP construct is anchored in the robust design paradigm first proposed by Taguchi (1986a). Using the cDSP construct several solutions are identified by carrying out tradeoffs among multiple conflicting goals. The obtained solutions are then evaluated by carrying out solution space exploration in order to identify the best solutions that satisfy the specific requirements identified. The cDSP is a hybrid formulation based on mathematical programming and goal programming. It also makes use of some new features. In goal programming (GP),[1] the target values for each goal are defined and the emphasis is on achieving the target for each goal as close as possible (Ignizio 1976, 1978, 1983, 1985). In cDSP, different weights are assigned to these goals and the compromised solutions obtained for different appropriate weights are explored. The obtained solutions are further evaluated by carrying out solution space exploration to identify solution regions that best satisfy the requirements identified. The generic formulation of cDSP is shown in Table 3.2.

The cDSP is similar to GP in that the multiple objectives are formulated as system goals, involving both system and deviation variables and that the deviation function is solely a function of the deviation variables. However, this is in contrast to traditional mathematical programming where multiple objectives are modeled as a weighted function of the system variables associated with the problem. From the traditional constrained optimization formulation, the cDSP retains the concept of system constraints. The cDSP places a special emphasis on the bounds of a system variable which is unlike traditional mathematical programming and GP. Contrary to GP formulation, the cDSP constraints and bounds are handled separately from system goals. In the cDSP, the feasible design space is defined by the set of system constraints and bounds. The aspiration space is defined by the set of system goals (see Fig. 3.13). For feasibility, the system constraints and bounds must be satisfied. A satisficing solution then is that feasible point that achieves the targets of the system goals as close as possible. The solution to this problem represents a tradeoff between that which is desired (as modeled by the aspiration space) and that which can be achieved (as modeled by the design space) (Mistree et al. 1993).

---

[1]GP was used as a short form for Gaussian Process in Chap. 2. The usage of GP in this section correspond to Goal Programming and is not related to the usage in Chap. 2.

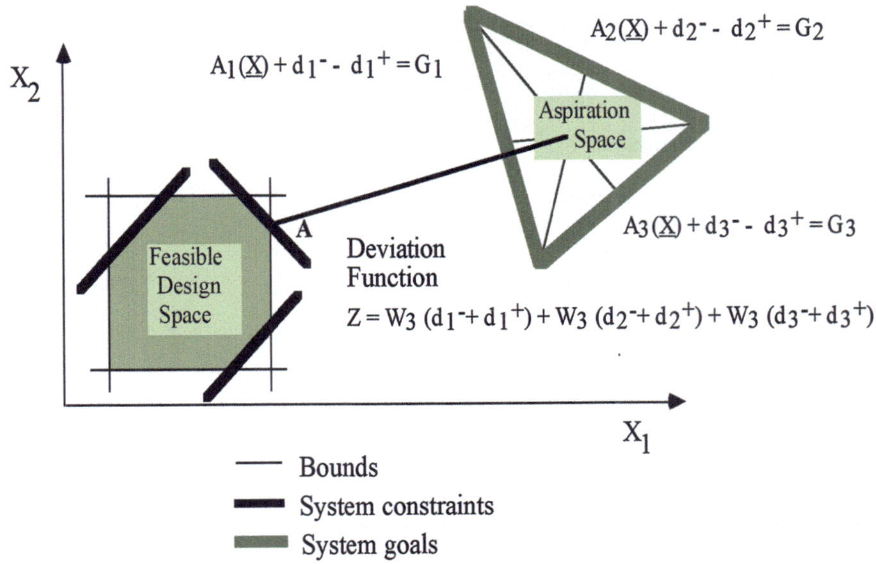

Fig. 3.13 Graphical representation of a two-dimensional compromise DSP, Archimedean formulation (Mistree et al. 1993)

There are four keywords in the cDSP formulation. All the information that is available for the designer to formulate the cDSP so as to make effective decisions are captured by the "Given" keyword. In the cDSP, for each objective an achievement function $A_i(X)$ is formulated and represents the achieved value of the *i*th objective as a function of a set of system variables, *X*. The deviation variables, d_i^- and d_i^+ represent the extent to which the goal target G_i is underachieved or overachieved with respect to the value of A_i(X). The information regarding the system variables and the deviation variables are embodied in the "Find" keyword. The information regarding system constraints, variable bounds, and system goals are captured by the "Satisfy" keyword to determine the feasible design space and the aspiration space. The "Minimize" keyword embodies the objective function which is formulated as a function of the deviation variables. The overall goal of the designer using the cDSP construct is to minimize the deviation function so that the target values specified for the system goals are attained as closely as possible by identifying the combination of design/system variables that best satisfy the conflicting requirements. The formulation of the deviation function is done in two ways: (a) as a pre-emptive (lexicographic) formulation or (b) as an Archimedean formulation—based on the manner in which importance is assigned to satisficing the goals. The most general form of the deviation function for "*m*" goals in the Archimedean formulation is:

$$Z = \sum_{i=1}^{m} W_i\left(d_i^- + d_i^+\right); \sum_{i=1}^{m} W_i = 1 \quad (3.4)$$

Table 3.3 The cDSP formulation (Mistree et al. 1993)

Given
An alternative to be improved, domain-dependent assumptions The system parameters: n number of system variables, q inequality constraints, $p + q$ number of system constraints, m number of system goals, $g_i(\mathrm{X})$ system constraint functions $f_k(d_i)$ function of deviation variables to be minimized at priority level k for the pre-emptive case
Find
System variables: The values of the independent *system* variables. X_i $i = 1, 2, \ldots, n$(They describe the physical attributes of an artifact.) *Deviation variables:* The values of the deviation variables. d_i^-, d_i^+ $i = 1, 2, \ldots, m$ (They indicate the extent to which the goals are achieved)
Satisfy
System constraints: These must be satisfied for the solution to be feasible (linear, nonlinear) $g_i(\mathbf{X}) = 0;\ i = 1 \ldots p$ $g_i(\mathbf{X}) \geq 0;\ i = p + 1 \ldots p + q$ *System goals:* These need to achieve a specified target value as far as possible (linear, nonlinear) $A_i(\mathrm{X}) + d_i^- - d_i^+ = G_i;\ i = 1 \ldots m$ *Bounds:* Lower and upper limits on the system variables. $X_i^{\min} \leq X_i \leq X_i^{\max};\ i = 1 \ldots n$ $d_i^-, d_i^+ \geq 0,\ d_i^- * d_i^+ = 0;\ i = 1 \ldots m$
Minimize
A deviation function: A function that quantifies the deviation of the system performance from that implied by the set of goals and their associated priority levels or relative weights. Case a: Pre-emptive formulation (lexicographic minimum) $\mathrm{Z} = [\mathrm{f}_1(\mathrm{d}_i^-, \mathrm{d}_i^+), \ldots\ldots, \mathrm{f}_k(\mathrm{d}_i^-, \mathrm{d}_i^+)]$ Case b: Archimedean $\mathrm{Z} = \sum_{i=1}^{m} \mathrm{W}_i(\mathrm{d}_i^- + \mathrm{d}_i^+);\ \sum_{i=1}^{m} \mathrm{W}_i = 1$

where the weights W_i reflect the level of desire to achieve each of the goals.

The details regarding formulating the cDSP and the associated rules are provided by Bras and Mistree (1993), Mistree et al. (1993) (Table 3.3).

The cDSP can be reformulated to carry out selection decisions also. Hence it is the principal mathematical DSPT formulation (Bascaran et al. 1989). The cDSP is used in this monograph and is the foundational mathematical construct to provide design decision support in identifying satisficing design solutions. In the next section, we discuss the philosophies of an "optimizer" and that of a "satisficer" (the designer's philosophy in this monograph).

3.3.5 *Optimizing Versus Satisficing Philosophy in This Monograph*

We believe there are two schools of thought for modeling design decisions—(i) that of an optimizer and (ii) that of a satisficer. The focus in this monograph is to share the observations with respect to modeling such decisions from the perspective of a satisficer and not that of an optimizer. The difference here is as follows. Consider a haystack with a number of needles hidden in it. An optimizer will continue searching the haystack until the last needle has been found. A satisficer, on the other hand, stops when he/she has found enough needles to proceed to the next step. We capture the perspective of the satisficer by using the compromise DSP. Our intention in solving the compromise DSP is to satisfice a set of goals. In our formulation the satisficing of goals solves the mathematical problem at hand; optimizing the numerical value of a goal function is not an issue. Let's discuss this in detail from the perspective of integrated materials and product design.

For complex systems design problems like the problem discussed in this monograph on the integrated realization of materials, products, and associated manufacturing processes, the following characteristics can be observed about the design problem and the information available (McDowell et al. 2009):

- Design problem could be loosely defined and open;
- Design information comes from different sources and disciplines;
- There will be multifunctional requirements in design and they are governed by multiple measures of merits and performances;
- All information required for design may not be available and thus the designer may have to work with incomplete, inaccurate, and infidel models and information;
- Design information may be hard (quantifiable) and some may be soft (qualitative).

Thus, such complex design problems are not characterized by utopian single-point unique solutions. The solutions due to these characteristics of complex design problems are less than optimal and therefore seek satisficing solutions. Simon (1996) coined the term "satisficing" to describe solutions that are "good enough to be acceptable but are neither exact nor optimal". Designer/decision-maker has two choices while formulating and solving such complex design problems:

- Solve the exact problem approximately, and
- Solve an approximation of the problem exactly.

In the first choice, the designer seeks an optimal solution using algorithms that are based on relatively simple models, by means of which an exact optimal solution can be found provided the assumptions on which the model is based can be satisfied exactly. However, only rarely does a solution that is optimal for a simple model is optimal in reality.

In the second choice, an approximate algorithm or heuristic, based on a relatively complex model, that can capture reality more closely than a simple model is used. The solutions thus obtained using an approximate algorithm are satisficing. The sequential linear programming approach is used for solving cDSP formulations because it has the highest potential for being used to develop a single algorithm for solving a range of DSPs in engineering design, as described by Mistree et al. (1993). Refinements to sequential linear programming (SLIP) and its multilevel version (SLIPML) have resulted in the adaptive linear programming (ALP) algorithm. The ALP algorithm with its multilevel, multigoal feature is incorporated in decision support in the design of engineering systems (DSIDES), a tailored computational infrastructure for formulating, solving, and analyzing decision support problems (Mistree and Kamal 1985; Reddy et al. 1996). Mistree et al. believe that three important features contribute to the success of the ALP algorithm, namely (Mistree et al. 1993):

- the use of second-order terms in linearization;
- the normalization of the constraints and goals and their transformation into generally well-behaved convex functions in the region of interest;
- an "intelligent" constraint suppression and accumulation scheme.

The approach followed in this monograph is based on the ***satisficing view of design*** embodied in the decision support problem technique proposed by Mistree et al..

3.3.6 Use of DBD and cDSP in This Monograph—Architecting Robust Materials, Product, and Manufacturing Process Networks

The primary goal in this monograph is to develop a systems-based design architecture for the integrated realization of materials, products, and processes and contribute the knowledge generated to advance the field of systems-based materials design. We adopt a decision-centric approach in this monograph because our end goal is to carry out decision-based meta-design and design of complex material and product systems. In this monograph, meta-design involves partitioning the complex material-product system based on functions into system-level function structures, partitioning the design process into decisions, and planning the sequence in which decisions are most appropriately made, following the procedures of DBD proposed by Mistree et al. (1990). A decision-centric view accommodates easily the other views of design processes like model-centric and tool-centric views, which is an added advantage. It also supports the domain-independent representation of design processes. Decision-based design is described as the seed that glues together the heretofore disparate engineering disciplines as well as economics, marketing, business, operations research, probability theory, optimization, and others (Hazelrigg 1998). Hazelrigg, therefore, describes DBD as omni-disciplinary. In the case of integrated materials and product design, we are dealing with multiple disciplines and a network of decisions to develop a product. The information flowing through the process chain serves as the source of the knowledge about the product which is captured in DBD.

In this monograph, ***decision-based design is used as a philosophical foundation for the systems-based design architecture for the integrated design of materials, products, and associated manufacturing processes***. Hence, the focus in this monograph is on making decisions that ***support architecting networked material-product systems***. The second, third, and fourth research questions (**RQ 2, 3, and 4**) addressed in the monograph are answered from a decision-based design perspective. The design of design processes (***designing design methods in this monograph***), hence, is equivalent to the configuration of **networked design decisions**—related to both materials/products and design processes. The **idea of robustness** in the network is key here and is reviewed in the next section and will be discussed in detail in Chap. 7. This leads to the key outcome in this monograph—***Product, and Manufacturing Process Networks***.

3.4 Robust Design of Materials and Products

In the design of complex networked systems, a very important factor that needs to be taken into account is uncertainty. As discussed in previous sections, in the model-based realization of complex systems, designers have to deal with models that are typically incomplete, inaccurate, and not of equal fidelity leading into uncertainty. The presence of uncertainty may lead to wrong decisions during the design of such networked systems. Now, the perfect approach will be to eliminate or mitigate this uncertainty. However, eliminating the uncertainty present in a system and its models is practically infeasible, time-consuming, and expensive. To address this challenge, a robust design paradigm was proposed where the focus is to **design systems to be insensitive to these uncertainties without removing their sources**. In this section, we carry out a review of robust design and investigate the progress made and their limitations in achieving robustness in design.

3.4.1 Uncertainty Classification

Uncertainty classification was discussed in brief in Chap. 1. In this section, we look at it in much greater detail. There are different views on the types of uncertainty that is present in engineering systems involving materials and products. From the ICME perspective the types of uncertainties are classified based on experimental (extrinsic) and modeling perspectives (intrinsic) (Horstemeyer 2012, see Fig. 3.14). The extrinsic uncertainty includes errors due to experimental setup, sensors, surroundings, and so on. Intrinsic uncertainty includes uncertainty due to modeling (model-related) and parametric (due to parameters involved in the model).

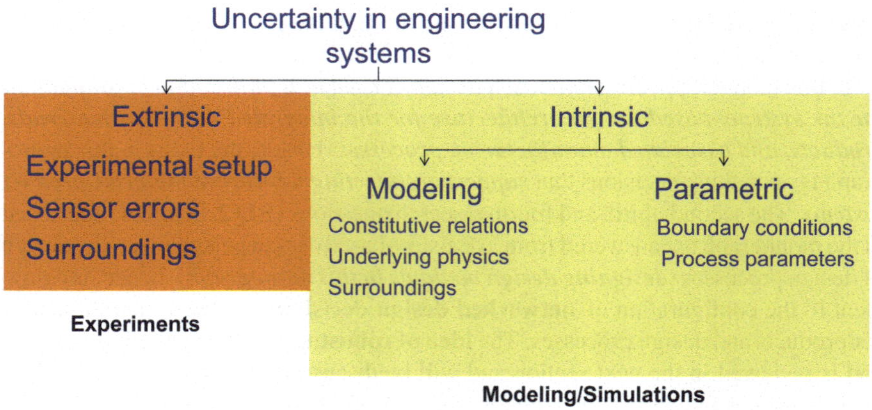

Fig. 3.14 Types of uncertainty from ICME perspective (Horstemeyer 2012)

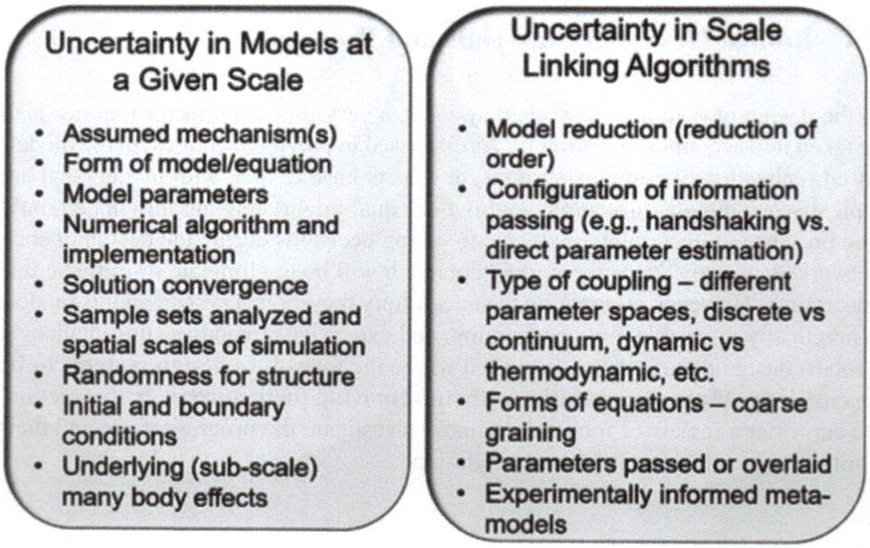

Fig. 3.15 Sources of uncertainty in models at a given scale in material structure hierarchy (left) and in scale linking or scale transition algorithms (right) (McDowell 2018)

The different sources of uncertainty that arise in multiscale modeling hierarchy, both at a given scale and scales linking different algorithms, are summarized by McDowell (see Fig. 3.15). McDowell (2018) observes that quantifying uncertainty in schemes for linking models at different length and time scale is still an immature field and formal mathematical approaches for doing this are largely undeveloped. The uncertainty in the coupling of models across length and time scales can compound

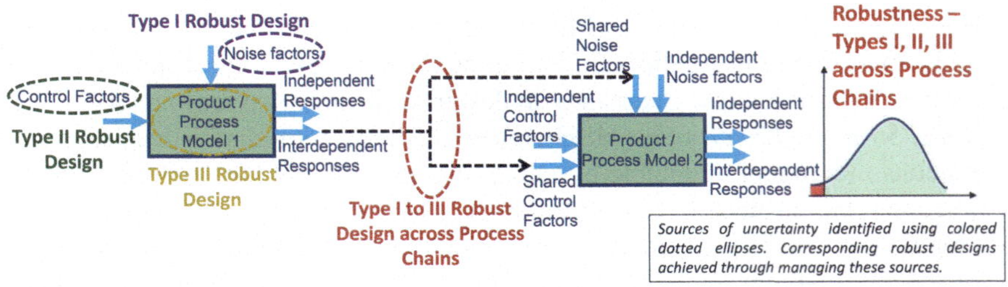

Fig. 3.16 A P-diagram showing the input and response in the design of product or process chains. Robust design is classified based on the source of variability

different other sources of uncertainty that are related to material model or material hierarchy at each scale, as summarized by McDowell in Fig. 3.15.

Since it's impractical to quantify uncertainty accurately in such multiscale models and systems, the recommendation made for simulation-assisted materials design is to focus on understanding the sensitivity of material properties to material microstructure and to capture dominant mechanisms and transitions that affect material responses or properties instead of focusing on accurately predict mean properties at higher scales. This is the philosophy of managing uncertainty. Sensitivity analysis of responses is important because of several reasons (McDowell 2018):

- It is challenging to isolate response sensitivity experimentally at specific scales in the material hierarchy,
- The identification of key design variables across material structure hierarchy is possible via sensitivity analysis, and
- Core to the concept of robust design is the sensitivity of process—structure and structure—property relations, where the focus is to explore a range of solutions that meet conflicting response requirements and identify satisficing design solutions that are relatively insensitive to uncertainty (McDowell et al. 2009).

Uncertainty Classification Adopted in this Monograph

In this monograph, we adopt a more general classification of uncertainty. We are not focused on the uncertainty due to experiments in this monograph. Uncertainty could be either *aleatory* (irreducible) or *epistemic* (reducible), depending on their causes. Improving the measurements and/or model formulation and/or increasing the accuracy are ways to diminish epistemic uncertainty. Aleatory uncertainty, however, is inherent in the physical system and can only be quantified in a statistical sense. Extending the classification by Isukapalli et al. (1998), the types of uncertainty in the simulation-based integrated design of the material, product, and processes are classified as (Choi et al. 2005; McDowell et al. 2009) (see Fig. 3.16):

- *Natural Uncertainty* (NU): uncertainty due to the inherent randomness or unpredictability of a physical system; aleatory in nature.

- *Model Parameter Uncertainty* (MPU): incomplete knowledge of model parameters/inputs due to insufficient or inaccurate data; reducible by sufficient data or accurate measurements; uncertainty in design variables or control factors.
- *Model Structure Uncertainty* (MSU): uncertain model formulation due to approximations in a model; reducible by improving model formulation; uncertainty in function relationship between control/noise and response.
- *Propagated Uncertainty* (PU): uncertainty compounded by the combination of all the above three types of uncertainty in a chain of models that are connected through input—output relations; interdependent responses, and shared control/noise factors as one model interacts with another (Allen et al. 2006).

All these types of uncertainty can exist in a system and it is very difficult to differentiate out which is dominant among them. It is not possible to eliminate all uncertainty. The possibility here is to manage them. The focus, therefore, is to manage uncertainty by designing the system to be insensitive to the sources without reducing or eliminating them. This is done by exploring the solution space and studying the sensitivity of responses to variations in noise, control factors, and models themselves and understanding the tradeoffs required with various compromises; this is called robust design (Chen et al. 1996a, b; Choi et al. 2005, 2008; Murphy et al. 2005; Allen et al. 2006; Nellippallil et al. 2017). There are several practical implications with this approach. Robust solutions are not focused on extensive optimization searches at individual levels and do not necessarily involve a large number of iterations (McDowell and Olson 2008). The practical interest here is for a ranged set of solutions that showcase good performance under variability rather than single-point solutions that are valid for a narrow range of conditions while performing poorly when the conditions are changed slightly. The human designer plays the role of an interpreter of the value of information in this approach. Concepts and mathematical constructs from information economics (Howard 1966; Panchal et al. 2006; Sinha et al. 2013) are used to manage uncertainty by studying the value of information for cost/benefit tradeoff to make design decisions in the presence of uncertainty. In the next sections, we review the robust design types and methods, frameworks that allow.

3.4.2 Robust Design Type I—Taguchi Method

In robust design (RD), the quality of products and processes is improved by reducing their sensitivity to variations without eliminating the sources (Taguchi 1986b; Taguchi and Clausing 1990; Nair et al. 1992; Tsui 1992). The robust design principles and methods are founded on the philosophy of Taguchi (1986a). Three categories of information interact with the system model in robust design (McDowell et al. 2009): (i) control factors, also known as design variables are parameters that the designer adjusts to move toward the desired product, (ii) noise factors, are exogenous parameters that affect the performance of product/process but cannot be controlled by the

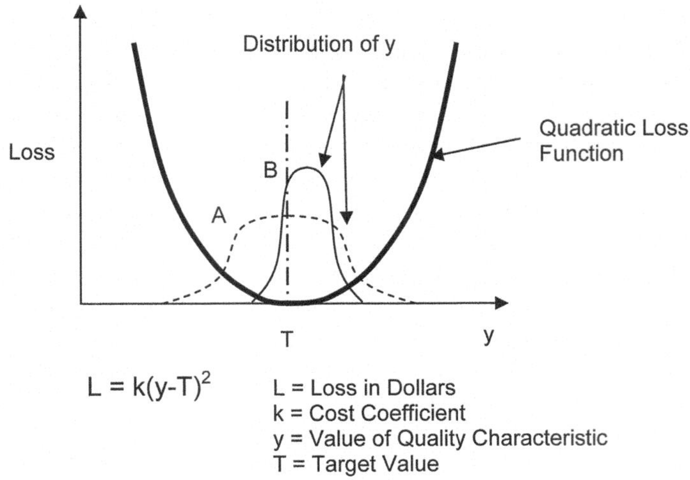

Fig. 3.17 Taguchi's quality loss function (adopted from Choi 2005)

designer, (iii) responses, are performance measures for the product or process. These categories of information are captured in Fig. 3.16.

As briefly described in Chap. 1, Type I robust design is to identify the control factor (design variable) values that satisfy a set of performance requirements despite variations in noise factors. Type I robust design was first proposed by Taguchi (1986a), Taguchi and Clausing (1990), Taguchi (1993). Taguchi's robust design principles and approach arobust designre focused on reducing the effects of variability without removing its sources and were foundational for product and process design in Japanese industries. Taguchi's approach uses experimental design (orthogonal arrays), quality loss function, and signal-to-noise (S/N) ratio. Taguchi proposed a quality loss function, and the objective here is to quantify the loss that is imparted by the product to the society from the time the product is shipped. In Taguchi's approach, the quality of the product is maximum when the loss imparted is minimum. Taguchi's quality loss function is specified as:

$$L = k(y - T)^2 \tag{3.5}$$

where L is the loss in dollars, k is the cost coefficient, y is the value of quality characteristic, and T is the target value. The quality loss function is illustrated in Fig. 3.17.

In Fig. 3.17, the two probability distributions A and B denote product outputs. The average loss of quality of B is smaller than that of A. This is predicted based on the deviation of the average value of y from the target, T and the mean squared deviation of y around its own mean, as described by Phadke (1995). Taguchi proposes three stages for engineering design and asserts that all three stages are important for achieving robust design. The three stages are system design, parameter design, and tolerance design. Taguchi specifically highlights the importance of parameter design stage to identify desirable parameters so as to minimize quality loss. The parameter design stage in Taguchi's robust design approach starts with clearly classifying parameters into control factors and noise factors. Control factors are design parameters that can be controlled while noise factors are uncontrollable or expensive to control. An orthogonal array as the experimental design is recommended. The control factors reside in an orthogonal array and the noise factors in an outer array. All combinations of control and noise factors are recorded in the experiment. Nominal responses are evaluated by identifying an average response by varying the noise factors for fixed control factor conditions. The signal-to-noise ratio proposed by Taguchi captures the sensitivity of responses to variations in noise factors. Based on the mean response and the signal-to-noise ratio, the best combination level of each control factor is selected by designers.

The approach by Taguchi is widely accepted in industry and academia. Many industrial problems are addressed from the perspective of robust design using Taguchi's approach.

Though Type I robust design principles as proposed by Taguchi are advocated widely, his statistical techniques that include orthogonal arrays and signal-to-noise ratio are widely criticized. Many researchers Box (1988), Vining and Myers (1990), Welch et al. (1990), Shoemaker et al. (1991), Tsui (1992), Parkinson et al. (1993), Sundaresan et al. (1995), Chen et al. (1996a, b) have actively worked on improving the statistical techniques in robust design and thus have over the years developed mathematical constructs that bring in robust design into a systematic framework. We discuss some of the criticisms about the Taguchi approach next.

Criticisms on the Taguchi Approach

Taguchi's experimental design and orthogonal arrays are widely criticized as computationally costly and inefficient as the approach requires a large number of experiments which are mostly unnecessary. There have been efforts to minimize the computational burden associated with this approach. Welch et al. (1990) and Shoemaker et al. (1991) address this issue and presents a combined single array for both control and noise factors and thereby reducing the associated computational cost. Also, approximations of the mean and variance using the response surface method were developed by Vining and Myer (1990) and Shoemaker et al. (1991).

The second major criticism of Taguchi's robust design is on the signal-to-noise ratio. This is criticized because there is a high chance for designers to miss useful information since the S/N ratio includes both mean and variance in its formulation. Since the effects on the mean are confounded with the effects on the variance, then the true factors affecting the mean cannot be separated. An alternative proposed here by statisticians (Box 1988; Vining and Myers 1990; Tsui 1992) is to model both the mean and variance directly instead of combining them into one signal-to-noise ratio function.

Another criticism is regarding the usage of Taguchi's robust design approach for unconstrained problems only. Parkinson et al. (1993) address this issue by proposing "feasibility robustness" for design. Feasibility robustness is considered by using first-order Taylor series expansion to calculate the amount of variation that needs to be considered in constraint function for variations in control and noise factors.

Even though the Taguchi method has been criticized and many researchers have developed alternatives for robust design, the philosophy of robust design by Taguchi has found a lot of applications in industry, leading to successful outcomes. Achieving design insensitive to the surroundings (noises) is a significant achievement by Taguchi's robust design approach. Taguchi's approach is invalid in situations where noise factors cannot be quantified as numeric parameters—a situation that can happen in integrated materials and product design.

3.4.3 *Suh's Axiomatic Design and Shannon's Information Theory in Robust Design*

Suh's axiomatic design facilitates robust design at the conceptual design phase, unlike Taguchi's robust design that is employed at the embodiment or detail design phase. Suh's axiomatic design is discussed in detail in Sect. 3.2.2. The two axioms by Suh, "The Independence Axiom" and "The Information Axiom" can be used to support robust design. As per Suh, a good design is one that satisfies both the axioms. Three types of designs are possible: (a) uncoupled design, (b) decoupled design, and (c) coupled design following the design equation matrix by Suh. The corresponding design equations following {FR} = [A]{DP} (Eq. 3.1) are shown in the following:

$$\begin{bmatrix} FR_1 \\ FR_2 \\ FR_3 \end{bmatrix} = \begin{bmatrix} X & 0 & 0 \\ 0 & X & 0 \\ 0 & 0 & X \end{bmatrix} \begin{bmatrix} DP_1 \\ DP_2 \\ DP_3 \end{bmatrix} \text{ Uncoupled Design} \tag{3.6}$$

$$\begin{bmatrix} FR_1 \\ FR_2 \\ FR_3 \end{bmatrix} = \begin{bmatrix} X & 0 & 0 \\ X & X & 0 \\ X & X & X \end{bmatrix} \begin{bmatrix} DP_1 \\ DP_2 \\ DP_3 \end{bmatrix} \text{ Decoupled Design} \tag{3.7}$$

$$\begin{bmatrix} FR_1 \\ FR_2 \\ FR_3 \end{bmatrix} = \begin{bmatrix} X & X & X \\ X & X & X \\ X & X & X \end{bmatrix} \begin{bmatrix} DP_1 \\ DP_2 \\ DP_3 \end{bmatrix} \text{ Coupled Design} \tag{3.8}$$

In Suh's axiomatic design, the uncoupled design is always considered the best design because it satisfies the independence axiom. The decoupled design follows the uncoupled design in terms of preference because it satisfies independence axiom by sequentially selecting design parameter values, thereby ensuring that the functional requirements are independent with each other with respect to the design parameters. However, the coupled design is not acceptable as per Suh's axiomatic design as it does not satisfy the independence axiom. Suh's suggestion to designers is to use the independence axiom first to select the best functional configuration of the system among all available design candidates.

Once the independence axiom is followed to select the system's functional configuration, the next step is to follow the information axiom and select design parameters so as to minimize information content. The axiom may look simple, and something that all designers understand—"that is keep everything simple". However, this has significant meanings in engineering design. After finding multiple designs using Axiom 1, the best one can be chosen based on Axiom 2. The best design has the minimum information content that is usually quantified by the probability of success. Now how can we quantitatively measure information? Usually, information is related to complexity. Then how is it possible to measure complexity? This leads us to the definition of information content and Shannon's Information Theory (Shannon 1949; Shannon and Weaver 1963).

In 1948, Claude Shannon published a paper called "A Mathematical Theory of Communication" (Shannon 1948). Stone (2015) reports "*This paper heralded a transformation in our understanding of information. Before Shannon's paper, information had been viewed as a kind of poorly defined miasmic fluid. But after Shannon's paper, it became apparent that information is a well-defined and, above all, measurable quantity.*"

The key point in Shannon's information is that ***Shannon information is a measure of surprise***. One way to express this is to define the amount of surprise of an outcome value x to be the 1/(the probability of x) or $1/p(x)$, so that the amount of surprise associated with the outcome value x increases as the probability of x decreases. Since information associated with a set of outcomes is obtained by adding the information of individual outcomes (Shannon's additivity condition), Shannon defines surprise as the logarithm of $1/p(x)$, see Fig. 3.18.

Using logarithms to the base 2, the Shannon information of a particular outcome measured in bits is written as:

$$\mathrm{h(x)} = \log_2 \frac{1}{\mathrm{p(x)}} \mathrm{bits} \tag{3.9}$$

where h stands for Shannon information. The average surprise is defined by the probability distribution of a random variable. The average surprise of a variable X which has probability distribution $p(X)$ is called the entropy of $p(X)$, represented as $H(X)$.

Now how does Shannon's information relate to Suh's axioms and robust design?

Let us look at a case of a uniform probability distribution of a design range. Let I be the information content measured in terms of entropy. The above equation can be rewritten as:

$$\mathrm{I} = \log_2 \frac{\text{System Range}}{\text{Common Range}} \tag{3.10}$$

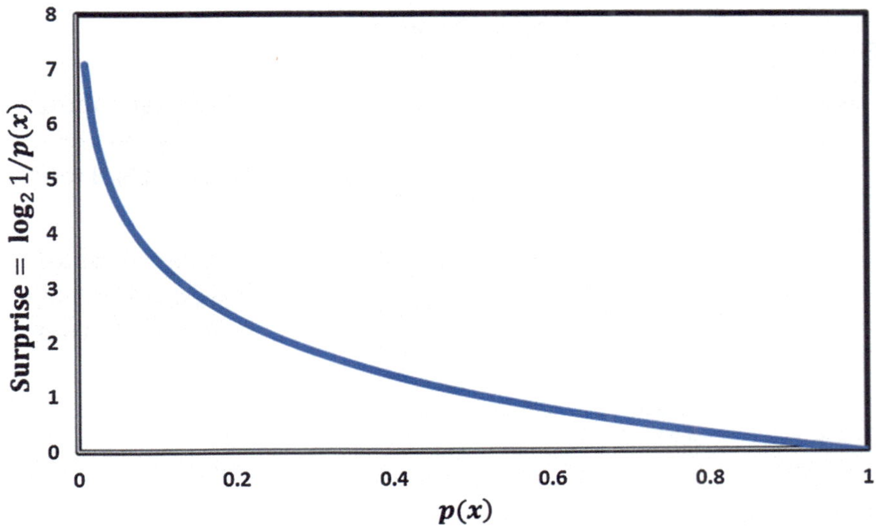

Fig. 3.18 Shannon information as surprise. Values of x that are less probable have larger values of surprise (Figure based on the work of Stone 2015)

Fig. 3.19 Design, system, and common range for calculating the probability of success (Choi 2005)

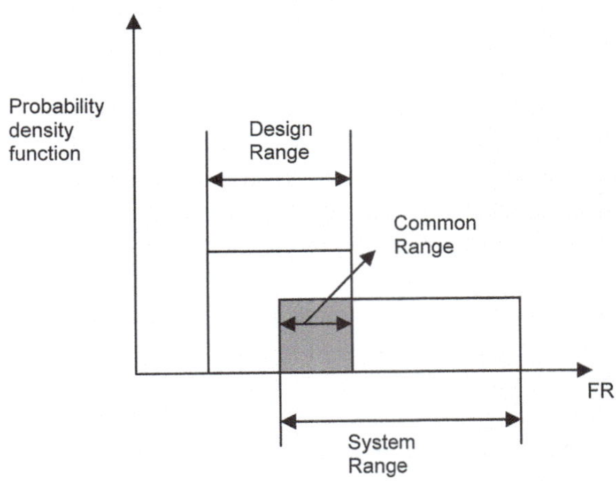

In Fig. 3.19, the desirable system range for meeting functional requirements is the design range, the system range defines the deviation of the functional requirement of a candidate, and the common range is the overlap between design range and system range. Now, the information content (I) is minimum when the probability of success is maximum. We achieve a maximum probability of success when the common range is maximized and/or when the system range is minimized. Therefore, in the perspective of Suh's information axiom, the designer should select a design candidate that has minimum information content based on the calculation of the probability of success.

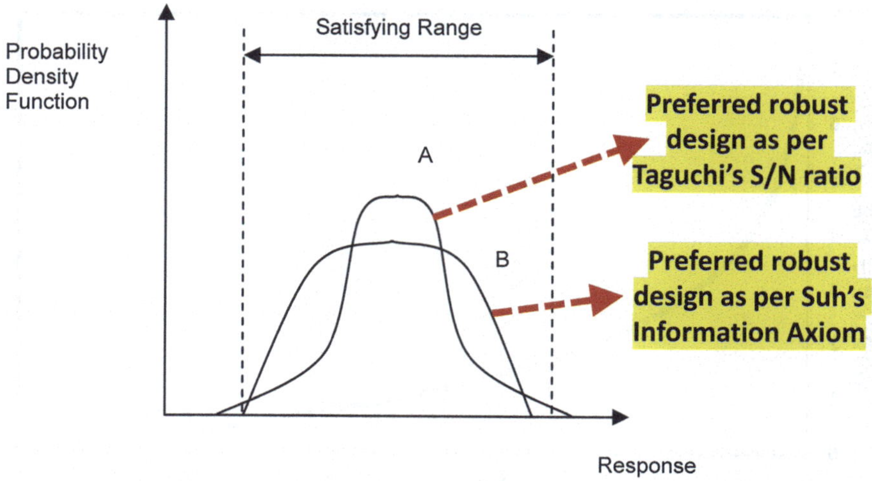

Fig. 3.20 Robust designs from the perspective of Taguchi and Suh

The philosophies in Taguchi's approach and Suh's axiomatic design on robust design are different. This can be illustrated using Fig. 3.20.

In Taguchi's method, a designer will select design A in Fig. 3.20 as design A results in minimum loss and maximum quality measured in terms of signal-to-noise ratio. However, in Suh's axiomatic design, a designer will prefer design B as it results in minimizing information content, thereby satisfying the information axiom. Now, this cannot let us conclude that one method is better than the other. The selection of these approaches should depend on the problem type and the requirement of the designer. The axiomatic design is preferred if the boundary of the satisfying range is strict. However, if the goal of the designer is to hit a target value and minimize the variation with respect to noise factors, then the Taguchi method has to be preferred. One remark that can be made based on this discussion is that Taguchi's approach is clearly a parametric design method meant for the detail design phase while Suh's axiomatic design is more like a decision-making tool that is more suited for the conceptual design stage.

One major drawback of Suh's axiomatic design for robust design is that it lacks a procedure to analyze the sensitivity of performance due to variations associated with the factors—control and noise factors .The information axiom can capture the designer's specific preference but is not a systematic procedure or method to be adopted for robust design. Also similar to the Taguchi method, Suh's axiomatic design cannot be applied to systems where we have unparameterizable variability—a situation that can happen in integrated materials and product design. Next, we discuss the robust concept exploration method for Type I and II robust design and the approaches followed in decision-based design (DBD) in accounting for robustness and managing uncertainty in the realization of complex systems.

3.4.4 Robust Design Type II—The Robust Concept Exploration Method (RCEM)

The focus of most of the work carried out on robust design is on the detailed design stage. The assumption in most of the work is that a preliminary design has already been established with a concrete layout and design specifications, which is mostly not the case. Some researchers have focused on infusing robustness at the early stages of design especially at the conceptual design stage. The decisions usually made at this stage have long-lasting and profound impact on the final product performance and quality. Now while exploring concepts designers are required to work with continuous design spaces. The efficacy of Taguchi's robust design method and Suh's axiomatic design is limited mostly to discrete design spaces or when we have a number of discrete design alternatives (for Suh's axiomatic design) and cannot be used to actively search a continuous design space. Along with this requirement, there is also a need for achieving designs insensitive to variations in not just the noise factors, but also the control factors. To address these needs, Chen et al. propose Type II robust design along with Type I for managing variations in control and noise factors (Chen et al. 1996a, b) and propose the robust concept exploration method (RCEM) to systematically explore robust solutions insensitive to variations in control and noise factors at the early stages of design (Chen et al. 1997).

Chen et al. categorize problems associated with simultaneously minimizing performance variations and bringing the mean on target based on their source of variation as (Chen et al. 1996a, b):

- Type I—minimizing variations in performance caused by variations in noise factors (uncontrollable parameters)
- Type II—minimizing variations in performance caused by variations in control factors (design variables).

The schematic of the concepts behind the two types of robust design is provided by Chen et al. in Fig. 3.21.

In Fig. 3.21, Chen et al. show the variation that happens in the performance function when there is variation in noise factors and control factors. In Type II robust design, we seek solutions in the nearly flat region where the variation in system performance is less for variation in control factors rather than the optimal solution point where even for a small variation in control factor, the system performance degrades significantly.

Chen et al. propose robust concept exploration method (RCEM), which is a domain-independent, systematic method that integrates statistical experimentation, approximate models (metamodels/response surface models), multiobjective decisions, and multidisciplinary analyses, to carry out robust design at early stages of design. The computing of RCEM is shown in Fig. 3.22.

In the RCEM, design parameters are categorized as noise factors and control factors in Processor A—Factors and Ranges. Simulation experiments are designed in Processor B—Point Generator to develop response surface models that capture the

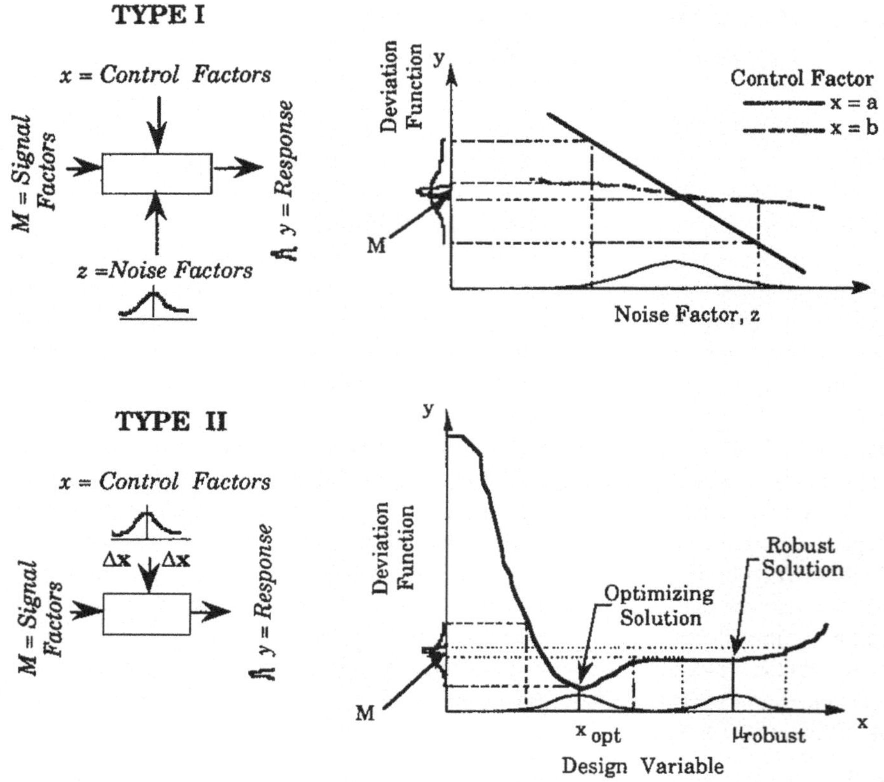

Fig. 3.21 Robust design for variations in noise factors (Type I) and control factors (Type II) (Chen et al. 1996a, b)

problem specifications. Using simulation program C, experiments that are designed are carried out, results are generated which are further analyzed and screened in Processor D—Experimental Analyzer. The unimportant factors are removed based on statistical analysis. The stages C and D are repeated sequentially until the best set of data is generated to build the response surface model (RCEM uses the response surface method to develop meta-models). The response surface models are build using Processor E. In Processor F, the cDSP construct is used to formulate the design problem and find a ranged set of design specifications that are robust against the variations associated with noise and control factors. The RCEM uses specific goals in cDSP formulation that are meant to bring the mean on performance target and minimize performance variation. The RCEM has been used for a variety of applications to design robust systems. This includes structural problem and design of a solar-powered irrigation system (Chen et al. 1996a, b), a high-speed civil transport (Chen et al. 1995), a general aviation aircraft (Chen et al. 1996a, b), product platforms (Simpson et al. 1996), and other applications (e.g. Chen et al. 2001). The RCEM finds applications in most common class of materials and product design problems. Using RCEM designers can formulate design problems to find ranges of material

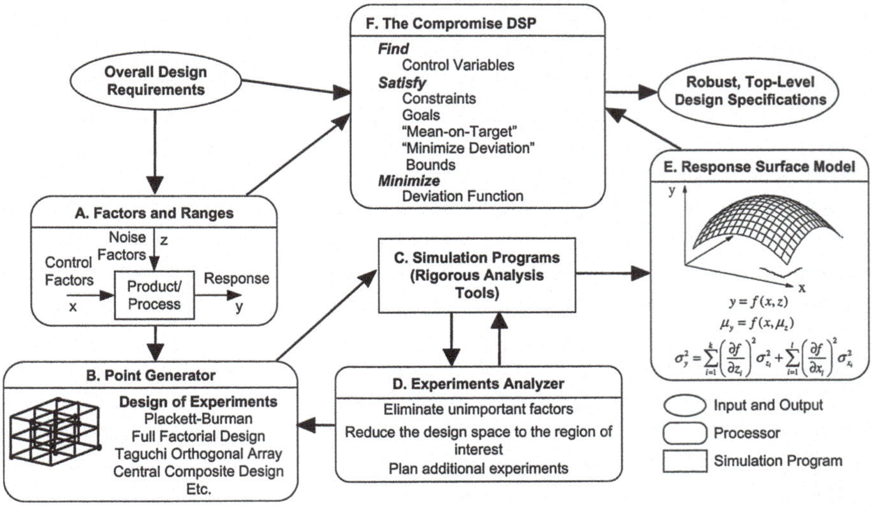

Fig. 3.22 Computational infrastructure of RCEM developed by Chen et al. (1997). In this figure, a modified version of the RCEM is shown; adopted from Allen et al. (2006)

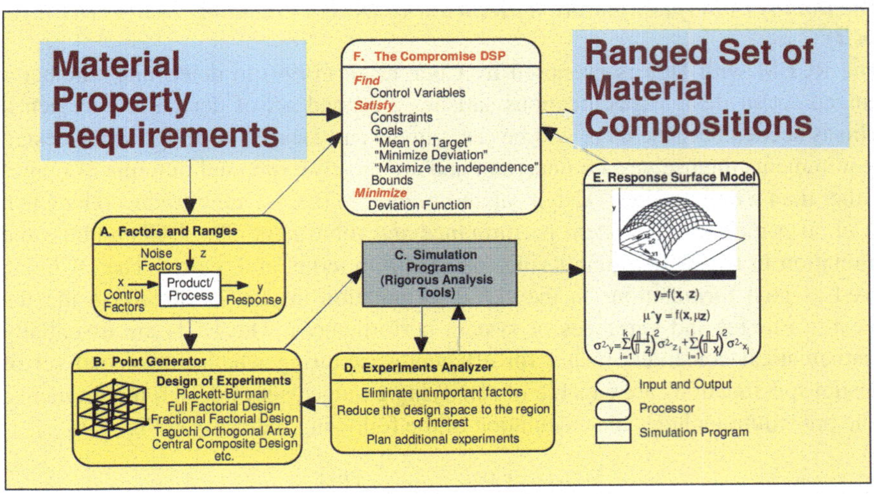

Fig. 3.23 Schematic showing application of RCEM to materials design problems; adopted from (Messer 2008)

structure and processing paths that satisfy specific material property or performance requirements. A schematic illustrating the application of RCEM in materials design is shown in Fig. 3.23.

RCEM, therefore, offers major advantages over other robust design approaches. It can be used to find a ranged set of robust design specification especially at the early stages of design, like the conceptual design stage. It also offers the ability to explore the entire continuous design space for finding robust solutions that are relatively insensitive to variations in noise as well as control factors. This is unlike other approaches like the Taguchi approach and Suh's axiomatic design. RCEM facilitates fast evaluation and reduces computational complexity by using approximate models after several rounds of experimental screening. This is a huge benefit as usually complex system simulations or analysis models consume time and are thus computationally expensive to perform. Another unique feature about RCEM is the fact that it uses the compromise decision support problem (cDSP) construct to formulate the design problem in terms of multiple goals that are meant to achieve a target mean and minimize performance variations with respect to control and noise factors.

There are also some limitations to RCEM and we discuss them here. The function evaluation in RCEM is not as accurate as actual analysis models or simulations due to the approximations. Thus, we have to deal with models that are incomplete and inaccurate. The performance variation estimation in RCEM is carried out using first-order Taylor series expansion, which could also add to the inaccuracy as we neglect the higher orders and also when we have problems with high order of nonlinearity. Also, RCEM in the form discussed in this section does not address robustness as a metric. The situation of having multiple conflicting goals that require different types of robust design is also not addressed in RCEM as presented by Chen et al.

The Robust Concept Exploration Method with Design Capability Indices (RCEM-DCI)

The RCEM with DCI is proposed by Chen et al. (1999) to determine whether a ranged set of design specifications satisfies a ranged set of design requirements. This is needed because there will be cases in the early stages of design when design requirements themselves are uncertain and are mostly expressed in terms of ranges rather than a target value. Design capability indices (DCIs) are introduced by Chen et al. as a measure of system performance and robustness. The DCIs address the limitation in RCEM by quantifying robustness in terms of a metric. The DCIs are used as goal formulations in the cDSP formulation instead of directly using the mean on target and variances of system performances. The DCIs are essentially mathematical constructs meant for efficiently capturing whether a ranged set of design specifications are capable of satisfying a ranged set of design requirements. The procedure includes the calculation of the following indices, see Fig. 3.24.

$$C_{dl} = \frac{\mu - LRL}{3\sigma}; \; C_{du} = \frac{URL - \mu}{3\sigma};$$
$$C_{dk} = \min\{C_{dl}, C_{du}\} \tag{3.11}$$

A negative value of DCI means that the system performance is currently outside of the system requirement range. A DCI value greater than unity means that the system is satisfying its requirement on the system performance. This is achieved by moving the mean of the performance away from the requirements limits and/or by reducing the deviation associated with the performance function. The designer's goal in a cDSP formulated using RCEM-DCI is to force the index to greater than or equal to one so that performance requirement lies within the required performance range. This is achieved by formulating DCI goals in the cDSP formulation. The mathematical aspects of DCI are fairly simple and easy to understand and compute for designers. A major assumption, however, made in RCEM-DCI is that the performance function is modeled accurately and that it has no uncertainty. This assumption is not valid for most complex problems as the models are typically incomplete and inaccurate and will have uncertainty associated with them.

3.4.5 Robust Design Type III

Another source of uncertainty that was not addressed by any of the methods discussed in previous sections is the uncertainty associated with models. Uncertainty in models is different from the uncertainty in control factors or noise factors. This uncertainty may be due to different reasons that include uncertain parameters (control and/or noise), model constraints, meta-models, approximations, model assumptions, functional relations, simulations, or analysis models. The presence of this uncertainty is high in materials and product design. This is due to the fact that material models are often nonlinear and include history effects associated with the processing of material resulting in uncertainty. Also, the assumptions made at different length and time scales; the boundary conditions used, and so on will all contribute to model uncertainty in material models. Therefore, robust design Type III is proposed to manage the uncertainty embedded within a model. In Fig. 3.25, a visual representation of robust design solutions for Types I and II and Types I, II, III together, along with the optimal solution is shown. The system response is shown by the solid curve. The two dotted curves represent the uncertainty limits for the system response. The Types I, II, and III robust solution identified has the least deviation in performance compared to Types I and II robust solution and optimal solution. Thus, the aim in

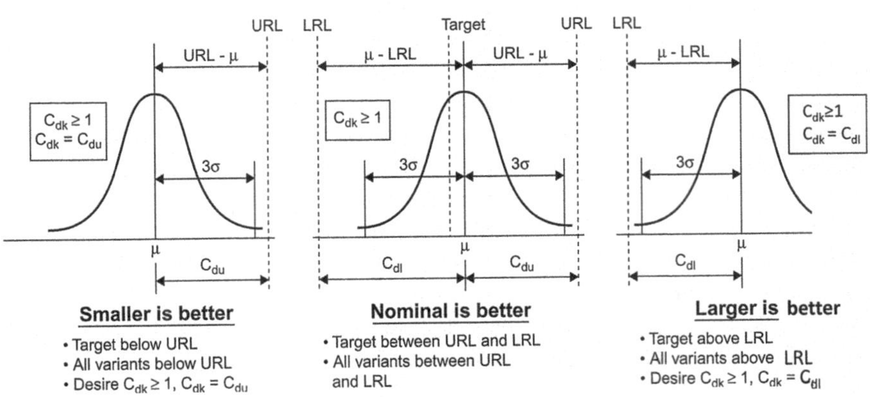

Fig. 3.24 Design capability indices (Chen et al. 1999)

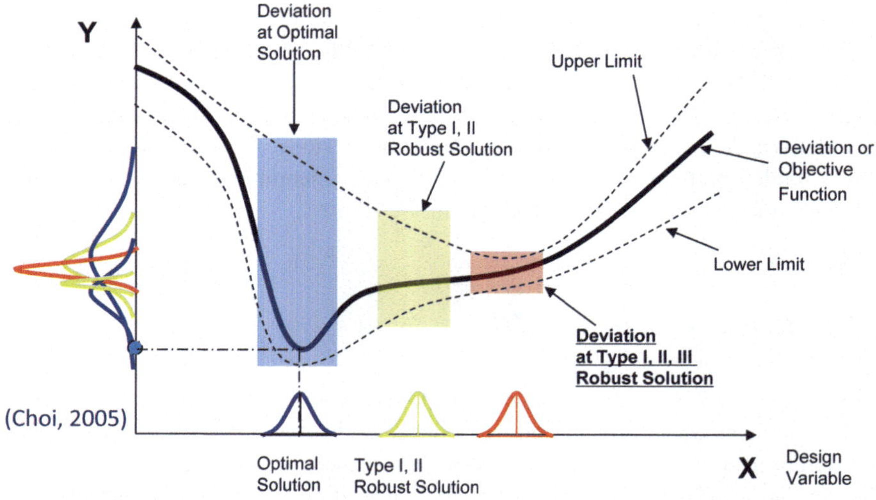

Fig. 3.25 Robust designs types I, II, and III (Choi 2005)

Type III robust design is to identify adjustable ranges for control factors (design variable) that satisfy a set of performance requirement targets and/or performance requirement ranges and are insensitive to the variability within the model.

Choi et al. (2005) propose the robust concept exploration method with error margin indices (RCEM-EMI) for Types I, II, and III robust designs. Error margin indices are mathematical constructs that indicate the location of the mean response and the spread of the response considering the variability associated with design variables and system models. EMIs represent the margin against failure due to uncertainty in both model and design variables. It is dimensionless just like DCI. The EMIs support Types I, II, and III robust designs. These are then incorporated as goals in the cDSP formulation to design the system under model structure and model parameter uncertainty. The RCEM-EMI procedure consists of (Choi et al. 2005): (a) clarification of the design task, (b) DOE and simulation, (c) integrated metamodel and prediction interval estimation, and (d) design space search using the cDSP for the RCEM-EMI. In the RCEM-EMI, the error margin indices (EMI) are metrics indicating the degree of reliability of a decision that satisfies system constraints and bounds. The entire procedure involved in RCEM-EMI is depicted in Fig. 3.26. The calculation of EMIs will be discussed in greater detail in Chap. 7.

The major advantage of RCEM-EMI over other approaches is that it can produce accurate results in design exploration. This is because the RCEM-EMI takes into consideration the uncertainty associated with noise factors, control factors, and the model itself. The RCEM-EMI helps a designer make decisions under a system's random variability and/or model parameter uncertainty in a model.

Some limitations associated with RCEM-EMI as presented by Choi et al. include the inability of RCEM-EMI to address multiple goals/performances that require different types of robust design. The RCEM-EMI in the form presented does not

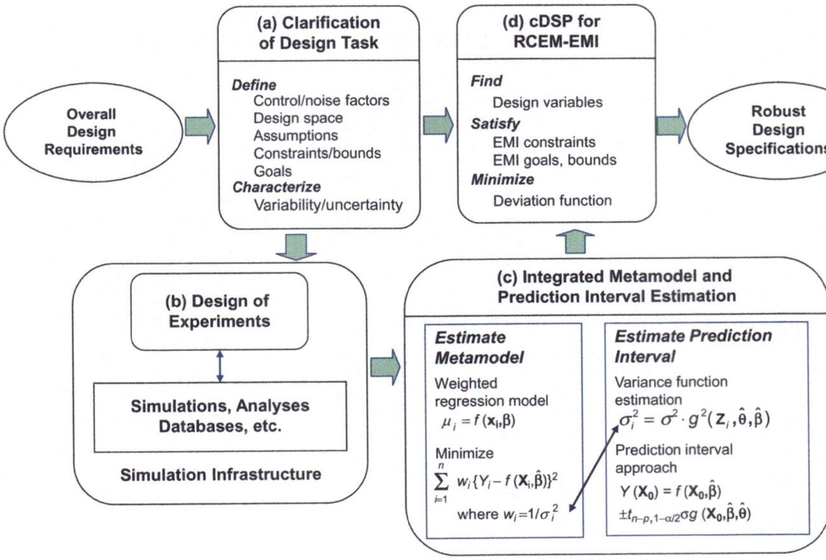

Fig. 3.26 The RCEM-EMI procedure (Choi et al. 2005)

support the management of propagation of all types of uncertainty across process chains. It also requires a large number of experiments for uncertainty analysis even in a single evaluation during design exploration, and thus is highly computationally expensive.

3.4.6 Managing Uncertainty Propagation Across Process Chains

Another challenge that needs to be addressed in materials and product design is the propagation of uncertainty across process chains. This is the uncertainty compounded by the combination of all the three types of uncertainty (natural, model parameter, and model structure) in a chain of models that are connected through input—output relations, interdependent responses, and shared control/noise factors as one model interacts with another, see Fig. 3.16 for the mode of interaction.

This model process chain could be a sequence of manufacturing processes or the processing—structure—property—performance mapping or even a multiscale model chain. In Fig. 3.16, as models interact, uncertainty may be accumulated and amplified through this sequential chain, resulting in a large variance in the final response (response of product/process model 2 of Fig. 3.16). This is an important issue and it leads to high levels of variability in the system response.

The Inductive Design Exploration Method (IDEM)

Choi (2005) developed the inductive design exploration method (IDEM) to achieve Type IV robust design specifically for the integrated multiscale design of material and

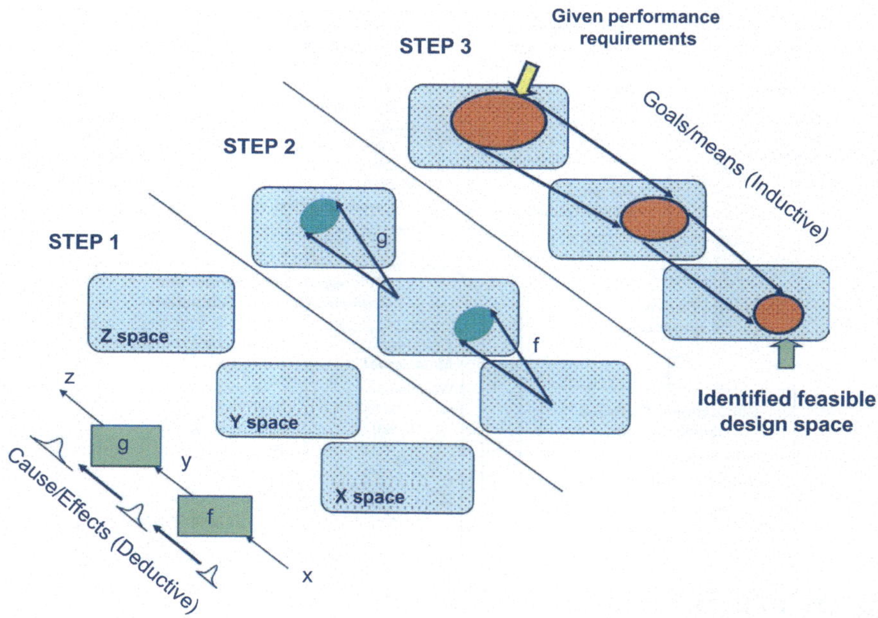

Fig. 3.27 Solution search procedure in IDEM (Choi et al. 2008a, b)

product. The hierarchical design of multiscale systems is facilitated by IDEM and accounts for NU, MPU, and MSU and the propagation of uncertainty (PU) through the scales (Choi et al. 2008). IDEM facilitates the exploration of robust solutions and uses a metric known as hyper-dimensional error margin index (HD_EMI) to assess the mapping across scales (Choi 2005). The higher HD_EMI value indicates that the mapped region is away from the boundary of the feasible region of interest and is less sensitive to any change. Thus, the HD_EMI value denotes the reliability of a chosen design variable, that the variable satisfies the constraints and bounds. IDEM is designed to provide a ranged set of robust solutions against propagated uncertainty (PU) and under model structure uncertainty (MSU) by passing feasible solution range in an inductive manner from the desired given performance range to the design space. The IDEM involves three steps as shown in Fig. 3.27. They are (Choi et al. 2008; Panchal et al. 2013):

(i) Parallel discrete function evaluation at each level of the design process. This step includes bottom-up simulations and experiments (STEPS 1 and 2 in Fig. 3.27).
(ii) Inductive discrete constraints evaluation (IDCE) is carried out in step where top-down feasible design space exploration is carried out using meta-models. This exploration uses the hyper-dimensional error margin index (HD_EMI) metric to assess mapping from higher space to lower space and to identify robust solution ranges from the feasible space (STEP 3 in Fig. 3.27).

(iii) The compromise decision support problem (cDSP) for finding the best solution under MSU. The cDSP facilitates the designer to identify the most desirable robust solution among the feasible range set of the solutions obtained. This is achieved by carrying out a tradeoff among the obtained HD_EMI values. The cDSP is the foundational computational construct in IDEM to carry out design decision-making (this step is not depicted in Fig. 3.27).

IDEM offers the following capabilities:

- Designers are able to identify robust solution ranges (multiple feasible solution ranges) with the consideration of uncertainty that is propagated across a process chain.
- None of the robust design approaches reviewed in the previous sections provided a range set of solutions—a characteristic specialty of IDEM. This can lead to a reduction in design iterations.
- The sequential uncertainty propagation analyses necessary to estimate final performance deviation are decoupled in IDEM and individual uncertainty analyses at each step.
- Designers can easily modify/change analysis models as there are no computational interfaces between models. The designer only needs to re-evaluate based on the altered model.

Some of the limitations of IDEM are highlighted in this section. This will be addressed in greater detail in Chap. 7. The limitations of IDEM include:

- Error due to discretization of design space—IDEM uses discretization of design space and further inductive discrete constraints evaluation for mapping from one space to another—this leads to discretization errors and also inability to capture the feasible boundary accurately—resulting in loss of information affecting system performance.
- Increasing accuracy by increasing the resolution of discrete points results in highly computationally expensive IDEM runs for evaluating feasible spaces.
- There is a limitation in terms of the number of design variables that can be used in IDEM for a design problem under study. The number of design variables increases the discrete points to be evaluated in the order of power—virtually impossible to evaluate beyond nine variables for an IDEM study.
- Limitation in terms of exploration and visualization—IDEM uses a three-dimensional visualization space using HD-EMI metric for exploration where only a maximum of three design variables can be studied at a time with the others variables taking defined values—these limits the scope of the simulation study and results.
- Issue of flexibility in design—IDEM does not allow designers to incorporate new goals or requirements at different levels during the process of design as the method is based on mapping to feasible spaces of "Y" and "X" for a given "Z" space.

3.4.7 Use of Robust Design in This Monograph

Robust design is used in this monograph for making material and product-related decisions. As discussed in the previous section, the decisions are formulated as compromise decision support problems. The design problems discussed in this monograph related to the integrated design of materials, products, and manufacturing processes are associated with different types of uncertainty that the system should be robust to. These include uncertainty inherent in the environment, uncertainty due to assumptions in the simulation models, uncertainty due to simplification of design processes (caused by ignoring dependencies in the design process), uncertainty propagated from one simulation model to another across process chains, and so on. Some of these aspects of uncertainty in materials design from different perspectives are discussed in Sect. 3.4. The robust formulation of compromise DSP in association with an inverse, decision-based design method is used in this monograph to make decisions that are robust to these uncertainties. The inverse decision-based design method developed in Chap. 4 is augmented to consider the different types of uncertainty across a process chain in Chap. 7 and consists of robust design techniques discussed in this section. Interested readers are also directed to the following articles for more information, see Nellippallil et al. (2017, 2018, 2020).

3.5 Foundations for a Platform for Decision Support—Template-Based Decision-Centric Perspective

In this monograph, a template-based decision-centric design is carried out to capture the knowledge associated with the design of a system. The fundamental assumption here, as discussed in previous sections, is that the role of a human designer is to make decisions given the information available and that decisions and information transformations are used to make satisficing decisions from a systems perspective. The need for a template-based approach modeling design decisions and interactions is to facilitate reusability, adaptability, extendibility, and modularity of design decision knowledge and support collaboration between distributed designers. Three key ideas are leveraged to achieve this: (i) hierarchical systems view of design processes as decision workflows, (ii) separation of declarative and procedural information, and (iii) design as a decision-centric activity. We discuss in brief about this in this section.

3.5.1 Hierarchical Systems View of Design Processes as Decision Workflows

Design processes can be progressively broken down into subprocesses from a hierarchical perspective. Looking at design from a decision-centric perspective, we can

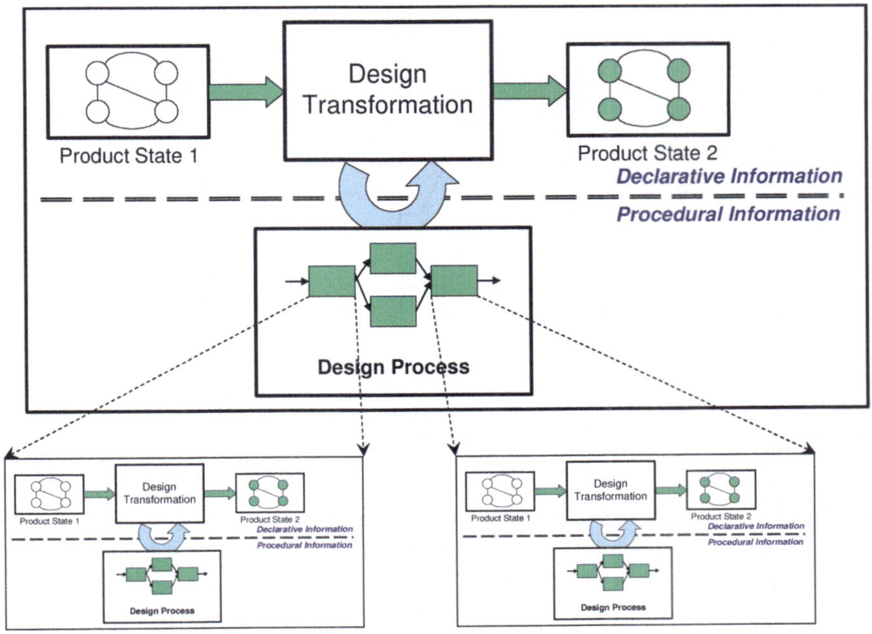

Fig. 3.28 Hierarchical systems view of design processes (Panchal et al. 2004)

model design processes as decision nodes interconnected with different other decision nodes forming a decision network or a workflow that supports decision-making and information transformations.

The transformations from the context of materials and product design involve an input material/product state, an output material/product state, and a design subprocess for execution—which is also a network of information transformations. In Fig. 3.28, we show the input, output, transformations, and design processes and their relations. These form the key elements of the hierarchical systems view of design processes as decision workflows.

The information related to the domain of the problem and the formulation of transformations, known as declarative information, and the information associated with the domain-independent procedure of solving the problem (procedural information) are separated clearly.

Modeling Design Processes Using Templates

The concept of modeling design processes using domain-independent decision templates is proposed by Panchal et al. (2004) and is illustrated using Fig. 3.29.

The design process shown in Fig. 3.30 has three information transformations: T1, T2, and T3. Now depending on the level of information available for a problem, the templates can be instantiated to support design decision-making. For example, in Fig. 3.30, T1 is a complete template, that is, it is completely instantiated as all the information available for executing it is available. T2 is a partially instantiated

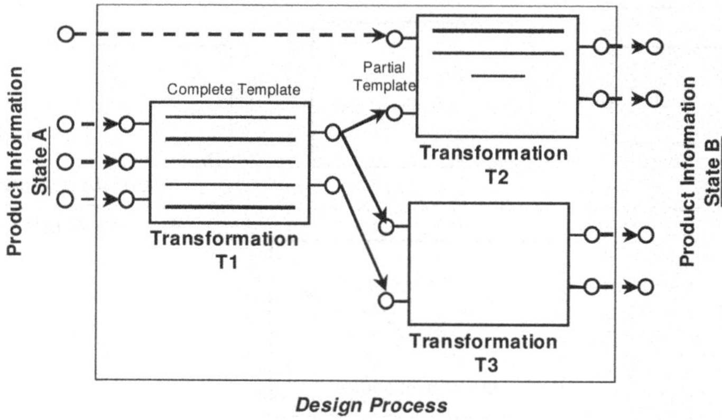

Fig. 3.29 Design process modeling procedure using templates (Panchal et al. 2005)

template and T3 is yet to be instantiated as the information on the problem for T3 may not be complete/available. However, the generic information transformation process remains the same as T1 or T2. This facilitates modular, reusable models for information transformation with a consistency in structure, thereby providing the ability to capture, archive, and reuse the design process knowledge at all levels of system hierarchy.

3.5.2 *Separation of Declarative and Procedural Information and Knowledge*

While developing templates, it is very important to separate the problem-formulated related information (declarative) and the process execution specific information (procedural). The declarative information is associated with design transformation and the product states. It represents what the designer does via information transformation and will be domain-specific. Procedural information is associated with how the information transformation is carried out and details how the transformation is executed via a decision workflow or decision network. It is domain-independent in nature. To separate out the declarative and procedural information, three levels are used—the product information level, the process information level (declarative), and the process execution information level (procedural), as shown in Fig. 3.31. More details on this are provided by Panchal et al. (2004, 2005), Panchal (2005), Messer (2008).

3.5.3 Design as a Decision-Centric Activity

From a decision-centric perspective, design is the process of transforming information that characterizes the needs and requirements for a product into knowledge of the product—as discussed in previous sections. The decision-centric templates are rooted in the decision support problem technique, specifically the compromise decision support problem construct. This was covered in detail in Sect. 3.3 and is not repeated here.

In this monograph, a knowledge-based platform for decision support in the design of engineering systems (PDSIDES) is presented. In PDSIDES decision-related knowledge is modeled with computational templates based on the decision support problem constructs using ontology to facilitate reuse and execution. In order to provide appropriate decision support for users of different knowledge levels, we define three types of users, namely template creators, template editors, and template implementers, who perform original design, adaptive design, and variant design, respectively. Using PDSIDES and its cloud-based version we facilitate the co-design of materials, products, and manufacturing processes. PDSIDES will be discussed in Chap. 8 of this monograph. Readers interested in this area are directed to the following article for more information, see Ming et al. (2018), Wang et al. (2018, 2019), Nellippallil et al. (2019).

3.6 Role of This Chapter in This Monograph

Our objective in this chapter is to introduce the design foundations and the fundamental constructs based on which the systems-based design architecture proposed in this monograph is developed. The design foundations reviewed and discussed in this monograph include: (i) Design as a goal-oriented activity, (ii) Pahl and Beitz design method, (iii) Suh's axiomatic design, (iv) Gero's model of design as a process, (v) decision-based design paradigm and DSP technique, (vi) robust design, and (vii) template-based decision-centric view of design. These foundations discussed in this chapter are used throughout this monograph. In Fig. 3.1, the utilization of these foundations in the development of the different components of the systems-based design architecture in the different chapters of this monograph is shown.

References

Allen, J. K., Seepersad, C., Choi, H., & Mistree, F. (2006). Robust design for multiscale and multidisciplinary applications. *Journal of Mechanical Design, 128*(4), 832–843.

Bascaran, E. (1991). A model for the conceptual design of thermal systems; concurrent decisions in designing for concept. Ph.D. Dissertation, Department of Mechanical Engineering, University of Houston, Houston, TX.

Bascaran, E., Bannerot, R. B., & Mistree, F. (1989). Hierarchical selection decision support problems in conceptual design. *Engineering Optimization, 14*(3), 207–238.

Beitz, W., Pahl, G., & Grote, K. (1996). Engineering design: a systematic approach, *MRS Bulletin*, vol. 71.

Box, G. (1988). Signal-to-noise ratios, performance criteria, and transformations. *Technometrics, 30*(1), 1–17.

Braha, D., & Maimon, O. (1997). The design process: properties, paradigms, and structure. *IEEE Transactions on Systems, Man, and Cybernetics-Part A: Systems and Humans, 27*(2), 146–166.

Bras, B., & Mistree, F. (1993). Robust design using compromise decision support problems. *Engineering Optimization, 21*(3), 213–239.

Bras, B. A. (1993). Foundations for designing decision-based design processes. Ph.D. Dissertation, Department of Industrial Engineering, University of Houston, Houston, TX.

Carroll, L., & Tenniel, J. (1865). Alice's adventures in wonderland and Through the looking-grass, Macmillan.

Chen, W., Allen, J. K., Mavris, D. N., & Mistree, F. (1996a). A concept exploration method for determining robust top-level specifications. *Engineering Optimization + A35, 26*(2), 137–158.

Chen, W., Allen, J. K., Tsui, K.-L., & Mistree, F. (1996b). A procedure for robust design: minimizing variations caused by noise factors and control factors. *Journal of Mechanical Design, 118*(4), 478–485.

Chen, W., Allen, J. K., & Mistree, F. (1997). A robust concept exploration method for enhancing productivity in concurrent systems design. *Concurrent Engineering, 5*(3), 203–217.

Chen, W., Garimella, R., & Michelena, N. (2001). Robust design for improved vehicle handling under a range of maneuver conditions. *Engineering Optimization, 33*(3), 303–326.

Chen, W., Simpson, T. W., Allen, J. K., & Mistree, F. (1999). Satisfying ranged sets of design requirements using design capability indices as metrics. *Engineering Optimization, 31*(5), 615–619.

Chen, W., Tsui, K.-L., Allen, J. K., & Mistree, F. (1995). Integration of the response surface methodology with the compromise decision support problem in developing a general robust design procedure. *ASME Design Automation Conference*, (pp. 485–492). Boston: Massachusetts.

Choi, H.-J. (2005). A robust design method for model and propagated uncertainty. Ph.D. Dissertation, The George W. Woodruff School of Mechanical Engineering, Georgia Institute of Technology.

Choi, H.-J., Austin, R., Allen, J. K., McDowell, D. L., Mistree, F., & Benson, D. J. (2005). An approach for robust design of reactive power metal mixtures based on non-deterministic micro-scale shock simulation. *Journal of Computer-Aided Materials Design, 12*(1), 57–85.

Choi, H.-J., Mcdowell, D. L., Allen, J. K., & Mistree, F. (2008a). An inductive design exploration method for hierarchical systems design under uncertainty. *Engineering Optimization, 40*(4), 287–307.

Choi, H., McDowell, D. L., Allen, J. K., Rosen, D., & Mistree, F. (2008b). An inductive design exploration method for robust multiscale materials design. *Journal of Mechanical Design, 130*(3), 031402.

Ebro, M., & Howard, T. J. (2016). Robust design principles for reducing variation in functional performance. *Journal of Engineering Design, 27*(1–3), 75–92.

Evbuomwan, N., Sivaloganathan, S., & Jebb, A. (1996). A survey of design philosophies, models, methods and systems. *Proceedings of the institution of mechanical engineers, part b: journal of engineering manufacture, 210*(4), 301–320.

Finger, S., & Dixon, J. R. (1989a). A review of research in mechanical engineering design. Part I: Descriptive, prescriptive, and computer-based models of design processes. *Research in Engineering Design, 1*(1), 51–67.

Finger, S., & Dixon, J. R. (1989b). A review of research in mechanical engineering design. Part II: Representations, analysis, and design for the life cycle. *Research in Engineering Design, 1*(2), 121–137.

Gero, J. S. (1990). Design prototypes: a knowledge representation schema for design. *AI magazine, 11*(4), 26.

Hazelrigg, G. A. (1996). *Systems engineering: an approach to information-based design*. NJ: Prentice Hall Upper Saddle River.

Hazelrigg, G. A. (1998). A framework for decision-based engineering design. *Journal of Mechanical Design, 120*(4), 653–658.

Horstemeyer, M. F. (2012). Integrated Computational Materials Engineering (ICME) for metals: using multiscale modeling to invigorate engineering design with science, Wiley.

Howard, R. A. (1966). Information value theory. *IEEE Transactions on systems science and cybernetics, 2*(1), 22–26.

Ignizio, J. P. (1976). Goal programming and extensions, Lexington Books.

Ignizio, J. P. (1978). A review of goal programming: A tool for multiobjective analysis. *Journal of the Operational Research Society, 29*(11), 1109–1119.

Ignizio, J. P. (1983). Generalized goal programming an overview. *Computers & Operations Research, 10*(4), 277–289.

Ignizio, J. P. (1985). Introduction to linear goal programming, Sage Publications.

Isukapalli, S., Roy, A., & Georgopoulos, P. (1998). Stochastic response surface methods (SRSMs) for uncertainty propagation: application to environmental and biological systems. *Risk Analysis, 18*(3), 351–363.

Kamal, S. Z., Garson, J., & Mistree, F. (1992). Heuristic Decision Support Problems. *Artificial Intelligence in Design'92*, Springer: 883–902.

Karandikar, H., & Mistree, F. (1993). Modelling concurrency in the design of composite structures. *Structural optimization*, vol. 150.

Krippendorff, K. (2007). Design research, an oxymoron? *Design research now*, pp. 67–80.

Kuppuraju, N., Ittimakin, P., & Mistree, F. (1985). Design through selection: a method that works. *Design Studies, 6*(2), 91–106.

McDowell, D. L. (2018). *Microstructure-Sensitive Computational Structure-Property Relations in Materials Design* (pp. 1–25). Computational Materials System Design: Springer.

McDowell, D. L., & Olson, G. (2008). *Concurrent design of hierarchical materials and structures*. Scientific Modeling and Simulations: Springer, pp. 207–240.

McDowell, D. L., Panchal, J., Choi, H.-J., Seepersad, C., Allen, J. & Mistree, F. (2009). Integrated design of multiscale, multifunctional materials and products, Butterworth-Heinemann.

Messer, M. (2008). A systematic approach for integrated product, materials, and design-process design, Ph.D. Dissertation, The George W. Woodruff School of Mechanical Engineering, Georgia Institute of Technology.

Miller, J. G. (1978). *"Living systems theory," Behavioral Sciences*. New York: McGraw Hill.

Ming, Z., Nellippallil, A. B., Yan, Y., Wang, G., Goh, C. H., Allen, J. K. & Mistree, F. (2018). PDSIDES—a knowledge-based platform for decision support in the design of engineering systems, *Journal of Computing and Information Science in Engineering, 18*(4).

Mistree, F., & Allen, J. K. (1997). Position paper Optimization in decision-based design, *Optimization in Industry, Palm Coast, FL, Mar*, pp. 23–27.

Mistree, F., Hughes, O. F. & Bras, B. (1993). Compromise decision support problem and the adaptive linear programming algorithm, *Progress in Astronautics and Aeronautics*, vol. 150, pp. 251–251.

Mistree, F., & Kamal, S. (1985). DSIDES: Decision support in the design of engineering systems, University of Houston.

Mistree, F., Marinopoulos, S., Jackson, D. M., & Shupe, J. A. (1988). The design of aircraft using the decision support problem technique. NASA Technical Report, Contract: NAS1-18117.

Mistree, F., & Muster, D. (1990). Conceptual models for decision-based concurrent engineering design for the life cycle. *Proceedings of the Second National Symposium on Concurrent Engineering*, Morgantown, WV.

Mistree, F., Smith, W., & Bras, B. (1993). A decision-based approach to concurrent design. Concurrent Engineering, Springer: 127–158.

Mistree, F., Smith, W., Bras, B., Allen, J. & Muster, D. (1990). Decision-based design: a contemporary paradigm for ship design, *Transactions, Society of Naval Architects and Marine Engineers*, vol. 98, pp. 565–597.

Mistree, F., Smith, W., Kamal, S., & Bras, B. (1991). Designing decisions: axioms, models and marine applications. *Fourth International Marine Systems Design Conference*.

Murphy, T. E., Tsui, K.-L., & Allen, J. K. (2005). A review of robust design methods for multiple responses. *Research in Engineering Design, 15*(4), 201–215.

Muster, D., & Mistree, F. (1988). The Decision-Support Problem Technique in Engineering Design. *The International Journal of Applied Engineering Education, 4*(1), 23–33.

Nair, V. N., Abraham, B., MacKay, J., Box, G., Kacker, R. N., Lorenzen, T. J., et al. (1992). Taguchi's parameter design: a panel discussion. *Technometrics, 34*(2), 127–161.

Nellippallil, A. B., Allen, J. K., Mistree, F., Vignesh, R., Gautham, B. P., & Singh, A. K. (2017). A Goal-Oriented, Inverse Decision-Based Design Method to Achieve the Vertical and Horizontal Integration of Models in a Hot-Rod Rolling Process Chain, *ASME Design Automation Conference*, Paper Number: DETC2017-67570.

Nellippallil, A. B., Ming, Z., Allen, J. K., & Mistree, F. (2019). Cloud-Based Materials and Product Realization—Fostering ICME Via Industry 4.0. *Integrating Materials and Manufacturing Innovation, 8*(2), 107–121.

Nellippallil, A. B., Mohan, P., Allen, J. K., & Mistree, F. (2020). An Inverse, Decision-Based Design Method for Robust Concept Exploration. *Journal of Mechanical Design*, https://doi.org/10.1115/1.4045877.

Nellippallil, A. B., Rangaraj, V., Gautham, B., Singh, A. K., Allen, J. K., & Mistree, F. (2018). An inverse, decision-based design method for integrated design exploration of materials, products, and manufacturing processes. *Journal of Mechanical Design, 140*(11), 111403-111403-17.

Nellippallil, A. B., Song, K. N., Goh, C.-H., Zagade, P., Gautham, B., Allen, J. K., & Mistree, F. (2017). A goal-oriented, sequential, inverse design method for the horizontal integration of a multistage hot rod rolling system. *Journal of Mechanical Design, 139*(3), 031403.

Olewnik, A. T., & Lewis, K. (2005). On validating engineering design decision support tools. *Concurrent Engineering, 13*(2), 111–122.

Olson, G. B. (1997). Computational design of hierarchically structured materials. *Science, 277*(5330), 1237–1242.

Pahl, G., & Beitz, W. (1996). Engineering design-A systematic approach, London: Springer-Verlag.

Panchal, J., Fernández, M., Paredis, C., & Mistree, F. (2004). Template-Based Design Process Modeling. *10th AIAA/ISSMO Multidisciplinary Analysis and Optimization Conference*.

Panchal, J. H. (2005). A framework for simulation-based integrated design of multiscale products and design processes, Ph.D. Dissertation, The George W. Woodruff School of Mechanical Engineering, Georgia Institute of Technology.

Panchal, J. H., Fernández, M. G., Allen, J. K., Paredis, C. J., & Mistree, F. (2005). Facilitating meta-design via separation of problem, product, and process information. *ASME 2005 International Mechanical Engineering Congress and Exposition*, Paper Number: IMECE2005- 80013.

Panchal, J. H., Kalidindi, S. R., & McDowell, D. L. (2013). Key computational modeling issues in integrated computational materials engineering. *Computer-Aided Design, 45*(1), 4–25.

Panchal, J. H., Paredis, C. J., Allen, J. K., & Mistree, F. (2006). Simulation model refinement for decision making via a value-of-information based metric. *ASME Design Automation Conference*, Paper Number: DETC2006-99433.

Park, G.-J. (2007). Analytic methods for design practice. Springer Science & Business Media.

Parkinson, A., Sorensen, C., & Pourhassan, N. (1993). A general approach for robust optimal design. *Journal of Mechanical Design, 115*(1), 74–80.

Phadke, M. S. (1995). Quality engineering using robust design. Prentice Hall PTR.

Reddy, R., Smith, W., Mistree, F., Bras, B., Chen, W., Malhotra, A., Badhrinath, K., Lautenschlager, U., Pakala, R., & Vadde, S. (1996). DSIDES User Manual. *Systems Realization Laboratory, Woodruff School of Mechanical Engineering, Georgia Institue of Technology*, Atlanta, Georgia.

Shannon, C. E. (1948). A Mathematical Theory of Communication. *The Bell System Technical Journal, 27*(3), 379–423.
Shannon, C. E. (1949). Communication Theory of Secrecy Systems. *Bell System Technical Journal, 28*(4), 656–715.
Shannon, C. E., & Weaver, W. (1963). *The mathematical theory of communication. 1949*. Urbana, IL: University of Illinois Press.
Shoemaker, A. C., Tsui, K.-L., & Wu, C. J. (1991). Economical experimentation methods for robust design. *Technometrics, 33*(4), 415–427.
Shupe, J. A. (1988). Decision-based design: Taxonomy and implementation. Ph.D. Dissertation, Department of Mechanical Engineering, University of Houston, Houston, Texas.
Simon, H. A. (1969). The sciences of the artificial MIT Press. Cambridge, MA.
Simon, H. A. (1996). The sciences of the artificial. MIT press.
Simpson, T. W., Chen, W., Allen, J. K., & Mistree, F. (1996). Conceptual design of a family of products through the use of the robust concept exploration method. *6th. AIAA/USAF/NASA/ISSMO Symposium on Multidisciplinary Analysis and Optimization*.
Sinha, A., Bera, N., Allen, J. K., Panchal, J. H., & Mistree, F. (2013). Uncertainty management in the design of multiscale systems. *Journal of Mechanical Design, 135*(1), 011008.
Stone, J. V. (2015). Information theory: a tutorial introduction. Sebtel Press.
Suh, N.P. (1990). The principles of design. Oxford University Press on Demand.
Sundaresan, S., Ishii, K., & Houser, D.R. (1995). A robust optimization procedure with variations on design variables and constraints. *Engineering Optimization+ A35, 24*(2), 101–117.
Taguchi, G. (1986a). Introduction to quality engineering, Asian Productivity Organization, Distributed by the American Supplier Institute, Inc., Dearborn, MI.
Taguchi, G. (1986b). Introduction to quality engineering: designing quality into products and processes. Quality Resources.
Taguchi, G. (1993). Robust technology development. *Mechanical Engineering-CIME, 115*(3), 60–63.
Taguchi, G., & Clausing, D. (1990). Robust quality. *Harvard Business Review, 68*(1), 65–75.
Tsui, K.-L. (1992). An overview of Taguchi method and newly developed statistical methods for robust design. *Iie Transactions, 24*(5), 44–57.
Vadde, S., Allen, J., & Mistree, F. (1994). Compromise decision support problems for hierarchical design involving uncertainty. *Computers & structures, 52*(4), 645–658.
Vadde, S., Allen, J., & Mistree, F. (1995). Selection using available assets catalog design: selection using available assets catalog design: selection using available assets. *Engineering Optimization+ A35, 25*(1), 45–64.
Vining, G. G., & Myers, R. H. (1990). Combining Taguchi and response surface philosophies: a dual response approach. *Journal of quality technology, 22*(1), 38–45.
Wang, R., Nellippallil, A. B., Wang, G., Yan, Y., Allen, J. K., & Mistree, F. (2018). Systematic design space exploration using a template-based ontological method. *Advanced Engineering Informatics, 36*, 163–177.
Wang, R., Nellippallil, A. B., Wang, G., Yan, Y., Allen, J. K., & Mistree, F. (2019). Ontology-based uncertainty management approach in designing of robust decision workflows. *Journal of Engineering Design, 30*(10–12), 726–757.
Welch, W., Yu, T., Kang, S. M., & Sacks, J. (1990). Computer experiments for quality control by parameter design. *Journal of Quality Technology, 22*(1), 15–22.

Chapter 4
Systems-Based Design Architecture for Integrated Design of Materials, Products, and Associated Manufacturing Processes

4.1 Addressing Systematic Model Integration and Information Flow

In this section, a function-based systematic approach is developed to achieve the integrated design of materials, products, and manufacturing processes from a systems perspective and to address systematic model integration and information flow. As discussed in Chap. 3, systematic approaches for the conceptual design of materials and products are not much addressed in the literature. A systematic design method/approach that involves strategic and tactically placed successive steps of information transformation that support designers to solve problems more efficiently and effectively is needed. Since changes at the detail design stage and at the product development stage are highly expensive, designers are required to make the right decisions at the early stages of design without overlooking or ignoring anything essential. To achieve this, a deliberate and systematic step-by-step procedure is required along with the designer's intuition and expertise support in the systematic approach.

As discussed in Chap. 3, after the planning and clarification of the task phase of a design process, the designer comes up with a requirements list and starts the conceptual design phase. In the conceptual design phase, the designer determines a principal solution or a concept. This is achieved by abstracting the essence of the problem, establishing function structures, searching suitable working inter-relationships (working principles, phenomenon, etc.), and then combining those (integration of information) to develop a system-level function structure or working structure or a workflow—from the context of this monograph. The working structure is then transformed into a more comprehensive representation to evaluate the essentials of the principal solution and review constraints, goals, and other design objectives. By oper-

A. B. Nellippallil et al., *Architecting Robust Co-Design of Materials, Products, and Manufacturing Processes*, https://doi.org/10.1007/978-3-030-45324-4_4

ating at the level of phenomena, and associated solution principles, the designer is able to go beyond a particular material/product system and will be able to adapt to changes, evolution, and other market cycles—resulting in dynamic behavior/response to market shifts and ability to provide that knowledge in classified form for easy retrieval. Thus, the question that we address in the first part of this chapter is:

What are the foundations needed for ***systematically identifying*** *and* ***integrating*** *material models with models of the rest of the system (product, manufacturing processes, and environment), so as to define the* ***processing—structure—property—performance*** *relationships and associated* ***information workflow*** *at early stages of design?*

Our hypotheses for addressing the question are:

- Through a systematic approach from a systems perspective, consisting of concept generation which includes:

(a) functional decomposition to generate multilevel function structures across the process chain for the end performance requirements, followed by
(b) identifying material and process phenomenon associated with function structures and systematically mapping them to solution principles (models identified from literature or developed through experiments),

- and framing the system structure for problem via,

(c) vertical integration of identified/developed material models and horizontal integration of identified/developed process models to systematically map material processing to material microstructure phenomena and next to macro-level properties and performances,

the design of product, process, and material concepts are integrated, and conceptual materials design is rendered more systematic.

To address this, the work carried out by Messer (2008) on function-based systematic design is leveraged and applied in the context of integrated design of materials, products, and manufacturing processes. An overview of the function-based approach to achieve systematic model integration and information workflow is presented in Fig. 4.1. A summary of elements for "systematic model integration and information flow" discussed and reviewed in this Chapter is included in Table 4.1.

4.2 Systematic Function-Based Conceptual Design

In Sects. 4.2.1–4.2.3, we discuss the systematic function-based conceptual design. We expand on some of the concepts in the early stages of design discussed in Chap. 3 and address it from the perspective of the question and hypotheses addressed.

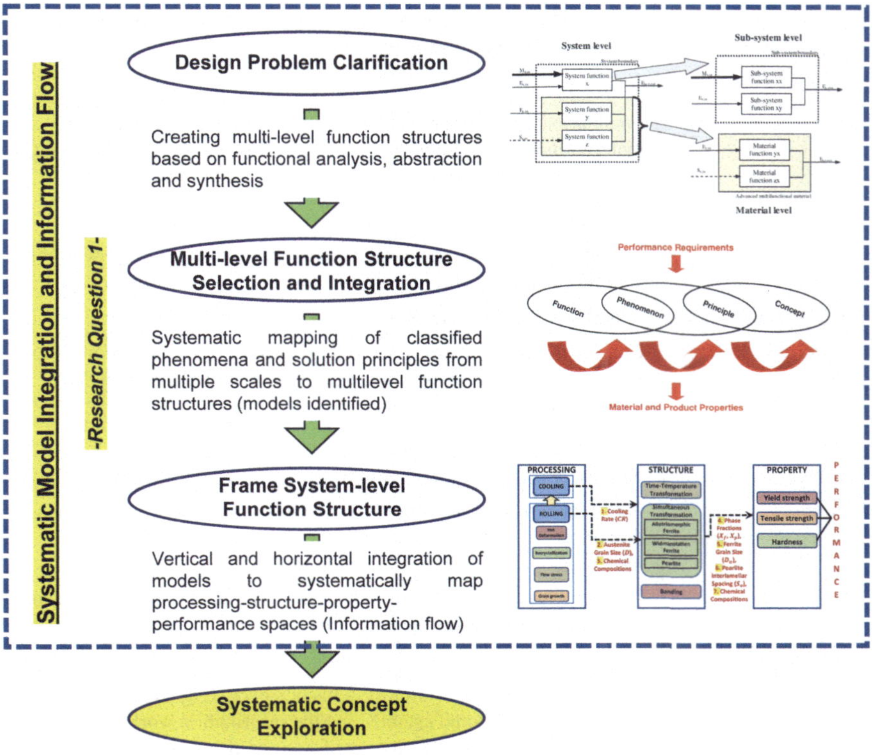

Fig. 4.1 Overview of function-based approach to achieve systematic model integration and information workflow

4.2.1 Systematic Conceptual Design Exploration

Conceptual design is considered as the most demanding activity in a design process. Let us look at Fig. 4.2. A change in design at later stages of design results in a cost of 1000×, while a design implemented at the conceptual level incurs only 1×. Traditionally, design is carried out using trial-and-error, and experiments resulting in design changes at later stages of design result in high design change cost at later stages, as can be seen from the dotted blue curve in Fig. 4.2.

Now, with the advent of simulations, digital twins, and digital threads, there has been a shift in design changes to early stages of design and thus a fall in associated design change cost, see the black curve in Fig. 4.2. What we intend to achieve in this monograph is to move all this process to the conceptual design phase. Hence, all the major design decisions need to be systematically made at the conceptual design phase leading to "***satisficing design concepts***" that result in fewer resolution costs and less design change costs at later stages of design, see the orange curve in Fig. 4.2. Systematic design as discussed in Chap. 3 involves discursive and intuitive thinking.

Table 4.1 Summary of elements for "systematic model integration and information flow" discussed and reviewed in this Chapter that form part of the architecture

Elements	*What?*
Processing elements	Foundations needed for systematically identifying and integrating material models with models of the rest of the system: (i) abstracting the essence of the problem, establishing function structures, (ii) searching suitable working inter-relationships (working principles, phenomenon, etc.), and (iii) combining those (integration of information) to develop a system-level function structure or working structure or a workflow
Data elements	Functional relationships and inter-relationships of a system
Connecting elements	Combination of working inter-relationships of the subfunctions into working structures, the solution principle for fulfilling the overall tasks for the system is established
Form	*How?*
Weights	Review of the foundations of function-based systematic design
Properties	Focus on current capabilities in function-based design, the associated limitations, and the research opportunities that are worthy of investigation
Relationship	Concepts discussed are used as theoretical foundations and as research philosophies to address the research gaps identified in the monograph
Rationale	*Why?*
Motivation	To generate concepts from a systems perspective by identifying and integrating models and establishing the information workflow that involves phenomena and associated solution principles for the integrated design of materials, products, and manufacturing processes
Characteristics/beliefs	• The design decisions made at the conceptual design phase are the most crucial ones that allocate a vast majority of the product's resources; • The system can be represented as a network of subsystems that are connected in terms of functions that they wish to satisfy and the information they share; • A subsystem can be considered separately to formulate a problem and its relations with other subproblems temporarily suspended; • Appropriate phenomena and solution principles are identified and the associated functional relationships/technology to develop functional relationships exist.
Interpretation	The need for function-based systematic design at the conceptual design phase

With both discursive thinking and intuitive thinking, a clarified problem and the associated information is systematically transferred from one state to the other to generate concepts.

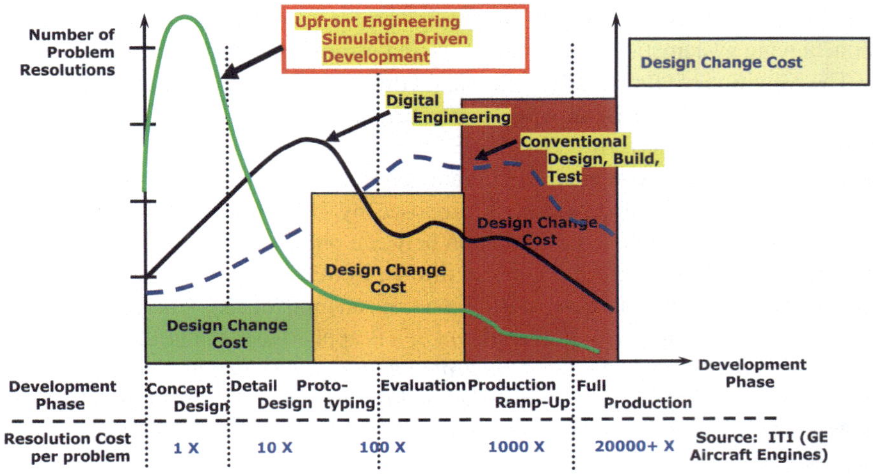

Fig. 4.2 Cost timeline across various design phases. *Source* ITI (GE Aircraft Engines)

4.2.2 *Function-Based Design*

Clarification of a design problem is followed by decomposing the system-level problem into subproblems by defining a boundary. Our aim in systematic conceptual design is to represent engineering systems in terms of their functions they must fulfill. Functional requirements are derived from customer needs and requirements in the case of Suh's axiomatic design or from desired product functions from the perspective of Gero's model of design as a process, as discussed in detail in Chap. 3. In this monograph, we use the approach of functional decomposition or function-based approach for modeling the system structure. The subparts of a complex system will perform specific subfunctions that contribute to the overall system-level function structure. The strategy here involves the functional decoupling of a coupled system to get the behavior of the system right and then focus on human decision-making via decision workflows.

The key to function-based design is to develop functional relationships and interrelationships of a system. This was discussed in detail in Chap. 3 while reviewing Pahl and Beitz design process and Suh's axiomatic design. Pahl and Beitz (1996) specify that the search for solutions is simplified and thus the discovery of design solutions is facilitated by establishing functional inter-relationships in a system. As per Suh (1990), the design process involves transforming customer needs into functional requirements and then interlinking the functional domain to the physical domain at every hierarchy level of the design process.

Simon (2013) notes that discovering viable ways of decomposing a complex system into semi-independent parts corresponding to the system's many functional parts is a powerful and well-respected technique. In this monograph, we focus on the use of functional decomposition of the design problem on multiple levels and scales for achieving systematic model integration and information flow.

The definition of "function" is different for different disciplines. The term function as defined by Pahl and Beitz is the general input/output relationship of a system whose purpose is to perform a task. Here "function" is an abstract formulation of the task to be performed. The input and output here refer to energy, material, or signals (information). A similar definition is proposed by Ullman (1992) as "the logical flow of energy, material, and information between objects or the change of state of an object caused by one or more of the flows". Both these definitions characterize function as involving certain transformation from an input of a system to an output of a system. In this monograph, the word *function* is applied to the *overall input/output relationship that is involved in the integrated material, product, and manufacturing process system*. Functions are represented in this monograph by leveraging the work by Messer (2008) where a solution-neutral natural language was used to represent functions by combining the nouns energy, material, and signal while taking into account input and output flows. This allows for the representation of a wider range of functions that can be used for functional representation in the materials and product domain.

Thus, the advantages associated with the function-based design are (Messer 2008):

- Domain-independent representation and solution exploration schemes;
- Generation of a broader solution field;
- Abstraction of essential problem characteristics;
- Allows for defining a problem boundary and problem formulation;
- Designers are more likely to explore new solutions than known solutions;
- Easy to identify non-intuitive solutions;
- Supports systematic problem formulation, design space generation and expansion, and exploration of design and solution space;
- A foundation for modular and reconfigurable design;
- Generation of rough idea of uncertainty and its propagation across process flow chain;
- Facilitation of planning and managing of design;
- Practical and easy to use for designers in any domain and can be interfaced with systems-based design exploration methods; and
- Emergence of innovative solutions and the logic underpinning the solutions can be clearly understood.

The major disadvantage associated with function-based design is that it does not provide systematic mappings to obtain concepts based on functional requirements.

4.2.3 Function-Based Systematic Conceptual Design—Pahl and Beitz Approach

The most well-known method for function-based systematic design is proposed by Pahl and Beitz (2013). The systematic approach proposed by Pahl and Beitz is based on best practices from industry and essentially includes the fundamentals of technical systems, systematic approach, and general problem formulation and solving process. The whole planning and design process includes planning and clarification of the task, through the identification of the required function structures, the systematic elaboration of solution principles, the construction of modular structures to generate system-level function structures, to the detailed documentation of the complete product (Messer 2008). The four main phases of planning and design process as proposed by Pahl and Beitz are discussed in great detail in Sect. 3.2.1, Chap. 3, see Fig. 3.3.

In this monograph, we leverage the function-based systematic approach by Messer (2008) which is strongly related to the working steps in the conceptual design phase of the (Pahl and Beitz 1996) design process. We start with the clarification of the problem and identification and documentation of system-level requirements in a solution-neutral form. The crux of the problem is then identified via analysis and further abstraction of problem-specific information. The system and subsystem level function structures are developed by iteratively analyzing, abstracting, and synthesizing. The individual subfunctions in the function structure are replaced by working solution principles. Working principles can be described quantitatively via physical laws governing the physical quantities involved. A functional relationship realized by the selected working principles and its characteristics results in a working inter-relationship that fulfills the function in accordance with the task to be performed. Through the combination of working inter-relationships of the subfunctions into working structures, the solution principle for fulfilling the overall tasks for the system is established.

To summarize the steps involved in generating a concept within the conceptual design phase of the Pahl and Beitz design process involves (Messer 2008):

- Abstraction to identify essential problems,
- Generation of system function structures,
- Search for working inter-relationships,
- Combination of working inter-relationships into working structures,
- Selection of suitable combinations,
- Refinement into principal solution variants, and
- Evaluation against technical and economic criteria to determine a system concept.

In this monograph, *functional decomposition, analysis, abstraction, synthesis, and systematic variation* are leveraged from Pahl and Beitz design process as core transformations for the function-based systematic approach.

4.2.4 Analysis of Research Gap

The application of function-based systematic approaches is traditionally carried out for embodiment and detail design phases of the design process and mostly for material selection. Generating concepts from a systems perspective to identify and integrate models and establish the information workflow that involves phenomena and associated solution principles for the integrated design of materials, products, and manufacturing processes is not addressed in the literature. The goal by addressing this research gap is to enhance the existing function-based design methods to increase design concept flexibility at the conceptual design phase in realizing the design of materials and products.

Systematic conceptual design is not typically addressed in current materials design. The design decisions made at the conceptual design phase are the most crucial ones that allocate a vast majority of the product's resources. A materials design revolution is underway in the recent past where the focus is to design the material microstructure and processing paths to satisfy multiple property or performance requirements. Recent advancements in computational modeling tools and frameworks that support simulation-based, integrated design exploration of materials, products, and the manufacturing processes through which they are made have resulted in the speeding up of the process of discovering new materials and has paved the way for rapid assessment of process–structure–property–performance relationships of materials, products, and processes. This has led to the simulation-based design of material microstructure (microstructure-mediated design) to satisfy multiple properties or performance goals of the product/process/system, thereby replacing the classical material design and selection approaches. This has resulted in a need to move the focus from embodiment and detail design phases to early stages of design like the conceptual design phase. A function-based systematic approach for conceptual design of materials and products is therefore proposed. This enables designers to identify the underlying phenomena and associated solution principles that embody the functional relationships. The limitations of existing approaches to the conceptual design of materials and products are summarized in Table 4.2.

An overview of the systematic function-based approach to the integrated design of materials, products, and manufacturing processes concepts and their subsequent exploration using a concept exploration framework is illustrated in Fig. 4.3. In Sect. 4.3, the systematic function-based conceptual design is proposed. Limitations and opportunities for future work are discussed in Sect. 4.5 along with verification and validation in Sect. 4.6. The systematic approach described in this section is tested with the hot rod rolling example problem in Chap. 6.

Table 4.2 Conceptual design efforts, limitations, and use in this monograph

Research effort	Limitations	Use in this monograph
Decision-based design	• Systematic concept generation at early stages of design not supported for materials and products • Limited to systematic selection at early stages and systematic design during embodiment and detail design	Used as foundational design philosophy in this monograph
Systems design	• The integrated design of materials, products, and associated manufacturing processes is not yet addressed from a systems perspective (goal-oriented)	The focus in this monograph is to achieve the systems-based design of materials, products, and manufacturing processes
Materials design	• Focus on intuitive methods for concept generation, followed by rigorous domain-dependent analysis using multiscale modeling efforts or material selection approaches • Highly domain-dependent • Systematic conceptual design at the early stages of design not addressed	Research efforts used to develop models that establish information linkage and workflow
Function-based design	• Focus on functional modeling • Do not support systematic mappings that allow concept generation	Foundation for addressing systematic model integration and information flow
Function-based systematic design	• Limited to material selection and not on early-stage concept generation	Foundation for addressing systematic model integration and information flow

4.3 Proposed Function-Based Approach for Systematic Conceptual Design

Leveraging from the work of Messer (2008) which is based on the Pahl and Beitz approach, the function-based approach for systematic conceptual design proposed in this monograph for the integrated design of materials, products, and associated manufacturing processes consists of Messer (2008):

(i) ***Functional decomposition*** of products and material systems into multilevel function structures through functional analysis, abstraction, and synthesis, and

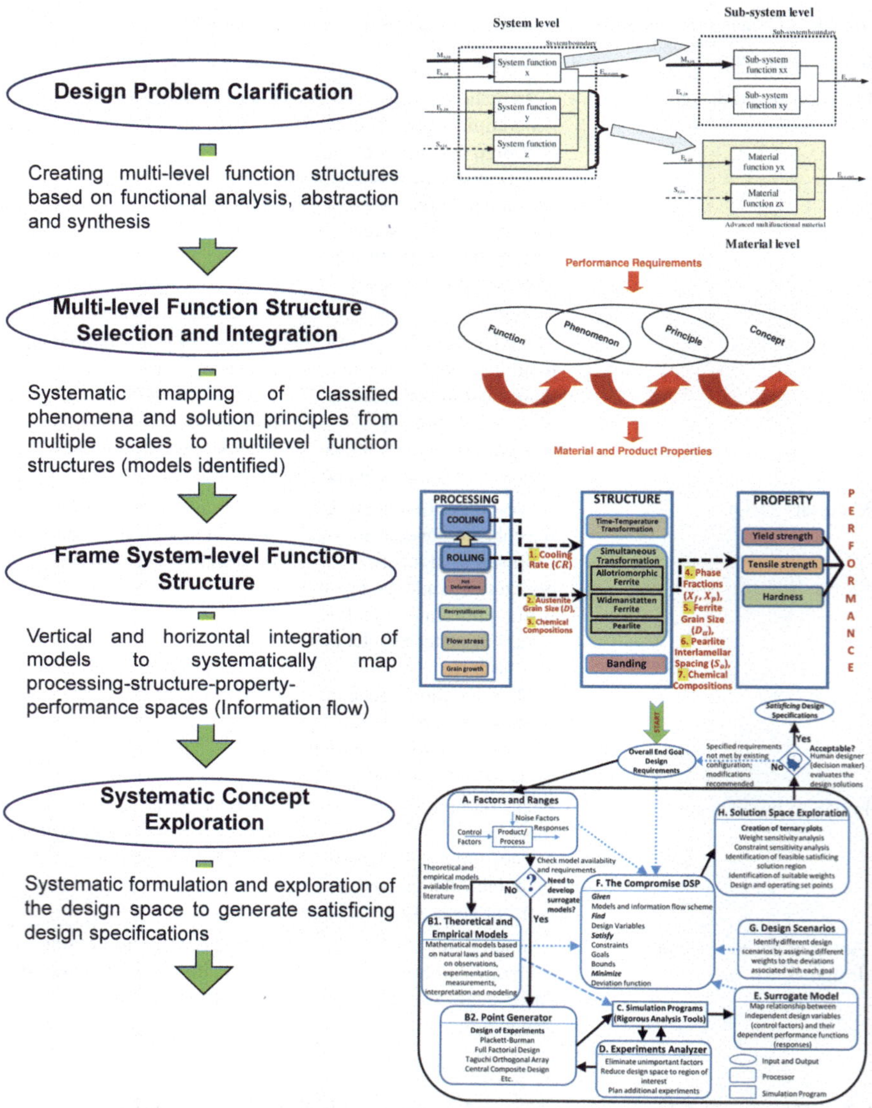

Fig. 4.3 Overview of systematic generation and exploration of concepts for the integrated design of materials, products, and manufacturing processes

(ii) ***Systematic mapping*** of phenomena and associated solution principles from multiple disciplines to multilevel function structures in order to develop principal material and product system solution alternatives and concepts.

The mappings in the systematic approach are shown in Fig. 4.4.

The key mapping includes (see Fig. 4.4):

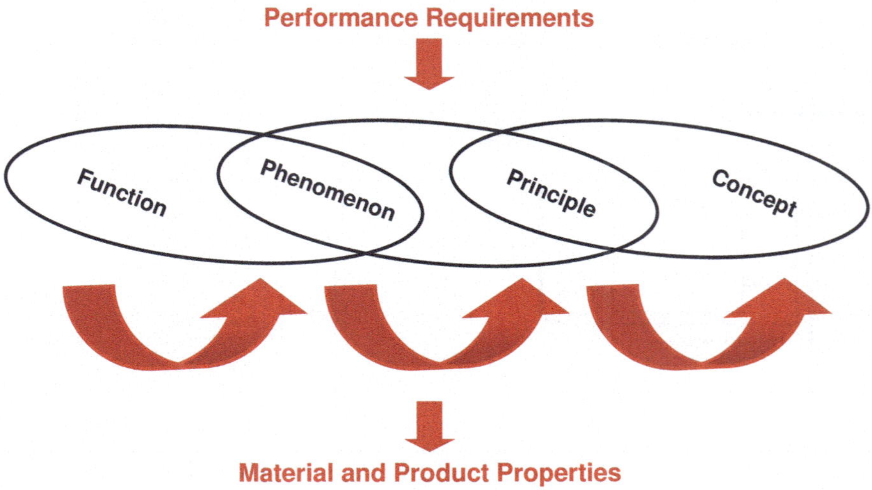

Fig. 4.4 Key mappings in systematic conceptual design (Messer 2008)

Clarified problem	→	Functional relationships
Functional relationships	→	Phenomena
Phenomena	→	Associated solution principles
Functional relationships, phenomena, and associated solution principles	→	Principal solution alternatives characterized by specific properties

The key mappings are described in the following sections.

4.3.1 Formulating Multilevel Function Structures

For the integrated design of materials, products, and manufacturing processes at the conceptual design phase, it is critical to take a systems perspective. This allows us to carry out the functional decomposition of the products and material systems into multilevel function structures through functional analysis, abstraction, and synthesis. Here, "function" is defined as the overall input/output relationship in the integrated material and product system. To address this in a domain-independent manner, functions here are represented in a solution-neutral way. Having defined and clarified a problem in a solution-neutral way, we specify an overall system function. "Function structure" here refers to a meaningful and compatible combination of subfunctions

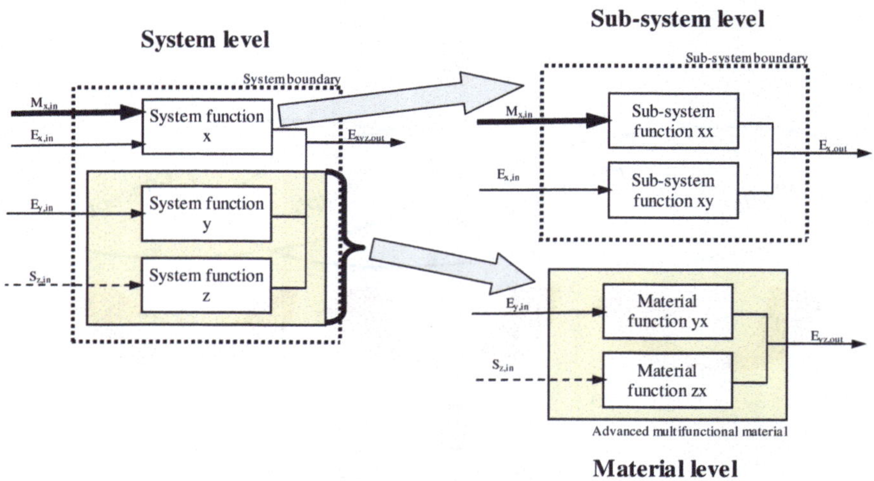

Fig. 4.5 Multilevel function structures. Adopted from (Messer 2008)

into the overall system function (Messer 2008). We create multilevel function structures by decomposing the overall system-level function into subfunctions on various levels of hierarchy like subsystem, assembly, component, part, materials, and so on (see Fig. 4.5).

In this monograph, multilevel function structures are proposed to include the material level at the conceptual design stage that spans the processing paths and material microstructures which further allows in making system-level connections with the product properties and performances. The path in the system configuration space made up of input/output relationships and their links can be mathematically varied. The subfunctions present can be mathematically varied and combined into function structure alternatives. ***Multilevel function structures support a designer in framing the system-level structure by identifying and combining functional relationships on different levels and scales.***

Function taxonomies are available for product domain in the literature. A comprehensive list of function taxonomies is provided by Szykman et al. (1999). A collection of verbs well-suited for function-based systematic conceptual design of materials is proposed by Messer (2008) and is reproduced in Table 4.3.

At the material level, the commonly used verbs could be "change" and transform". "Change" refers to changing material/energy from one state to another while transform refers to transforming from one form to another.

The key to the systematic approach is a function-based abstraction, through which the problem complexity is reduced and the essential characteristics of the problem are emphasized. This supports the discovery of non-intuitive solutions. The function structure thus developed consists of several linked subfunctions that represent the flow of information, energy, and material. Individual subfunctions are first represented as "black-boxes" and are further replaced by more concrete statements

Table 4.3 Verbs frequently used to define functional relationships (Messer 2008)

Sink	Source	Store	Combine	Modify	Convey	Control
Absorb Dissipate Receive	Create Emit Supply Release	Accumulate Deposit Record	Mix Couple Link (sequential) Constrain Separate Compare	Amplify Attenuate Prepare Filter Shape Transform Change	Channel Conduct Direct Divert Guide Transmit Transport	Actuate Detect Sense Display Maintain Resist (deformation) Support (loading) Stabilize (buckling)

that are in mathematical form to allow for simulation-based material and product design. Phenomena and associated solution principles are identified to develop a wide range of principal solution alternatives. Phenomena are described quantitatively by means of laws (laws of physics and mathematics) governing the quantities involved. Solution principles refer to means to embody phenomena. These means are physical, quantitative descriptions in terms of laws of physics and mathematics, to fulfill functional relationships. From the standpoint of integrated materials, product, and manufacturing process design, the phenomena and associated solution principles for property–structure relationships are not yet addressed.

It is very essential to identify phenomena and associated solution principles for property–structure relationships on multiple scales for effective and efficient integrated design of material and product concepts. To facilitate a function-based systematic approach to the integrated design of the product and material concepts, design catalogs are proposed for phenomena and associated solution principles that represent property–structure relations on multiple scales. In this monograph, we are interested in identifying material models and integrating them so that we can establish the process–structure–property–performance relationships given the performance requirements. For achieving that, we focus on identifying phenomena and corresponding solution principles (models in our case) that embody the phenomena. One approach to systematically identifying solution principles is by using design catalogs. We discuss the design catalogs and their application in the next section.

4.3.2 *Design Catalogs*

Design catalogs are classification schemes that are used to provide a classified collection of known and proven solutions (knowledge) for easy retrieval. They are characterized by the systematic presentation of information from which the required knowledge can be retrieved. It also facilitates identification and combination of essential solution characteristics that meet the requirements. Design catalogs are therefore knowledge base in a classified form. It provides designers with an overview of a certain domain or subarea. Design catalogs are therefore intended to provide (Messer 2008):

- knowledge, and experiences of different designers are captured in a single location,
- quicker, more problem-oriented access to the accumulated solutions or data,
- self-explanatory classification for easy retrieval,
- a comprehensive range of possible solutions or at the very least the most essential ones which can be extended later, and
- the greatest possible range of interdisciplinary applications.

From the perspective of integrated materials and product design, the requirement is a knowledge base (an open-ended map) that enables designers to identify the structure–property relationships—which in terms of systematic conceptual design is the identification of underlying phenomena and associated solution principles that cause a certain behavior and thereby provide solutions for a variety of problems or cases.

An example of the mapping and the design catalog is adopted from Matthias Messer's doctoral dissertation (see Fig. 4.6; Messer 2008). The system-level function is to dissipate energy. The identified phenomenon is "inelastic deformation" that embodies the system-level function "dissipate energy". The most promising solution

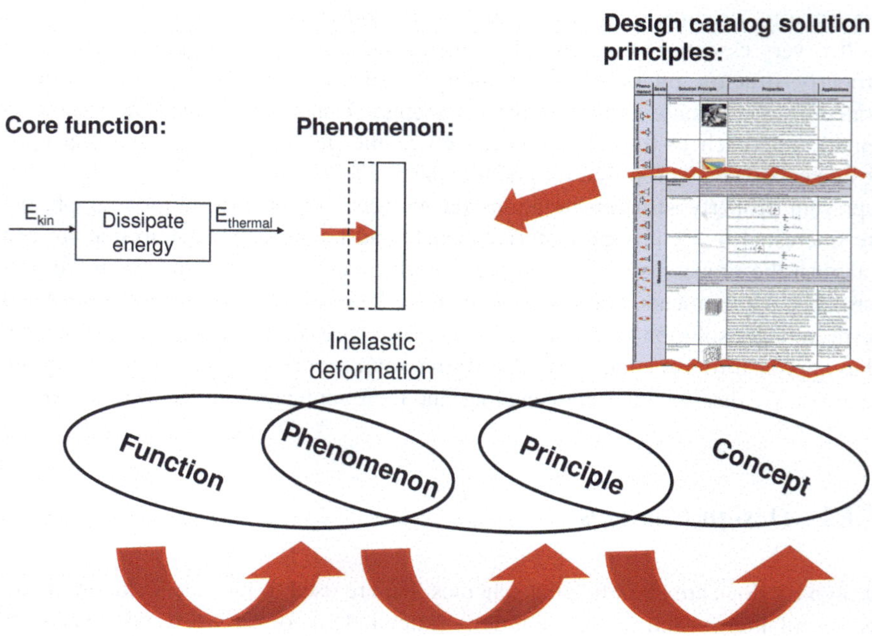

Fig. 4.6 Mapping solution principles to phenomenon inelastic deformation to embody system-level function "dissipate energy". Adopted from Messer (2008)

principles are selected and evaluated for feasibility based on the given performance requirements.

A design catalog showing solution principles is adopted from Matthias Messer's monograph (Messer 2008) and is shown in Fig. 4.7.

Alternatives

Solution Principles	Acronym	Illustration	Remarks	Feasibility
Compression of structures	C	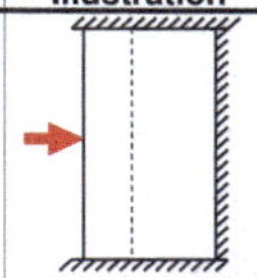	Ultimate tensile strength is an excellent indicator of the ability of a material to absorb energy. The uniform elongation parameter is unimportant provided that the ductility is sufficient to allow the requisite plastic deformation. However, compression of structures usually suffers form the limited zone of plastic deformation and the unavailability of lightweight materials that combine high ultimate tensile strength and sufficient plastic deformation in this configuration.	🚫
Tension of structures / Nanotubes	T(N)	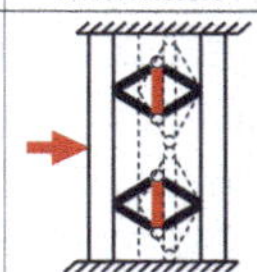	The maximum energy that can be absorbed per unit weight before tensile instability supervenes depends upon the ultimate tensile strength and strain. Structures with incorporated carbon nanotubes could be substantially stronger and lighter than conventional, but, besides the stroke (maximum displacement limitation), cost and manufacturability are still major problems with nantobes.	🚫
Torsion of structures	T	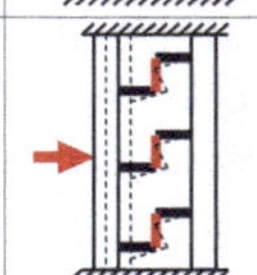	The maximum energy that can be absorbed per unit weight before torsional instability supervenes depends upon the ultimate torsional strength and strain. Besides the maximum stroke limitation, energy absorbers based on torsion suffers from the unavailability of lightweight materials that combine high ultimate torsional strength and strain.	🚫
Bending of Plate structures	BP	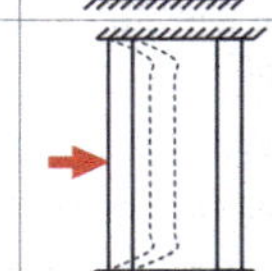	Kinetic energy of an incoming blast impinging on a plate can be dissipated by bending and stretching. The superior performance of sandwich plates relative to solid plates of equal mass for shock loading is due (i) the substantial bending strength of a sandwich plate as described above, and (ii) energy absorption by the sandwich core.	✓
Bending of Shell structures	BS	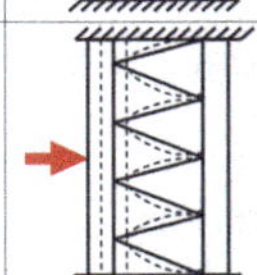	The failure mode of laterally loaded shell structures starts as local denting followed by global bending collapse. For shell structures such as tubes, the axial buckling mode has a specific energy absorption capacity which is approximately 10 times that of the same tube in lateral bending. Moreover, in the bending of shell structures, usually not all material participates in the absorption of energy by plastic work.	🚫
Inversion of structures	I	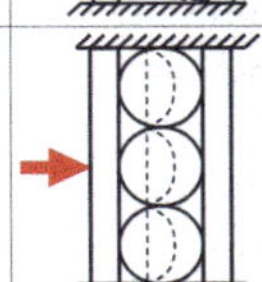	Inversion of structures involves for example the turning inside or outside in of a thin circular tube made of ductile material. Another example is the inversion of spherical shells where material passes into the central dimple region through a circular knuckle whose radius increases with deformation. In general, practical shortcomings, such as difficulties in securing axial loading, did not yet proove inversion to be a feasible energy dissipation mechanism.	🚫
Splitting of structures	S	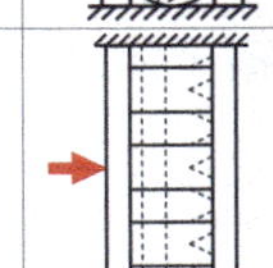	Tube splitting is a special case of tube inversion where the die radius is large enough to cause splitting instead of inversion. Again, practical shortcomings, such as difficulties in securing axial loading, did not yet prove inversion to be a feasible energy dissipation mechanism.	🚫
Progressive Buckling of Open-top tubes and frustra	BO	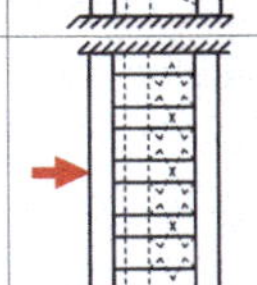	Circular tubes under axial compression provide one of the best energy absorption devices since they feature "optimal" energy dissipation characteristics. They provide a high stroke length per unit mass and a reasonably constant operating force. Best energy dissipation is obtained through progressive plastic buckling via a series of ringlike folds which avoids overall elastic buckling.	✓

Fig. 4.7 Evaluation of solution principles for the function "dissipate energy". Adopted from Messer (2008)

4.4 Discussion on Function-Based Systematic Approach and Use of Design Catalogs to the Integrated Design of Materials, Products, and Manufacturing Processes

Using the function-based systematic approach, an abstract solution to the given design problem is developed by representing systems (product and material systems) in terms of functions to be fulfilled. Functions are represented in this work as solution-neutral using taxonomies. Using functional representations at a material level for structure–property relations will bridge the gap between product design and materials design and support the integrated design of materials and products. Design catalogs facilitate access to classified and reusable knowledge and expertise in both materials and product domains. Functional relationships are used as a common interface here and this leads to the increased synergy between the two domains leading to integration, transparency, modularity, and reconfigurability of design processes. This encourages designers to think more deeply about fundamental phenomena and solution principles. Phenomena are physical effects that can be described in terms of mathematical equations. Solution principles associated with certain phenomena are further embodying phenomena. The approach followed in this monograph is based on systematic identification and variation of functional relationships, phenomena, and associated solution principles. This leads to a comprehensive design space exploration. It is reported that more innovative solutions emerge from function-based systematic approaches and the logic behind the solutions is clearly understood and evident. Thus the result of applying the function-based systematic approach to the integrated design of the product and material concepts is (Messer 2008):

(i) a more thorough search through,
(ii) an informed decision-making, as well as
(iii) an expansion of the design space and hence an increase in concept flexibility, which is crucial when facing dynamic demands in a global marketplace.

4.4.1 *Discussion on Design Catalogs*

Design catalogs support the function-based systematic design and help a rational designer by extending cognitive abilities and increasing the designer's concept flexibility. Design catalogs, however, are not static. They must be considered as a living document that is updated continuously, extended and maintained to keep up with expanding body of knowledge. Design catalogs are intended for multidisciplinary exchange of knowledge, exploration of the relationships between phenomena and associated solution principles at multiple scales, and their differing behaviors. Since designers are operating at a phenomenal level, it helps them to view the problem from a systems perspective. Once the specific phenomena and associated solution principles are decided, designers focus on material properties and start the design activities from there for conceptual design.

Design catalogs thus provide knowledge in a form classified for easy retrieval for different applications. They are essentially an open-ended map that enables a designer to identify underlying phenomena and associated solution principles. Focusing on phenomena and associated solution principles, that is, property–structure relations a designer is able to step out of the technological cycle of obsolescence and evolution. To convert design catalogs into "living" documents, web-based tools can be used/developed that support live editing and updating of catalogs.

4.5 Addressing the Requirements of Systematic Concept Exploration and Inverse Design Exploration

Having addressed the systematic function-based design of materials and products, the requirements of systematic concept exploration and inverse design exploration are addressed next. In Sect. 4.6, the concept exploration framework (CEF) is presented and introduced as a step-by-step approach for formulating design problems and quickly evaluating design alternatives to generate satisficing design specifications. The CEF is developed to support systematic and rapid concept exploration of complex engineered systems involving the material, product, and manufacturing processes. In Sect. 4.7, we present the goal-oriented, inverse design (GoID) method that uses the concept exploration framework to facilitate the systems-based inverse design exploration of material microstructures and processing paths to meet multiple product-level performance/property requirements. Both CEF and GoID together address the requirements of systematic concept exploration and inverse design exploration.

The question, therefore, addressed here is:

What are the computational foundations needed for performing the ***systematic*** and ***rapid concept exploration*** of complex engineered systems involving the material, product, and manufacturing processes ***satisfying certain end performance requirements***, when simulation models are typically ***incomplete, inaccurate, and not of equal fidelity***?

Our hypotheses for addressing the question are:

- Developing a concept exploration framework anchored in decision-based design-construct—the cDSP can support the designer in formulating the design problem systematically and exploring the solution space to generate satisficing design specifications (To address **G3**).
- Developing a goal-oriented, inverse design method that uses the concept exploration framework to facilitate the systems-based inverse design exploration of material microstructures and processing paths to meet multiple product-level performance/property requirements (To address **G4**).

In this chapter, the focus is to develop a concept exploration framework that is based on a simulation-based design approach for designing complex systems. An ideal concept exploration framework should support both the activities of a priori analysis and *a posterior* evaluation/synthesis in concept exploration. The focus, however, in this monograph is to use the concept exploration framework to support systems-based top-down design to generate "***satisficing design specifications***". The inverse method is supported by the CEF to systematically explore design alternatives and generate "***satisficing***" design solutions across process chains. In Table 4.4, we summarize the architecture component constituents for systematic concept exploration and inverse design exploration as discussed in this chapter.

4.6 The Concept Exploration Framework (CEF)

The CEF is introduced in this monograph as a general framework that includes systematic steps to identify design alternatives and generate *satisficing* design solutions. The CEF is inspired by the RCEM (Chen et al. 1997) with the addition of features (processors) to consider different material and product models and options to explore the solution space for different design scenarios. Core to the CEF is the foundational mathematical construct—the compromise decision support problem (cDSP) (Mistree et al. 1993). The cDSP construct used here is anchored in the robust design paradigm first proposed by Taguchi. The fundamental assumption is that the models are not complete, accurate, and of equal fidelity (Taguchi 1986; Bras and Mistree 1993). The cDSP is a hybrid of mathematical programming and goal programming. Target values for each goal are defined in a cDSP and the emphasis of the designer is to satisfy these target goals as closely as possible. This is achieved by seeking multiple solutions through tradeoffs among multiple conflicting goals. The solutions obtained are further evaluated by solution space exploration to identify solution regions that best satisfy the requirements identified. There are four keywords in the cDSP—*Given*, *Find*, *Satisfy*, and *Minimize*. The overall goal of the designer using the cDSP is to minimize a deviation function—a function formulated using the deviations (captured using *deviation variables*) that exists from the goal targets.

The details regarding formulating and solving the cDSP are available (Bras and Mistree 1993; Mistree et al. 1993) and explained in detail in Chap. 3, and hence are not explained here.

Next, we explain the concept exploration framework (CEF). In Fig. 4.8, the computing infrastructure for the CEF is shown. The computing infrastructure for CEF includes eight processors (A, B1, B2, D, E, F, G, and H) and simulation programs (C). The application of the CEF begins with the designer identifying the overall end goal design requirements for the problem under study. The further steps in the CEF are below. The solid arrows in Fig. 4.8 are used to highlight the steps of CEF in sequence. The dotted lines and dashed lines are used to represent information sharing within the framework.

Table 4.4 Summary of elements for "systematic concept exploration and inverse design exploration" discussed and reviewed in this chapter that form part of the architecture

Architecture component constituents	Summary interpretation for the systems-based design architecture
Elements	*What?*
Processing elements	Foundations needed for systematic concept exploration and inverse design exploration
Connecting elements	Decision-based design constructs
Form	*How?*
Weights	Concept exploration framework (CEF) and goal-oriented, inverse design (GoID) method
Properties	Focus on presenting the steps involved in CEF and GoID and highlighting the associated functionalities
Relationship	GoID together with CEF facilitate systematic concept exploration and inverse design exploration
Rationale	*Why?*
Motivation	To develop a concept exploration framework that supports a designer in systematically formulating a design problem and exploring the solution space to identify satisficing design solutions and a goal-oriented, inverse design method to support a designer in designing the system starting from the end goals and requirements
Characteristics/beliefs	• Design exploration has to be systematic and need to take into account end requirements/goals • Decisions can be formulated mathematically, and analysis models are available for design decision-making • Design process starts with an end goal in mind—could be an end product that needs to be supplied to customers • There exist models or computational tools or data that can establish the forward information workflow for the process/product that is being considered • Analysis models for problems are typically incomplete, inaccurate, and not of equal fidelity • Designer's interest is in identifying satisficing robust solutions given the uncertainty in the problem and not single-point optimum solutions • Design is a goal-oriented activity with design requirements subject to change at any time for the problems considered
Interpretation	For finding satisficing solutions in a multilevel, multistage process chain that involves the processing–structure (PS), structure–property (SP) relations are passing down the satisficing solution ranges in an inverse manner, from given final performance range to the design space of the previous space (defined by model input and output) with designer having the flexibility to choose solution of preference

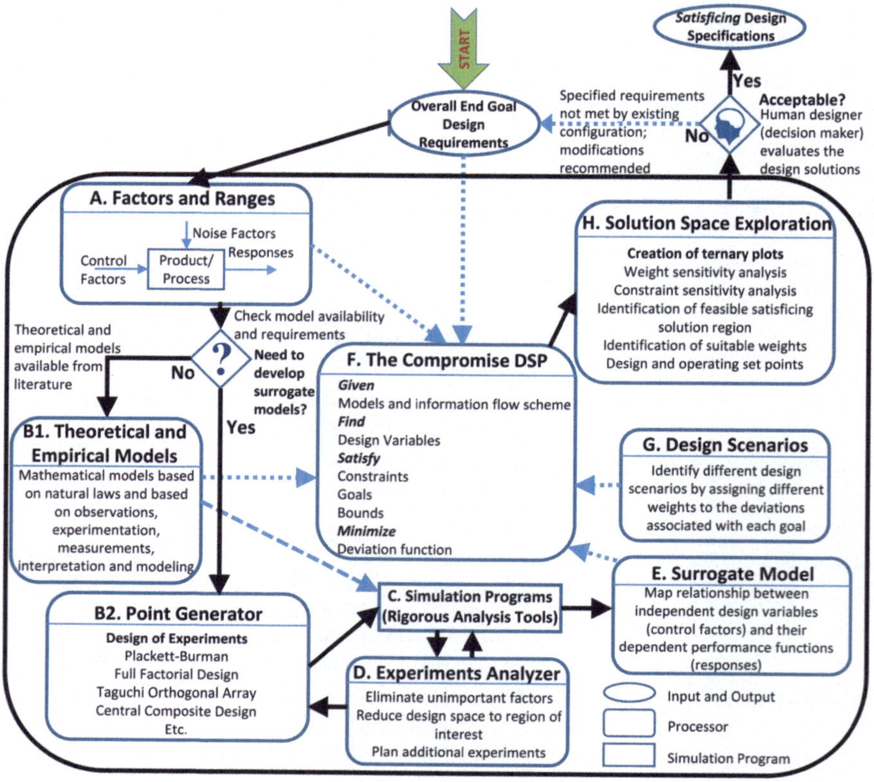

Fig. 4.8 The computing infrastructure for concept exploration framework (CEF)

Step 1 using Processors A and B1

In this step, the initial concept exploration space is defined and the cDSP is formulated. For the requirements identified for the problem, the control factors (factors that the designer can control), noise factors (factors that the designer cannot control), and the responses (the performance goals identified) and their ranges are identified in processor A. This information is input to the foundational mathematical construct—the cDSP, processor F. In parallel with the identification of factors, ranges, and responses, a designer identifies the models available/required. For problems related to manufacturing processes such as hot rolling and cooling, several different models defining material/process behavior are available in the literature (Hodgson and Gibbs 1992; Majta et al. 1996; Kuziak et al. 1997; Pietrzyk et al. 1999; Phadke et al. 2004). Such available theoretical and empirical models are identified in processor B1 and are communicated to the cDSP.

Step 2 using Processors B2, D, E, and Simulation Program C

In Step 2, the designer carries out low-order screening experiments. If models for the problem are not available or if there is a need to develop reduced order or surrogate models so as to reduce the size of the problem, an experiment is designed to develop them. The point generator, processor B2 is used to design the experiments. The simulation program (C) is used to run the experiments. The simulation programs for manufacturing-related problems may use some of the theoretical and empirical models from processor B1. This information flow is shown using the dotted arrow in Fig. 4.8. An example of this is a finite-element simulation (simulation program, C) for rolling that uses a constitutive model (empirical model, processor B1) to define the flow behavior of the material. The experiments analyzer, processor D evaluates the simulation results and recommends additional experiments if needed. Regression analysis and ANOVA are used to evaluate the significance of the results. Processor E is used to create the surrogate models using the simulation program results that are acceptable to the designer.

Step 3 using Processors F, G, and H

All models are communicated to the cDSP, processor F. The cDSP is then exercised for different design scenarios as specified by processor G. These scenarios which are identified by assigning different weights to the deviations associated with the goals define a solution space. This solution space is then explored using processor H. Ternary plots are generated to visualize and explore the solution space to identify feasible solution regions that satisfy the requirements. A human designer evaluates the design solutions, checks feasibility, and satisficing solution regions. If the overall end goal requirements are not satisfied or there are no feasible satisficing regions, then the overall end goal requirements may be modified as in Fig. 3. In such a situation, a designer can also make use of the ternary plots to carry out design tradeoffs to identify regions that satisfy the modified end goal requirements instead of repeating the CEF.

Thus the generic functionalities offered by CEF, in summary, includes: (i) identification of end goals and requirements for a problem, (ii) systematic identification of control factors, noise factors that influence the responses of the goals and requirements, (iii) systematic identification of mathematical models—theoretical, empirical models available from literature on the problem domain and systematic development of surrogate models using simulation programs and design of experiments, (iv) systematic formulation of the design problem using the cDSP construct for the given information available for the problem, (v) systematic planning of the design scenarios to be explored for the problem, (vi) exercising the problem formulated for the design scenarios, and (vii) systematic analyzing of the solution space with the opportunity for the human designer to visualize the solution space and make design decisions. These functionalities can be used to formulate and execute any complex systems problem in a systematic fashion and provide decision support, provided there is availability of required information (see Fonville et al. 2019; Nellippallil et al. 2016, 2017a, b, c, 2018a, b, 2019, 2020). To facilitate the generic applicability of the CEF and extend the designer's abilities in making design decisions that are robust, flexible, and modifiable, particularly in the early stages of design, an ontology for design space exploration and a template-based ontological method that supports systematic design space exploration using CEF is proposed (see Ming et al. 2018; Wang et al. 2018).

The concept exploration framework along with its features of multigoal decision support can be readily incorporated into a design method that supports the design of the material and product (processing, composition, and microstructure) as part of a larger overall system design process. The framework can embody the hierarchy of process–structure–property–performance proposed by Olson (1997) by systematically accounting the information flow and mappings across these spaces and transforming overall design requirements into a set of satisficing design specifications for the material/product and manufacturing process system of interest.

In Sect. 4.7, we describe the goal-oriented, inverse method in its generic form and the application of the method to explore the design space for the hot rod rolling process chain problem.

4.7 The Goal-Oriented, Inverse Method

4.7.1 Generic Form of the Goal-Oriented, Inverse Method

The basic idea of our method for finding satisficing solutions in a multilevel, multistage process chain that involves the processing–structure (PS), structure–property (SP) relations are passing down the satisficing solution ranges in an inverse manner, from given final performance range to the design space of the previous space (defined by model input and output) with designer having the flexibility to choose solution of preference. The method will be explained using the information flow diagram shown in Fig. 4.9. It is a *goal-oriented* method because we start with the end goals that need to be realized for the product as well as process and then design the preceding stages to satisfy these end goals as closely as possible by exploring the design space. Then the design decisions that are made for the end requirements of the product/process after exploration are communicated to the stages that precede them to make logical decisions at those stages to satisfy the requirements identified, thereby carrying out a design space exploration process in an inverse manner, as described by Steps 2.1–2.3 (Fig. 4.9). To demonstrate the generic nature of the method, we call the different sequential processes as "n" to "$n + 2$" and the decision support constructs as "i" to "$i + 2$".

Step 1: Establish forward modeling and information flow across the process chain (forward material workflow)

Step 1 of the proposed method involves establishing forward modeling and information flow across models. In Step 1, the designer makes sure that there is a proper flow of information as models are connected across different "Processes". These processes could be different manufacturing processes that are sequentially connected to produce the product with information passing across processing–microstructure–property–performance spaces. Mathematical models are either identified or developed to establish the information flow. Steps 1 and 2 of the concept exploration framework are used to identify factors, ranges, responses, and models for the specific materials

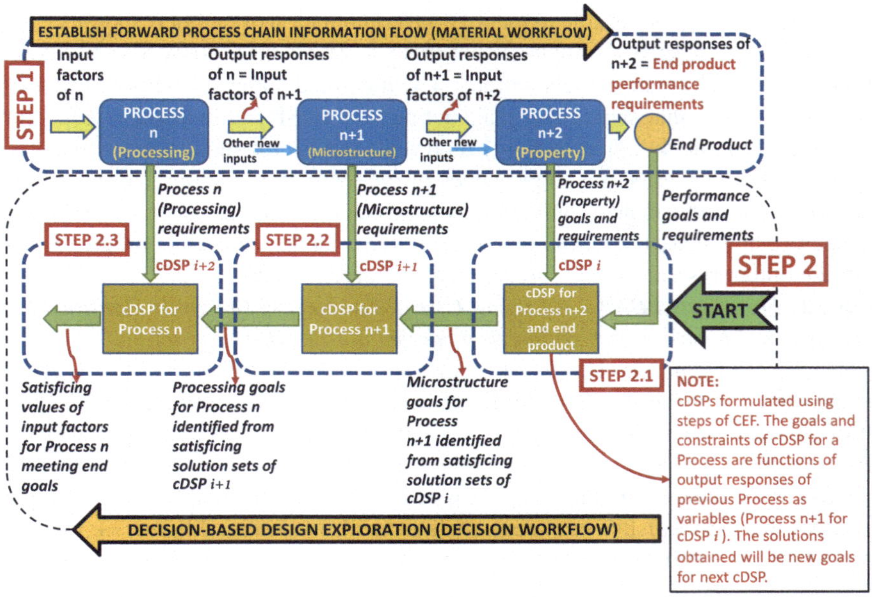

Fig. 4.9 Generic form of the goal-oriented, inverse method illustrated using Steps 1 and 2

design problem under study. In Fig. 4.9, Step 1, we see that the output of a process serves as the input for the next process along with other new inputs specific to the next process with the final output being the end product. We can imagine these "processes n, $n+1$ and $n+2$" as processing, microstructure, and property spaces, respectively, as shown in Fig. 4, to understand the method clearly. Thus, process n (processing space) generates output that serves as input for process $n+1$ (the microstructure space). The output of process $n+1$ (the microstructure identified) serves as the input for process $n+2$. The output of process $n+2$ defines the property space and this directly defines the final performance characteristics of the end product. From a design standpoint, the input to a process are design variables and the output response from the process serves as input variables to the next process.

Step 2: Carry out decision-based design exploration starting from performance space and sequentially identifying satisficing regions of interest in previous spaces in an inverse manner

Step 2.1: Formulate cDSP i using CEF for achieving the desired end product properties and performances

In Step 2, we start the decision-based design exploration starting from the end goals and requirements and have Steps 2.1–2.3 to complete the process chain in Fig. 4.9. cDSP i is formulated for process $n+2$ in Step 2.1. The design variables of this cDSP will be the output responses from process $n+1$ that serves as input to process $n+2$. The property and performance goals that are desired are defined in this cDSP. On executing the cDSP for different design scenarios and exploring the solution space using CEF, the designer is able to identify the combination of output responses from process $n + 1$ (that serves as input for process $n + 2$) that best satisfy the conflicting property and performance goals defined. The identified values of output responses for process $n + 1$ that satisfy the goals defined for cDSP i are passed as goals for cDSP $i + 1$. In Fig. 4.9, process $n + 1$ represents the microstructure space, then the output of cDSP i will be the target values of microstructure factors that satisfy the properties and performances defined for the product. In Step 2.2, using cDSP $i + 1$ we analyze how these target microstructure values can be achieved in process $n + 1$ with the output responses from process n as the input variables.

Step 2.2: Formulate cDSP i + 1 using CEF for achieving the goals identified for process n + 1 based on the exploration carried out in cDSP

In Step 2.2, we formulate cDSP $i + 1$ for process $n + 1$. The target goals in this cDSP are the values of the design variables for cDSP i identified after solution space exploration in Step 2.1. The design variables for cDSP $i + 1$ are the output responses from process n that serves as input to process $n + 1$. Executing this cDSP and exploring the solution space using CEF, the designer is able to identify the combination of input variables that best satisfies the target goals defined. From Fig. 4.9, we see that the output will the combination of processing variables that best satisfy the microstructure targets defined in cDSP $i + 1$. Again, we pass these identified values of design variables from cDSP $i + 1$ that satisfy requirements to the next cDSP $i + 2$ as target goals.

Step 2.3: Formulate cDSP i+ 2 using CEF for achieving the goals identified for process n based on the exploration carried out in cDSP i+ 1

In Step 2.3, in a similar fashion to previous steps, the designer formulates cDSP $i + 2$ for process n with target goals being the design variable values identified from cDSP $i + 1$. On exploration of solution space, the designer is able to identify the combination of input factors of process n that best satisfies the target performance goals identified for cDSP $i + 2$.

Thus, using this proposed method, the designer is able to carry out the top-down driven, simulation-supported, decision-based design of processing paths and material microstructure to satisfy a ranged set of product-level performance requirements. The method is generic and can be applied to similar problems with information flow from one process to another as shown in Fig. 4.9. The method supports the coordination of information and human decision-making and is suited for problems involving a network of forward, sequential information flow. Given any complex system that involves the sequential flow of information across processes/levels, the proposed method has the potential to be applied to support information flow by making effective decisions across the processes/levels in order to realize an end goal.

4.8 Discussion on Concept Exploration Framework and Inverse Design Exploration Method

In this chapter, we present a goal-oriented, inverse method supported by the concept exploration framework (CEF) to achieve the integrated design exploration of the material, product, and manufacturing processes. The method is goal-oriented and inverse because we start with the end mechanical properties of the product and inversely maps the requirements to microstructure and processing spaces of the material to identify multiple solutions that satisfy the requirements. The utility of the proposed method is demonstrated by carrying out the integrated solution space exploration of the processing and microstructure spaces of the rolling and cooling processes to identify *satisficing solutions* that realize the end mechanical properties of the rod product. The method and its application are characterized by a confluence of different disciplines, like engineering mechanics, materials science, manufacturing, and systems engineering. The functionalities offered by the method supported by CEF include:

- The method is based on requirements-driven, "top-down" design of the system and associated subsystems by taking a goal-oriented approach which is different to the standard practice of bottom-up modeling and design of material and product systems,
- There is a perception of obtaining a satisficing design space across process chains; augmenting the human ability to make design decisions—visualizing a solution space and making logical judgments through tradeoffs to identify satisficing solution regions of interest,
- There is the capability to handling "n" number of design variables—this is an advantage over other design exploration methods like IDEM where there is a limitation on the number of design variables,
- Propagation of end goal requirements (product performance or properties) across a process chain with the designer having the capability to check whether the end goals are actually achievable at previous spaces in their current configuration or not—the designer can recommend adjustments in the design space if needed,
- Offers flexibility in design: The capability to define new goals and requirements at each level as the method uses individual cDSPs to facilitate information flow allowing to formulate a design space at each level—advantage over other design exploration methods like IDEM and pyDEM where the design space is defined by mapping from previous spaces (Choi et al. 2008; Kern et al. 2017),
- The capability to carry out rapid, integrated design exploration of material and products using simulation models that we accept is typically incomplete and inaccurate,
- The capability to coordinate information and human decision making,
- The CEF offers the capability to prioritize models, input factors, output responses, and computational tools in terms of their value in design, and

- Ensuring feasible design solutions that allow investing in new complex material systems with confidence.

The proposed method and the concept exploration framework are generic and support the integrated decision-based design of similar manufacturing processes involving the material and product. Given any complex systems problem that involves the sequential flow of information across processes/levels, the proposed method has the potential to be applied to support information flow and human decision-making across the processes/levels in order to realize an end goal. Through the proposed method an approach is proposed for *microstructure-mediated design* by *integrating the design of the material, product, and associated manufacturing processes* involved.

4.9 On Verification and Validation—Theoretical Structural Validity (TSV)

4.9.1 TSV of Function-Based Design

Theoretical structural validation refers to accepting the validity of individual constructs used in the systematic function-based approach and accepting the internal consistency of the way the constructs are put together. Theoretical structural validation involves systematically identifying the scope of the proposed approach's application, reviewing relevant literature and identifying the research gaps that are existing, identifying the strengths and limitations of the constructs used based on literature review, determining the constructs and approaches that can be leveraged for the systematic function-based approach while reviewing literature on the advantages, disadvantages and accepted domains of application, and checking the internal consistency of the constructs both individually and when integrated. In this chapter, we establish the generic nature of the systematic approach and why the approach is appropriate for concept generation during the early stages of design for the integrated design materials, products, and associated manufacturing processes. By carrying out a literature search, it is shown that the systematic function-based approach and the associated constructs have been previously applied for problems in various domains in a successful manner and are verified and validated. The use of these generic systematic approaches for the integrated design exploration of materials, products, and associated manufacturing processes to establish systematic model integration and establishment of information workflow is not addressed in the past literature.

Based on the critical review of literature in this chapter, it is inferred that the application of the function-based systematic method is mostly on areas related to mechanical, control, software, and process engineering and is mostly applied for selection of materials for different applications from existing classes of solutions. The focus is also more on product design by developing concepts at the early stages of design. Our focus by using the function-based design is in establishing model integration and information flow chain to facilitate systematic problem-oriented conceptual design via functional decomposition and representation of the problem in solution-neutral natural language taking into account the input and output flows. This allows us to establish the integrated conceptual design of materials and products in a more systematic and domain-independent manner which helps in increasing the designer's flexibility and easy establishment of the information workflow for material/product system.

Once the phenomena and associated solution principles (models in our case) are identified, design catalogs are used to facilitate function-based systematic material and product design. Based on the literature review in this chapter, it is established that design catalogs are previously used and validated for facilitating function-based systematic design in different domains successfully. However, the use of design catalogs for identifying and capturing material and product models to facilitate integrated materials and product design is not addressed in the literature. The determination of phenomena and associated solution principles on multiple system levels is crucial and this allows for developing a wide range of principal solution alternatives and increase a designer's concept flexibility.

The use of design catalogs in past literature has been confined to mechanical components like gearboxes, bearings, connections, and so on. The use of design catalogs for defining processing–structure–property–performance relationships via material models at multiple levels/scales using information generated through the integration of such models is not addressed as of now.

In this monograph, the past efforts on function-based design and identification of phenomena and associated solution principles are leveraged to achieve the integrated design of materials, products, and manufacturing processes. The focus in this monograph is to establish processing–structure–property relations from a systems perspective by addressing phenomena and associate solution principles, thereby integrating the conceptual design of materials and products in a systematic and domain-independent manner. To facilitate function-based systematic design at the level of phenomena and solution principles, the functionalities associated with design catalogs are leveraged to support a designer in designing material and product concepts in an integrated fashion.

The systematic approach followed is shown as a flowchart in Fig. 4.3. The details are provided with the description of each task in a step-by-step manner in this chapter. The input needed, and the output generated is clarified, the internal information flow is checked to ensure sufficient information availability to execute the next steps. Through a critical evaluation of each step and the way individual constructs are put together, internal consistency of the systematic approach is verified and accepted.

The theoretical structural validity of the function-based systematic approach for conceptual materials and product design to achieve systematic model integration and information workflow is accepted by the logical procedure of literature review, gap analysis, and development and evaluation of individual and integrated constructs. Empirical studies need to be carried out to establish the usefulness and effectiveness of the approach, and they are addressed in Chap. 6.

4.9.2 TSV of Concept Exploration Framework and Goal-Oriented, Inverse Design Method

Theoretical structural validation refers to accepting the validity of individual constructs used in the concept exploration framework and the goal-oriented, inverse design method and accepting the internal consistency of the way the constructs are put together. Theoretical structural validation involves systematically identifying the scope of the proposed framework's and design method's application, reviewing relevant literature and identifying the research gaps that is existing, identifying the strengths and limitations of the constructs used based on literature review, determining the constructs and approaches that can be leveraged for the concept exploration framework and inverse design method while reviewing literature on the advantages, disadvantages, and accepted domains of application, and checking the internal consistency of the constructs both individually and when integrated.

In Chap. 1, the need for a concept exploration framework for the systematic concept exploration of materials and products is established. The CEF is inspired by the RCEM. The RCEM is critically reviewed in Chap. 3 and the functionalities and limitations associated with the method are established. The limitations of RCEM in terms of the following are discussed and are summarized into the following points: (i) RCEM does not take into account already available material and product models and relationships and focuses on establishing reduced-order meta-models/surrogate models, (ii) RCEM has limitations in terms of exploration of solution space and does not have processors for establishing design scenarios for exercising the cDSP, (iii) RCEM also lacks visualization tools and constructs for solutions space exploration and carry out design tradeoffs, (iv) RCEM cannot be individually used to support design exploration of process chains and thus needs support from a design method to achieve the same, (v) RCEM in terms of EMI and DCI does not address robust design of a system having multiple conflicting goals that require different types of robust design across process chains. Based on these limitations, the requirements for an improved framework and a design method that facilitates the integrated design exploration of materials, products, and associated manufacturing processes are established. To address the need for the inverse design method, the current research efforts focusing on inverse design exploration of material hierarchies are reviewed in detail in Chaps. 1 and 2. The existing challenges and limitations are addressed and the need for a systems-based, top-down design exploration method is established in Chaps. 1 and 2. In this chapter, the goal-oriented, inverse design method is proposed. Several challenges associated with similar inverse design exploration methods like the IDEM are highlighted in Chap. 3 and some of these challenges are addressed by the inverse design exploration method proposed in this monograph. A detailed analysis of the functionalities offered compared to methods like IDEM is provided in Chap. 7. The basic idea of the method proposed in this chapter for finding satisficing solutions in a multilevel, multistage process chain that involves the processing–structure (PS), structure–property (SP) relations is passing down the satisficing solution ranges in an inverse manner, from given final performance range to the design space of the previous space (defined by model input and output) with designer having the flexibility to choose solution of preference. The method is goal-oriented because the designer starts with the end goals that need to be realized for the product as well as process and then design the preceding stages to satisfy these end goals as closely as possible by exploring the design space. Then the design decisions that are made for the end requirements of the product/process after exploration are communicated to the stages that precede them to make logical decisions at those stages to satisfy the requirements identified, thereby carrying out a design space exploration process in an inverse manner. The proposed concept exploration framework is shown in Fig. 4.8, and the goal-oriented, inverse design method along with the associated steps is shown in Fig. 4.9. The details of the framework and the design method with a description of each step to be performed to formulate, exercise, and explore a complex systems design problem are provided in this chapter. The input needed and the output generated are clarified; the internal information flow is checked to ensure the availability of sufficient information to execute the next steps. Through a critical evaluation of each step and the way individual constructs are put together, internal consistency of the concept exploration framework and the inverse design method is verified and accepted.

The theoretical structural validity of the concept exploration framework and the goal-oriented, inverse design method to achieve inverse decision-based design exploration of process chains from a systems perspective is accepted by the logical procedure of literature review, gap analysis, and development and evaluation of individual and integrated constructs like the cDSP, surrogate modeling techniques, ternary analysis and plots, the inverse design method, and so on. Empirical studies need to be carried out to establish the usefulness and effectiveness of the framework and the method addressed in Chaps. 5–7.

4.10 Role of This Chapter and Connection with Other Chapters in This Monograph

In this chapter, the focus is on developing a systematic function-based approach for model integration and establishing information workflow. The function-based approach supports a designer in establishing the forward material workflow of the material/product system. Also, in this chapter, the focus is on developing a concept exploration framework that supports a designer in systematically formulating a design problem and exploring the solution space to identify satisficing design solutions. A goal-oriented, inverse design method is proposed to support a designer in designing the system starting from the end goals and requirements. The method is proposed as a generic one with implications in several fields given there are models that establish the forward information workflow. The relationship of this chapter with other chapters is shown in Fig. 4.10. The utility of the systems-based design architecture developed in this chapter is tested using example problems in Chap. 5 and a

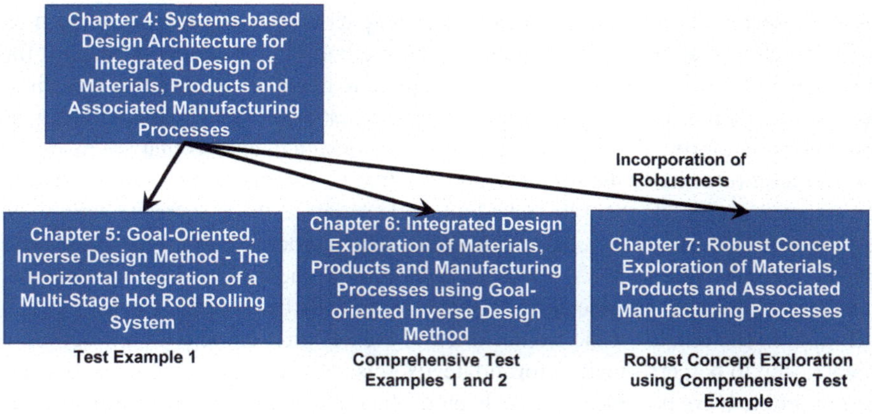

Fig. 4.10 Relationship of this chapter with other monograph chapters

comprehensive steel manufacturing example problem in Chap. 6. In Chap. 7, the concept exploration framework and inverse design exploration method proposed in this chapter are updated to include robustness. The interest in Chap. 7 for the designer is to identify satisficing robust solutions for multiple conflicting goals across a process chain.

References

Bras, B., & Mistree, F. (1993). Robust design using compromise decision support problems. *Engineering Optimization, 21*(3), 213–239.

Chen, W., Allen, J. K., & Mistree, F. (1997). A robust concept exploration method for enhancing productivity in concurrent systems design. *Concurrent Engineering, 5*(3), 203–217.

Choi, H., McDowell, D. L., Allen, J. K., Rosen, D., & Mistree, F. (2008). An inductive design exploration method for robust multiscale materials design. *Journal of Mechanical Design, 130*(3), 031402.

Fonville, T. R., Nellippallil, A. B., Horstemeyer, M., Allen, J. K., & Mistree, F. (2019). A goal-oriented, inverse decision-based method for an American Football Helmet. *ASME Design Automation Conference*, Paper Number: IDETC2019-97388.

Hodgson, P., & Gibbs, R. (1992). A mathematical model to predict the mechanical properties of hot rolled C-Mn and microalloyed steels. *ISIJ International, 32*(12), 1329–1338.

Kern, P. C., Priddy, M. W., Ellis, B. D., & McDowell, D. L. (2017). pyDEM: A generalized implementation of the inductive design exploration method. *Materials & Design, 134*, 293–300.

Kuziak, R., Cheng, Y.-W., Glowacki, M., & Pietrzyk, M. (1997). Modeling of the microstructure and mechanical properties of steels during thermomechanical processing. *NIST Technical Note (USA), 1393*, 72.

Majta, J., Kuziak, R., Pietrzyk, M., & Krzton, H. (1996). Use of the computer simulation to predict mechanical properties of C-Mn steel, after thermomechanical processing. *Journal of Materials Processing Technology, 60*(1–4), 581–588.

Messer, M. (2008). *A systematic approach for integrated product, materials, and design-process design*. Ph.D. Dissertation, The George W. Woodruff School of Mechanical Engineering, Georgia Institute of Technology.

Ming, Z., Nellippallil, A. B., Yan, Y., Wang, G., Goh, C. H., Allen, J. K., & Mistree, F. (2018). PDSIDES—A knowledge-based platform for decision support in the design of engineering systems. *Journal of Computing and Information Science in Engineering, 18*(4).

Mistree, F., Hughes, O. F., & Bras, B. (1993). Compromise decision support problem and the adaptive linear programming algorithm. *Progress in Astronautics and Aeronautics, 150*, 251–251.

Nellippallil, A. B., Song, K. N., Goh, C.-H., Zagade, P., Gautham, B., Allen, J. K., & Mistree, F. (2016). A goal oriented, sequential process design of a multi-stage hot rod rolling system. *ASME Design Automation Conference*, Paper Number: DETC2016-59402.

Nellippallil, A. B., Rangaraj, V., Allen, J. K., Mistree, F., Gautham, B., & Singh, A. K. (2017a). A decision-based design method to explore the solution space for microstructure after cooling stage to realize the end mechanical properties of hot rolled product. In *Proceedings of the 4th World Congress on Integrated Computational Materials Engineering (ICME 2017)*. Springer.

Nellippallil, A. B., Rangaraj, V., Gautham, B., Singh, A. K., Allen, J. K., & Mistree, F. (2017b). A goal-oriented, inverse decision-based design method to achieve the vertical and horizontal integration of models in a hot rod rolling process chain. *ASME Design Automation Conference*, Paper Number: DETC2017-67570.

Nellippallil, A. B., Song, K. N., Goh, C.-H., Zagade, P., Gautham, B., Allen, J. K., & Mistree, F. (2017c). A goal-oriented, sequential, inverse design method for the horizontal integration of a multistage hot rod rolling system. *Journal of Mechanical Design, 139*(3), 031403.

Nellippallil, A. B., Mohan, P., Allen, J. K., & Mistree, F. (2018a). Robust concept exploration of materials, products and associated manufacturing processes. *ASME Design Automation Conference*, Paper Number: DETC2018-85913.

Nellippallil, A. B., Rangaraj, V., Gautham, B., Singh, A. K., Allen, J. K., & Mistree, F. (2018b). An inverse, decision-based design method for integrated design exploration of materials, products, and manufacturing processes. *Journal of Mechanical Design, 140*(11), 111403-111403-17.

Nellippallil, A. B., Mohan, P., Allen, J. K., & Mistree, F. (2019). Inverse Thermo-Mechanical Processing (ITMP) design of a steel rod during hot rolling process. *ASME Design Automation Conference 2019*, Paper Number: IDETC2019-97390.

Nellippallil, A. B., Mohan, P., Allen, J. K., & Mistree, F. (2020). An inverse, decision-based design method for robust concept exploration. *Journal of Mechanical Design*, https://doi.org/10.1115/1.4045877.

Olson, G. B. (1997). Computational design of hierarchically structured materials. *Science, 277*(5330), 1237–1242.

Pahl, G., & Beitz, W. (1996). *Engineering design: A systematic approach*. London: Springer-Verlag.

Pahl, G., & Beitz, W. (2013). *Engineering design: A systematic approach*. Springer Science & Business Media.

Phadke, S., Pauskar, P., & Shivpuri, R. (2004). Computational modeling of phase transformations and mechanical properties during the cooling of hot rolled rod. *Journal of Materials Processing Technology, 150*(1), 107–115.

Pietrzyk, M., Cser, L., & Lenard, J. (1999). *Mathematical and physical simulation of the properties of hot rolled products*. Elsevier.

Simon, H. A. (2013). *Administrative behavior*. Simon and Schuster.

Suh, N. P. (1990). *The principles of design*. Oxford University Press on Demand.

Szykman, S., Racz, J. W., & Sriram, R. D. (1999). The representation of function in computer-based design. In *ASME Design Engineering Technical Conferences, 11th International Conference on Design Theory and Methodology*.

Taguchi, G. (1986). *Introduction to Quality Engineering*. Dearborn, MI: Asian Productivity Organization, Distributed by the American Supplier Institute, Inc.

Ullman, D. G. (1992). A taxonomy for mechanical design. *Research in Engineering Design, 3*(3), 179–189.

Wang, R., Nellippallil, A. B., Wang, G., Yan, Y., Allen, J. K., & Mistree, F. (2018). Systematic design space exploration using a template-based ontological method. *Advanced Engineering Informatics, 36*, 163–177.

Chapter 5
Goal-Oriented, Inverse Design Method—The Horizontal Integration of a Multistage Hot Rod Rolling System

5.1 Model-Based Horizontal Integration of Process Chains

Steel mills are involved in the production of semi-products such as sheets or rods with certain grades of steel. Process designers are very much aware of the operating constraints and process requirements for each of the operations as they are involved in the whole process day-in and day-out. Due to the advancements in material technology, new improved materials with enhanced properties are introduced to market posing a serious challenge to steel manufacturers. Suppose that owing to the changing properties and performance requirements, manufacturers must produce a semi-product such as a rod with a newer grade of steel. This new steel grade can be used at the laboratory scale to produce a rod, but the challenge posed to a steel manufacturer is to scale-up production. This requires the exploration of the ***design setpoints*** for each unit operation in the plant scale production of the rod (Tennyson et al. 2015). Plant trials are one way of achieving this. However, plant trials usually take a lot of time and are expensive. Another option is to use computational models for exploring the design setpoints for these operations and thereby reduce the time and cost. However, these models are for specific phenomena that occur during an integrated process. Isolated models for individual processes will not give a true representation of the whole system and the desired solution. In this context, we define horizontal integration of processes as the facilitation of information flow from one process stage to another, thereby establishing the integration of manufacturing stages to realize an end product. For exploring the design setpoints to achieve an end product, knowledge of the operational constraints and requirements are necessary for the newer grades of steel. However, this information is not available. Therefore, the first task is to identify operating constraints and design information for each unit operation. These operating constraints are imposed by the previous and subsequent unit operations as each process step is connected and information flows from one operation to another. Such process design problems are characterized by their complexity due to a large number of variables and their relationships at multiple stages. Two types of associations are possible for such problems—sequential and non-sequential. In the case of

A. B. Nellippallil et al., *Architecting Robust Co-Design of Materials, Products, and Manufacturing Processes*, https://doi.org/10.1007/978-3-030-45324-4_5

a non-sequential association, there is no definite order among the subsystems and most network problems fall under this category (Tiwari et al. 2008).

In this chapter, we focus on demonstrating the inverse method for designing a multistage hot rod rolling system for manufacturing a rod which is one of the semi-products in a steel manufacturing process chain. We view design as a decision-making process and believe that the fundamental role of a human designer is to make decisions. The hot rod rolling problem is sequential in which information flows from the first rolling stage/pass to the last rolling pass and the decisions made at the first pass influence the decisions that must be made at later passes (Tiwari et al. 2008). We carry out the design process by means of a goal-oriented method that uses well-established empirical models, response surface models along with the compromise decision support problem (cDSP) construct (Reddy et al. 1992; Mistree et al. 1993; Allen et al. 2006) to support integrated information flow across different stages of rolling process. The method is goal-oriented because the decisions are first made based on the end requirements identified for the product and the process at the last rolling stage, and these decisions are then passed to the preceding stages. Thus, the decisions at the first rolling stages are influenced by the decisions made at the last rolling stage, thereby making this an inverse design scheme based on end goals. The cDSP is formulated using empirical models and the response surface models developed using simulation experiments and are then exercised for different design scenarios to explore the design space and to identify the best set of variables (design and operating setpoints) that meets the conflicting goals. Ternary plots are used to visualize these scenarios and to identify the appropriate feasible design space. The design of the multistaged rolling process is carried out using the setpoints identified. The entire goal-oriented, inverse design method is generic and has the potential to be applied to design any set of manufacturing processes where there is a sequential flow of information (material) in order to realize an end product with specified target goals.

In Sect. 5.2, we describe the hot rod rolling process and the challenges associated with the modeling and design exploration of the process. In Sect. 5.3, we describe the problem in terms of the boundary defined and parameters considered in this study. The solution strategy (in terms of process design scheme) and the method adopted for this problem are also described in this section. In Sect. 5.4, we describe the empirical models and the response surface models developed. The mathematical formulation of the rod rolling problem using the cDSP construct is also presented in this section. The ternary analysis for visualizing and exploring the solution space is covered in Sect. 5.5. The key findings and closing remarks are presented in Sect. 5.6. We showcase the design calculations in Sect. 5.7. In Table 5.1, we summarize the testing of the architecture constituents carried out in this chapter.

Table 5.1 Summary of the testing carried out in this chapter

Elements	*What?*
Processing elements	The functionalities of CEF and GoID—carry out systematic goal-oriented, inverse design exploration of horizontal manufacturing process chains (horizontal integration of information flow)
Data elements	Horizontal integration of hot rolling process chain problem
Connecting elements	The GoID method
Form	*How?*
Weights	Testing the utility of CEF and GoID using a test example
Properties	Focus on macrostructure-level hot rolling process requirements
Relationship	Theoretical and empirical performance validation using the test example
Rationale	*Why?*
Motivation	To validate the utility of CEF and GoID functionalities in providing design decision support in the horizontal integration of process chains
Characteristics/beliefs	• Design is a goal-oriented synthesis activity and essentially a decision-making process—the manufacturing process chain is therefore represented in terms of a horizontal decision workflow • Decisions can be formulated mathematically, and analysis models are available for design decision-making • Design process starts with an end goal in mind—could be an end product that needs to be supplied to customers • There exist models or computational tools or data that can establish the forward information workflow for the process/product that is being considered • Analysis models for problems are typically incomplete, inaccurate, and not of equal fidelity • Designer's interest is in identifying satisficing robust solutions given the uncertainty in the problem and not single-point optimum solutions • Design is a goal-oriented activity with design requirements subject to change at any time for the problems considered
Interpretation	Understanding the efficacy of the design constructs and broader impacts

5.2 The Hot Rod Rolling Process

Hot rod rolling is a complex, multistage manufacturing process that plays a critical role in producing specific grades of steel with specified target properties. The complexity in the process arises not only from the high working temperatures but also because of the requirement to precisely control the process parameters to obtain the desired microstructure and properties. Due to increasing competition facing steel and aluminum manufacturers, there is an increasing need to make this process more flexible, agile, and energy-efficient. Process designers must determine cost-effective solutions to assist in decision-making and to improve efficiency. Multipass rolling systems design (RSD) is the preparation of a set of rolls that are laid in series in the right sequence for different rolling passes to achieve the desired profile (Oduguwa and Roy 2006). RSD helps in producing workpieces with the desired work profile subject to the constraints of the mill with acceptable quality, minimum cost, and maximum output. This is equivalent to a search problem where the design space is explored to satisfy the requirements in order to determine the required number of passes to achieve a product of the required dimensions with minimum defects by controlling design variables. This requires considering different behaviors of the material during rolling, including geometrical, mechanical, thermal, thermo-mechanical, and metallurgical behaviors at multiple scales. Rolling is a multidisciplinary process involving reheating, interstand operations, mill engineering, roll pass design, metallurgical transformations, and so on (Oduguwa and Roy 2006).

The challenges associated with the design of a rolling system arise from the complex nature of the process due to a large number of process parameters, constraints, bounds, and so on, the multistaged nature of process involving handshakes,[1] the hierarchical nature in terms of process–structure–property–performance relationships, multidisciplinary nature requiring knowledge and expertise from different fields, and complex relationships between stress/strain, temperature, and microstructure that requires model coupling at different scales (Roberts 1983; Lapovok and Thompson 1994, 1997; Michalewicz 1995; Shin 1995; Roy et al. 2000; Jupp 2001; Oduguwa and Roy 2001, 2002a, b, 2006; Oduguwa et al. 2004). The challenges are addressed in detail below.

The challenges associated with the design of a rolling system are listed below in detail (Oduguwa and Roy 2006):

(i) *Complex search space*: Rolling is a complex deformation process involving several process parameters, hard constraints, and bounds, so it is difficult to define the search space for rolling systems design (Michalewicz 1995; Roy et al. 2000). The functions involved in the process are nonlinear, discontinuous, and require coupling with models that are at different scales. Functions relating to all the influencing parameters involved are not directly available for this problem in order to carry out design exploration.
(ii) *Multistaged process*: Hot rolling is a multistage problem requiring multiple passes to produce the end product. Hence the sequential nature of linked unit operations that can be seen in a steelmaking process chain can be attributed to the hot rolling process. Each of these passes has essential requirements for achieving the desired property/performance of the end product. The output of one pass becomes the input to the next pass and there is a handshake between each pass which must be designed (Oduguwa et al. 2004).
(iii) *Hierarchical nature*: The hierarchical nature of the rolling process arises not only from process–structure–property–performance relationship in the material system but also from the way levels are defined in the system in terms of objectives to be met (Oduguwa and Roy 2002a, b). For example, at the top level, an engineer is interested in the productivity and quality of output

[1]Handshake, the flow of information between passes as the output of one pass is the input to the next. Thus, the passes are linked by the relationships that exist when material flows between them.

from the system. At a lower level, an engineer is interested in the number of passes required to achieve a target and the form of roll grooves for a particular operation, and so on (Lapovok and Thompson 1994, 1997; Shin 1995).

(iv) *Multidisciplinary problem*: The various aspects of the rolling process, such as manufacturing, mill design, thermal aspects, material aspects, computational aspects, metallurgical phenomenon, modeling and simulation, process design, and quality control, require knowledge and expertise from a wide range of engineering fields, making the entire process highly complex.

(v) *Complex relationships*: Stress/strain, temperature, and microstructure have such a complex relationship that it is essential that they are analyzed for modeling the process (Jupp 2001). Accurate measures of each of them are required to fully describe the behavior of the material. For example, the temperature developed in the material while rolling involves the heat that is developed due to plastic deformation. This temperature developed influences the microstructure during rolling. Stress/strain that occurs in rolling affects the stored energy in the material and influences the recrystallization which changes the microstructure. Similarly, the deformation temperature, strain, and strain rate affect the stress developed in the material while rolling (Jupp 2001). All these complex relationships need to be modeled precisely. This is a huge challenge from a design perspective.

(vi) *Multidimensional*: Rolling involves various product and process parameters, such as rolling speed, rolling temperature, reduction ratio, the geometry/work profile, spread of the workpiece in rolling, temperature distribution, friction conditions, heat transfer, cooling conditions, and thermal conditions of the roll (Roberts 1983). The challenge of bringing these varied parameters into a single formulation is a complex task.

(vii) *Knowledge-driven process*: Rolling system design is a knowledge-driven process and is mostly qualitative in nature (Oduguwa and Roy 2001). The traditional way of rolling system design involved some expert personnel who have sound knowledge in the mixture of engineering fields that are involved, having many years of experience in practically handling the system. However, with the evolving trends in the market and rise in competitions in the industry, the requirement is to design the system as fast as possible in the best way with changing demands. This requires people with expertise in multidisciplinary areas to work together and share knowledge in order to effectively add value to the system design. The capability to do that is defined as co-design in this monograph, see Chaps. 1 and 8. Also, the rise in computational tools and techniques has to be exploited in the best possible way to bring in new design changes and modifications so as to improve the design process.

In this chapter, we address some of these challenges by developing a design method using simulation models along with the compromise decision support problem construct and solution space exploration techniques to design the multiple stages of a rolling system ensuring information flow to support horizontal integration of stages in order to realize an end product. The complex search space is managed by framing a proper boundary for the problem formulated and will be explained in Sect. 5.3. Well-established empirical models along with a finite-element model developed for rolling are used to define the complex relationships. The academic and industrial collaboration involved in this work between people from mechanical, design, material science, and metallurgy domains helped to deal with the multi-disciplinary nature of the problem. The decision support problem constructs along with the solution space exploration techniques support a designer to manage the uncertainty associated with the models and address a way of handling such complex problems from a systems design perspective. In the next section, we describe the foundational construct for our work—the compromise decision support problem (cDSP) construct.

5.3 Foundational Constructs—The cDSP and Solution Space Exploration

In the model-based realization of complex systems, we have to deal with models that are typically incomplete, inaccurate, and not of equal fidelity. This brings into the design process different types of uncertainties associated with the system, the parameters considered, the models considered, and the uncertainties due to their interactions (McDowell et al. 2009). From the decision-based design perspective, the fundamental role of a human designer is to make decisions given the uncertainties associated. In this regard, we define robust design as a design that is relatively insensitive to changes. This involves achieving a desired performance for the system while the sensitivity of the performance objectives with respect to the system variables is minimized (Ebro and Howard 2016). Thus, the designer's objective here is to find "*satisficing*" solutions that showcase good performance given the presence of uncertainties and not optimum solutions that are valid for a narrow range of conditions while performing poorly when the conditions are changed slightly. The cDSP is proposed by Mistree et al. for robust design with multiple goals (Bras and Mistree 1993; Mistree et al. 1993). The fundamental assumption here is that the models are not complete and accurate; opposed to the fundamental assumption in optimization where the models are complete and accurate, and the objective function can be modeled accurately so that solution obtained is implementable. Hence the cDSP construct is anchored in the robust design paradigm first proposed by Taguchi (1986). Using the cDSP construct several solutions are identified by carrying out tradeoffs among multiple conflicting goals. The obtained solutions are then evaluated by carrying out solution space exploration in order to identify the best solutions that satisfy the specific requirements identified. The cDSP is a hybrid formulation based on mathematical programming and goal programming. In goal programming, the target values for each goal are defined and the emphasis is on achieving the target for each goal as close as possible. In cDSP, different weights are assigned to these goals and the compromised solutions obtained for different appropriate weights are explored.

The formulation and solving of the cDSP followed by the exploration of solution space for any problem are carried out using the steps in the concept exploration framework (CEF), see Chap. 4, Sect. 4.6. A generalized four-step method is illustrated in Fig. 5.1 that captures the overall steps of CEF in a simplified manner (Shukla et al. 2014, 2015; Nellippallil et al. 2016). After having defined the problem and requirements, Step 1 is to identify the theoretical and empirical models and relationships that exist for the process/problem of interest. Response surface models are developed to represent certain parameters as a function of the process variables. These response surface models are developed by carrying out simulation experiments, which could be finite-element model-based experiments or other similar experiments depending on the problem of interest. The response surface models developed through simulation experiments along with the theoretical and empirical models and relationships available are used to formulate the cDSP for the process/problem that is under study (Step 2).

In Step 3, we exercise the cDSP for different design scenarios and the results are recorded for each scenario. These scenarios are identified by assigning different weights to the goals of the cDSP formulated. The collective design results for different scenarios are visualized using ternary plots and the feasible design space that satisfies the design requirements in the best possible manner is identified (Step 4). Multiple solutions that satisfy the design requirements are identified from the feasible design space. The designer makes design decisions from the set of solutions depending upon the preferences set for the problem under study. For the manufacturing problem under consideration, these identified design solutions are the design and operating setpoints.

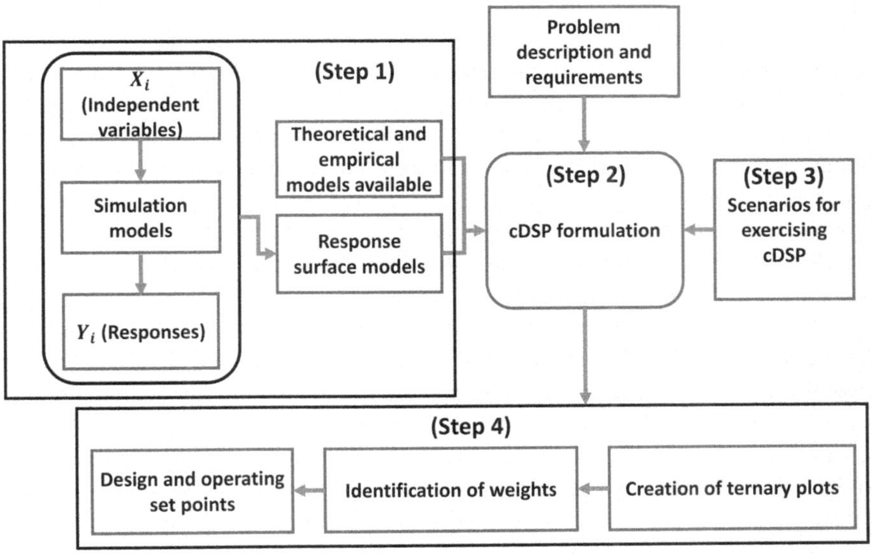

Fig. 5.1 cDSP-based steps to predict setpoints (includes key steps of CEF—a simplified form)

In the next section, we describe the goal-oriented method for carrying out sequential process design of manufacturing stages utilizing the cDSP construct and solution space exploration techniques to achieve the integrated design of the product and the processes. We use the hot rod rolling system design problem as an example to illustrate the efficacy of the method presented.

5.4 Problem Description and Application of Goal-Oriented, Inverse Design Method

Rod *quality* depends on many factors, starting from the material microstructure to the macrostructure. Key factors influencing quality include steel composition, segregation of alloying elements, distribution of inclusions, microstructure, and rod geometry. Ovality is one such geometrical property that is defined as the difference between the height of the rod section and the width from the center of the rod (Oduguwa and Roy 2002a, b). Ovality is desirable in the initial roll passes as it helps to reduce the geometry of the square billet. However, it is not desired in the end rod product as the output requirement is for a round/circular rod. Thus, there is a need to minimize/control the ovality induced at the last rolling stage. One way of minimizing ovality is to ensure high contact between the workpiece and the roll. However, this requires high rolling loads and thus minimization of ovality is possible at the expense of a high rolling load. Rolling load influences the overall functioning of the process and is representative of the overall process performance (Tiwari et al. 2008). Rolling load ensures the flow of material across passes. Higher rolling loads require increased rolling power requirements and can also yield deflections in the rolling system which is detrimental to the rolls themselves. This adds to the costs of the process. Hence maintaining the rolling load within a target value in an acceptable range is necessary but conflicts with the objective of minimizing ovality. Excessive rolling load resulting in roll breakage and wear are detrimental to production efficiency as it conflicts with rolling process *productivity* which is expressed in terms of throughput (Tiwari et al. 2008). Therefore, this is a multiobjective design problem with three objectives: minimize ovality, maximize throughput, and minimize rolling load subject to the rolling constraints.

In this process, the output of one stand is input for the next and there are successive reductions of the billet at each rolling stand. Therefore, modeling this process demands information exchange between these stands as the intermediate product developed in one stand will affect the form, properties, and performance of the product developed at consecutive stands that follow, which results in an impact on the end product. Therefore, a method to ensure the determination of the right combination of design variables to meet the constraints for each rolling pass and thereby meet the overall performance requirement is essential.

We have developed a computational method for this sequential problem that has information exchange between rolling passes and is used to identify the setpoints of the various rolling passes involved. For this example, we assume that there are four passes that follow a square-oval, oval-round, round-oval, and oval-round sequence moving from Pass 1 to Pass 4. The final requirement of the product of Pass 4 is to have minimum ovality, maximum throughput, and a minimum rolling load value within an acceptable range. The different sequential relationships that exist among passes define the problem. The constraints for the process include the range for rolling load, the range for throughput, the maximum value of rolling wear, minimum and maximum values of elongation and spread for each pass. The cDSPs for two passes—Pass 2 and Pass 4—are formulated. The cDSP for Pass 4 takes into account the end goals identified for the problem in terms of ovality, throughput, and rolling load. The cDSP for Pass 2 is developed to support information flow across passes and perform the design of other passes. The goals for the cDSP for Pass 2 are maximization of throughput (maintaining target throughput values achieved for Passes 3 and 4) and achieving a target value of rolling load within a defined range. The ovality goal is an end goal for the rod produced after Pass 4 and is not required for Pass 2 as the material is again subjected to deformation to oval shape in Pass 3 to facilitate the progressive breakdown of geometry.

This goal-oriented, inverse design method proposed to design the rolling system will be explained using the information flow diagram shown in Fig. 5.2. In order to generalize the method, we are naming the four stages of rolling passes as "manufacturing stages" which are numbered from "n" to "$n + 3$". We will be using the term "end product" for the rod developed after rolling and the term "input material" to refer to the billet that comes from the continuous casting stage of the steel manufacturing process chain. The arrows that denote the flow of information need to be

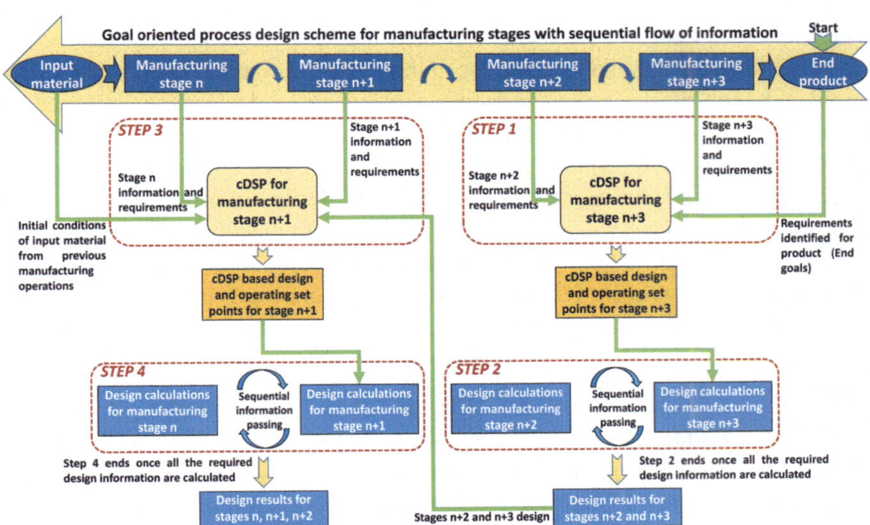

Fig. 5.2 Goal-oriented, inverse design method for manufacturing stages having a sequential flow of information

followed to visualize the design process. There are four steps in the design method for designing these four manufacturing stages to realize the end product.

Step 1: Formulation of cDSP for the last manufacturing stage (n + 3) using the information from the end product to be realized and the sequential relationship existing between stages n + 2 and n + 3

The whole design process starts with the identification of requirements for the end product to be produced after the manufacturing stage $n + 3$, as shown in Fig. 5.2. In Step 1, the cDSP for manufacturing stage $n + 3$ is formulated. The cDSP is formulated using the information available on manufacturing stage $n + 3$ and by incorporating the sequential relationship that the stage $n + 3$ has with manufacturing stage $n + 2$. The requirements identified for both the end product and for the manufacturing stage $n + 3$ are embodied in this cDSP as goals. The requirements from the manufacturing stage $n + 2$ along with the sequential relationships that exist are captured by the "Given" and "Satisfy" keywords of the cDSP formulated. The cDSP is exercised for different identified scenarios by assigning different weights for each goal, and the scenarios that suit the design requirements the most are selected after carrying out solution space exploration using ternary plots. The system variables identified are basically the design and operating setpoints for manufacturing stage $n + 3$.

Step 2: Design of stages n + 3 and n + 2 using the design and operating setpoints identified and the information available from end product requirements

In Step 2, the design and operating setpoints generated for manufacturing stage n + 3 from Step 1 are used to design the stage by carrying out design calculations to determine information. Design calculations essentially involve analysis to check the achievement of goals and using the design and operating setpoints generated to calculate the values of parameters of both the manufacturing stages using the sequential relationships that exist between them that was incorporated in the cDSP formulated. First, the design and operating setpoints are used to generate information for stage $n + 3$. The new design information generated for stage $n + 3$ has a sequential relationship with manufacturing stage $n + 2$ and hence they are passed to carry out the design of manufacturing stage $n + 2$. Once new design information is generated for manufacturing stage $n + 2$, they are again passed to manufacturing stage $n + 3$ to come up with information which was unknown before. Thus, a cyclic process of information exchange is carried out at this step to generate new information for both the manufacturing stages using the design and operating setpoints identified in Step 1. Step 2 ends once all the required design information for the problem formulated is identified.

Step 3: Formulation of* cDSP *for manufacturing stage n + 1 using the design information generated for stages n + 2 and n + 3; and the sequential information existing between stages n and n + 1; along with information on input material

In Step 3, the cDSP for manufacturing stage $n + 1$ is formulated. The design information generated for stages $n + 2$ and $n + 3$ are communicated to the cDSP for manufacturing stage $n + 1$. The "Given" keyword of this cDSP captures the design information from stages $n + 2$ and $n + 3$. Along with that, sequential information related to stages n and $n + 1$ and the initial conditions of input material are also captured during the formulation of this cDSP using the "Given" and "Satisfy" keywords. Specific requirements identified for manufacturing stage $n + 1$ are formulated as system goals. The cDSP formulated is exercised for different scenarios to find design and operating setpoints for manufacturing stage $n + 1$ that satisfy the requirements identified for the stage as well as the end requirements of the product.

Step 4: Design of manufacturing stages n + 1 and n using the design and operating setpoints identified; the information available from input material and the information from stages n + 2 and n + 3

In a similar fashion to Step 2, the design and operating setpoints identified for manufacturing stage $n + 1$ are used to design the stage by carrying out design calculations. The design information generated for stage $n + 1$ is passed to the design manufacturing stage n using the sequential relationships that exist. The information available from the input material is also used at this stage to carry out the design of the manufacturing stage n. The new design information generated for stage n is then communicated back to stage $n + 1$ to determine stage $n + 1$ information that was unknown before. The sequential information passing is carried out until the required design information for the problem formulated is identified. The design information generated for stages n and $n + 1$ are also used to carry out design calculations for stages $n + 2$ and $n + 3$ as the information from those stages are available in the cDSP formulated for stage $n + 1$. Hence the final result obtained using this goal-oriented, sequential method is the design information for all the four stages n, $n + 1$, $n + 2$, and $n + 3$ in order to realize the requirements identified for the process as well as the end product.

The proposed four-step method using the cDSP construct is generic and the method can be used for the design of other such unit operations where there is a sequential flow of information by identifying the design and operating setpoints that satisfy certain system goals and then design the entire system using these identified setpoints.

In Sect. 5.5, we describe the empirical models and theoretical models as well as the important relations that exist for the rod rolling problem under study. We also describe the response surface models that are developed as a part of the study here in this section. In Sect. 5.5.2, we explain the cDSP formulation for the Pass 4 (stage $n + 3$) of the hot rod rolling problem. The cDSP for Pass 2 (stage $n + 1$) which follows a similar pattern to that of Pass 4 will be explained in Appendix 1. In Sect. 5.6, we explain the scenarios identified for the cDSP for Pass 4 and visualization of the scenarios using ternary plots to identify the design and operating setpoints.

5.5 Designing a Multipass Rolling System

The purpose of a roll pass design is (Wusatowski 2013): (a) To ensure the production of a correct profile within permissible dimensional limits and with a good surface finish, free of surface defects, but at the same time keeping the internal stress in the section being rolled to a minimum, (b) to ensure the maximum output at minimum cost, (c) to ease the working conditions of the rolling crew, (d) to reduce roll wear to a minimum. For our hot rod rolling example problem, the design requirements are:

- Achieve a round profile by minimizing the ovality at the end of the fourth rolling pass.
- Maximize throughput while ensuring that product quality is not reduced.
- Maintain a minimum rolling load within a specified range and ensuring that it never exceeds the maximum.
- Control the elongation and spread during the rolling process within specified limits.
- Control the entry and exit speeds of the stock within specified limits.
- Ensure that the wear on the rolls is within an acceptable limit.
- Obey the sequential relationships between the different rolling passes (in terms of geometry and workpiece profile, etc.)

First, a process model for a rolling system is developed that ensures the flow of information through the sequential relationships between rolling passes as shown in Fig. 5.2. In Figs. 5.3a, b, we represent the geometry for the oval and round passes with key dimensions of interest for the rolling problem. The entire breakdown sequence consists of two more such passes in a cascaded fashion where the output of an oval pass is the input for the round pass. The rolls are laid horizontally and vertically for the oval and round passes, respectively. Therefore, the horizontal major axis of the oval stock in Fig. 5.3a coincides with the vertical axis of a round pass as in Fig. 5.3b. A detailed description of the models along with the mathematical expressions related to the goals identified is provided in the following sections.

5.5.1 *Major Relations and Calculations for the Rolling Pass Design Study*

Condition of Constant Volume

This condition requires that the volume of the material rolled remains the same after each pass.

$$V = Fl = F_j l_j = V_j \tag{5.1}$$

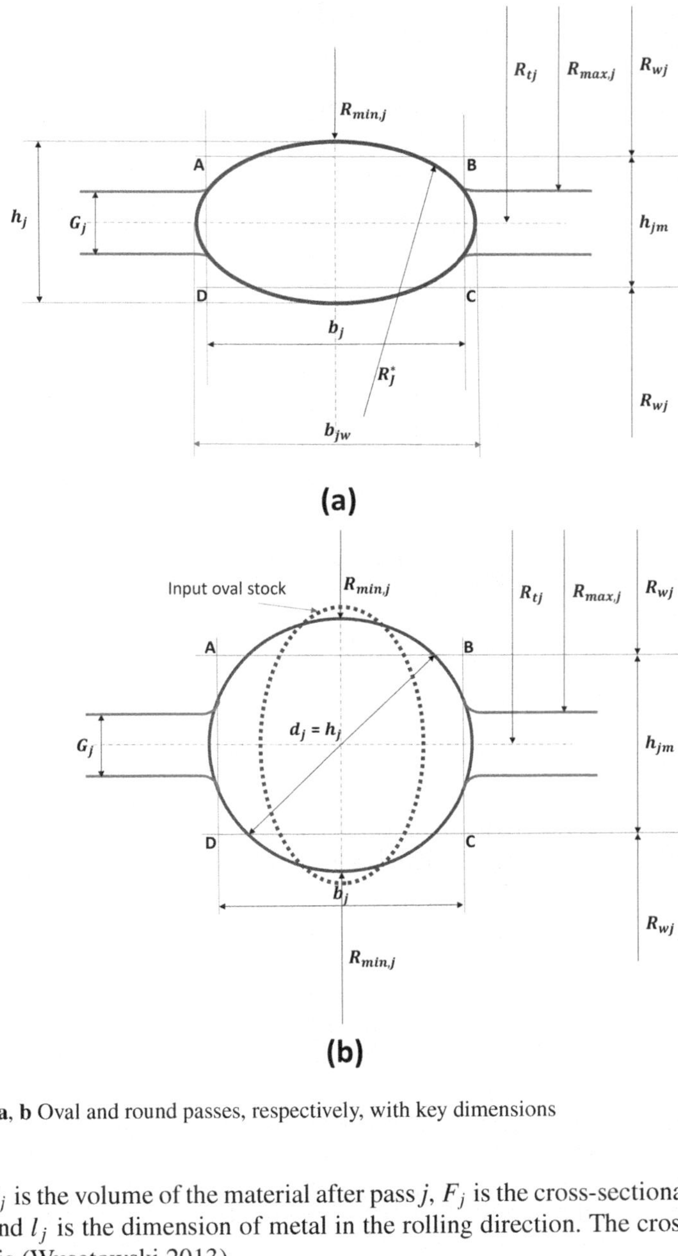

Fig. 5.3 **a**, **b** Oval and round passes, respectively, with key dimensions

where V_j is the volume of the material after pass j, F_j is the cross-sectional area after pass j, and l_j is the dimension of metal in the rolling direction. The cross-sectional area F_j is (Wusatowski 2013)

$$F_j = h_j b_j \tag{5.2}$$

This expression is valid for the rolling of rectangular cross-sections. For the rolling of non-rectangular cross-sections such as bars, shapes, and rails, an additional term, the mean height of stock is introduced and this is expressed as (Wusatowski 2013)

$$h_{jm} = \frac{F_j}{b_j} \tag{5.3}$$

It is calculated by dividing the cross-sectional area F_j by the maximum breadth b_j of the filled section for a particular pass *j*.

Thus the condition of constant volume during rolling is (Wusatowski 2013)

$$\begin{aligned} V_0 = F_0 l_0 = h_{0m} b_0 l_0 = V_1 = F_1 l_1 = h_{1m} b_1 l_1 \\ = V_n = F_n l_n = h_{nm} b_n l_n \end{aligned} \tag{5.4}$$

On dividing these relations (Wusatowski 2013)

$$\frac{h_{2m} b_2 l_2}{h_{1m} b_1 l_1} = \gamma_m \beta \lambda = 1 \tag{5.5}$$

where

$$\gamma_m = \frac{h_{2m}}{h_{1m}} = \text{mean coefficient of draught} \tag{5.6}$$

$$\beta = \frac{b_2}{b_1} = \text{spread in rolling} \tag{5.7}$$

$$\lambda = \frac{F_1}{F_2} = \frac{h_{1m} b_1}{h_{2m} b_2} = \frac{l_2}{l_1} = \frac{w_2}{v_1} = \text{coefficient of elongation} \tag{5.8}$$

where v_1 is the entry speed during a rolling pass, w_2 is the exit speed during the same pass. For round-oval rolling for rod production, an equivalent rectangle approximation (shown by ABCD in Figs. 5.3a, b) is carried out and the geometrical parameters are identified during the design process.

Rod Ovality

The ovality of the final rod product is a serious concern for manufacturers. It is mainly due to: (i) geometric factors such as the incoming width and height of the workpiece, radius of the roll, and the roll gap, (ii) metallurgical parameters such as strain values, stress developed, temperature of material during rolling, and (iii) rolling process parameter such as rolling speed (Oduguwa and Roy 2002a, b).

The geometric factors like incoming height (h_{j-1}) and width (b_{j-1w}) of the workpiece will define the amount of elongation and spread that occurs while rolling. This helps to determine the ovality of the rod produced. The roll radius ($R_{\max}$) and the roll gap (G_j) are critical parameters defining rolling contact and output size. Both of these parameters affect the ovality induced. The temperature (T_j) during rolling is also critical and determines the material flow. Higher temperature favors flow and thus plays a role in defining ovality. Also, the rolling speed (N_j, measured in rpm) affects the geometry formed.

Table 5.2 Factors and factor levels for design simulation

Level	h_{j-1} mm	b_{j-1w} mm	G_j mm	$R_{\max,j}$ mm	T_j K	N_j rpm
1	22	55	5.5	200	1280	20
−1	18	52	3.5	155	1270	10

Although these variables are known to influence the ovality during rolling, the exact relationships with respect to these variables are not available, and therefore simulation experiments using finite-element (FE)-based rolling model are carried out to determine models to predict ovality as a function of the variables identified. Appropriate ranges for the variables of interest are identified and a two-level fractional factorial design of experiments (DoE) is carried out. The steps associated with the same are (Oduguwa and Roy 2002a, b; Montgomery 2008):

Step 1: Fractional factorial design

The factors and factor levels for the simulations are depicted in Table 5.2.

A two-level six-factor fractional factorial design is used for the DoE. Finite-element (FE) simulations are carried out using the experimental design for the different runs of DoE. The coupled temperature–displacement finite-element model developed for the fourth oval to round rolling pass in ABAQUS is shown in Figs. 5.4 and 5.5.

The material being rolled is modeled as a deformable body of oval shape and is meshed using C3D8RT, an eight-node thermally coupled brick element. The material properties, for example, conductivity as a function of temperature, elastic properties, and so on, for steel are assigned to the billet. The plastic behavior of the material is described by assigning yield stress values for steel at different plastic strains. The rollers with round grooves are modeled as a discrete rigid body and are meshed using R3D4 elements. The surface profile of the rolls is modeled using the analytical models developed by Lee et al. (2000). The oval-shaped billet is constrained to move along the rolling direction. The rollers are constrained to only rotate along the axis of rotation. An initial temperature is defined for the billet before rolling which serves as the temperature for rolling. A surface-to-surface contact is defined between the billet

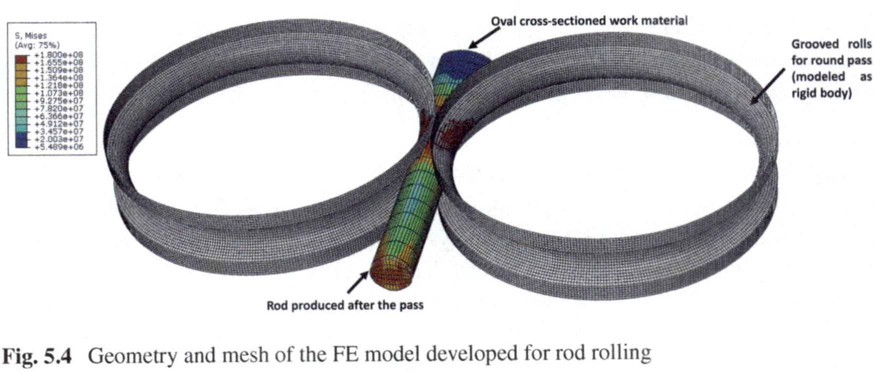

Fig. 5.4 Geometry and mesh of the FE model developed for rod rolling

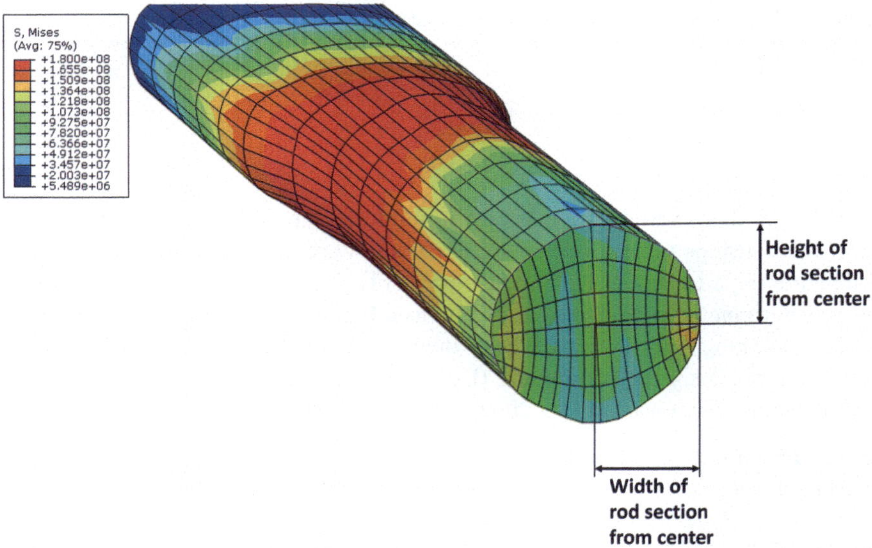

Fig. 5.5 Cross-section of rod produced using FE simulation showing the stress contours and the geometrical variables measured for calculating ovality

surface and the grooves of the rollers. The kinematic contact method is selected for mechanical constraint formulation. The heat transfer coefficient is defined between the roll gap and to air and the reported values from literature are selected (Phaniraj et al. 2005). The coefficient of friction value is set to 0.3 for the rolling simulations to develop the response surfaces for ovality. In preliminary studies, the coefficient of friction was shown to have a negligible effect on ovality; however, it does have an effect on roll wear as discussed later. The heat due to plastic deformation is defined and a value of 0.9 is used (Galantucci and Tricarico 1999). The angular velocity of the roll is applied based on the average strain rate associated with the rolling pass schedule (Lee et al. 2002). The developed FE model is validated for temperature predictions at billet center and surface, stresses developed, and geometry such as the final area of the rod produced following a similar pattern as in our previous works (Nellippallil et al. 2017a; Goh et al. 2014). The value of ovality in the rods is measured for each run and is recorded from the FE results as the absolute difference between the height and width of the rod section from the center.

Step 2: Model fitting

In Step 2, we develop response surface models for ovality by fitting the results obtained with a second-order polynomial. We carry out ANOVA and find that the effect of roll radius is negligible by analyzing the *p*-values obtained and thus roll radius is eliminated from the list of factors. The parameters of the second-order polynomial are determined using least squares regression analysis by fitting FE responses to input data. More detailed descriptions of RSM techniques and tools can be found in Myers and Montgomery (1995) and Simpson et al. (2001).

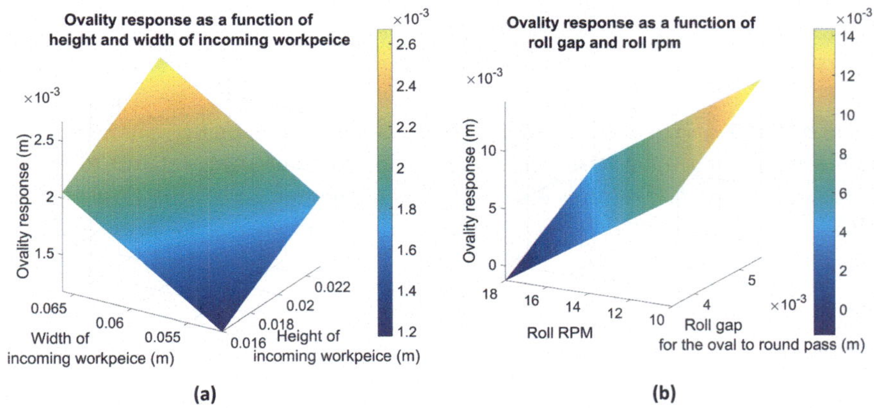

Fig. 5.6 a, b Ovality responses for different variables considered

The response surface model thus developed for ovality with an R^2 value of 0.99 is

$$O_{vj} = 8.6153 \times G_j + 27.539 \times b_{j-1w} - 0.0009 \times N_j + 0.0001 \times h_{j-1} \times T_j - 0.0023 \times h_{j-1} \times N_j - 0.0041 \times G_j \times T_j - 0.0269 \times G_j \times N_j - 0.0216 \times b_{j-1w} \times T_j - 0.0026 \times b_{j-1w} \times N_j \quad (5.9)$$

The response surface of the ovality model as a function of height and width of the incoming workpiece with fixed values of other variables is shown in Fig. 5.6a. In Fig. 5.6b, we show the response of the ovality model as a function of the roll gap and roll rpm.

Throughput

Throughput defines process productivity. Throughput is expressed as a function of exit speed during rolling (w_j) and the final stock cross-sectional area (F_j) that leaves the roll (Wusatowski 2013). The subscript *j* refers to the pass number.

$$T_{pj} = F_j w_j \quad (5.10)$$

F_j, the area of cross-section for the round pass is

$$F_j = \left(\frac{\pi h_j^2}{4} \right) \quad (5.11)$$

where $h_j = d_j$ = diameter of rod as shown in Fig. 5.3b.

For an oval sectional with a defined (*b/h*) ratio, the cross-section area is (Wusatowski 2013)

$$F_j = \left(\frac{\left(\frac{b}{h}\right)^2 h_j^2}{4.35} \right) \tag{5.12}$$

where b/h is a ratio defined for pass j. The equation is based on the values obtained from a nomogram for determining $h_m/h_{\max}$ for common ovals relative to s (roll clearance), h and b (Wusatowski 2013). The expression for b_{jw} is (Wusatowski 2013).

$$b_{jw} = (b/h)h_j \tag{5.13}$$

Hence from Eqs. 5.13 to 5.14,

$$b_{jw} = \sqrt{4.35\, F_j} \tag{5.14}$$

Rolling load

Excessive rolling load in various passes can affect productivity while minimum ovality is achieved through high contact and higher loads. Shinokira and Takai (Shinokura and Takai 1982) introduce a method for calculating the effective roll radius, the projected contact area, the non-dimensional roll force, and the torque arm coefficient expressed as simple functions of the geometry of the deformation zone. The rolling load (P) is defined as a function of a multiplier (Q_s), projected contact area (F_p), and mean flow strength of material ($2k$).

$$P = Q_s F_p(2k) \tag{5.15}$$

The mean flow strength of the material ($2k$) in the pass is approximated as the yield stress of material under plain compression as expressed in Sim's model (Sims 1954). The projected contact area is given by (Said et al. 1999),

$$F_p = \frac{2}{\pi}(0.9b_j)L_d = 0.573b_j L_d \tag{5.16}$$

where b_j is the final width after a pass. The projected length of contact in the deformation zone is (Lee et al. 2002),

$$L_d = \sqrt{\left(R_{\max} - \frac{h_o - G}{2}\right)(h_i - h_o)} \tag{5.17}$$

where $R_{\max}$ is the radius of the roll, G is the roll gap, h_i and h_o are the heights of the incoming and outgoing workpiece, respectively. Since there is a 90° rotation from an oval to a round pass, the incoming height of the workpiece for a round pass will be the width from the oval pass that precedes it. For a typical round pass j, the formula becomes

$$L_d = \sqrt{\left(R_{\max,j} - \frac{h_j - G_j}{2}\right)\left(b_{j-1w} - h_j\right)} \tag{5.18}$$

The multiplier Q_s is given by (Said et al. 1999)

$$Q_s = -0.731 + 0.771M + \frac{1.61}{M} \tag{5.19}$$

where M depends on the projected contact area, F_p, and the initial and final cross-sections, F_i and F_o, respectively (Said et al. 1999).

$$M = \frac{2F_p}{F_i + F_o} \tag{5.20}$$

For a typical pass j, $F_o = F_j$ and $F_i = F_{j-1}$.

Roll wear during rolling

Reducing the wear during rolling is important. To estimate it, we use an expression that estimates the change in the radius of a work roll due to wear during rolling (Roberts 1983). Roll wear is expressed as (Pietrzyk et al. 1999).

$$\frac{\Delta R}{l_j} = \frac{K\mu_j L_d^2 r\bar{\sigma}\exp\left[\frac{\mu_j L_d}{h_i(2-r)}\right]}{D^2\sigma_{\text{roll}}} \tag{5.21}$$

where ΔR is the change in roll radius, l_j is the rolled length, K is the wear constant, μ_j is the coefficient of friction, L_d is the projected contact length, r is the reduction during rolling, $\bar{\sigma}$ is the flow strength of the material rolled, σ_{roll} is the flow strength of roll, and D is the roll diameter. Here, we use $K = 8 \times 10^{-5}$, $\bar{\sigma} = 250$ MPa for the material rolled and $\sigma_{\text{roll}} = 600$ MPa (Roberts 1983; Pietrzyk et al. 1999). The rolled length l_j is

$$l_j = \lambda_j \times l_{j-1} \tag{5.22}$$

The value of l_{j-1} is assumed to be 3 m. The coefficient of friction, μ_j, is a system variable in this study and is between 0.3 and 0.45.

5.5.2 The cDSP for Roll Pass 4 (Step 1 of Method Proposed)

In this section we describe the mathematical formulation of the compromise decision support problem (cDSP) for Pass 4 of rod rolling. The cDSP for Pass 4 incorporates the end requirements identified for the rolling process. The cDSP is:

Given:

(1) End requirements identified for the rod rolling process

- Minimize ovality
- Maximize throughput
- Minimize rolling load
- Minimum limit of rolling load, $P_{\text{min}} = 28\,t$ (metric)
- Maximum limit of rolling load, $P_{\text{max}} = 35\,t$ (metric)
- Minimum limit of throughput, $T_{\text{pmin}} = 0.0001\,\text{m}^3/\text{s}$
- Maximum limit of throughput, $T_{\text{pmax}} = 0.0008\,\text{m}^3/\text{s}$
- Target value for ovality, $O_{v,\text{Target}} = 0.001 \pm 0.001\,\text{m}$
- Target value for rolling load, $P_{\text{Target}} = 28\,\text{t}$
- Target value for throughput, $T_{p4,\text{Target}} = 0.0006\,\text{m}^3/\text{s}$

(2) Number of passes = 4
(3) Initial billet size = 42 × 42 mm
(4) Pass sequence = Square–oval–round–oval–round
(5) Other parameter values for passes
(6) The RSMs and well-established empirical and theoretical correlations for the oval to round pass

- Area of round section obtained after Pass 4

$$F_4 = \left(\frac{\pi h_4^2}{4}\right) \tag{5.23}$$

- Coefficient of elongation for Pass 4

$$\lambda_4 = \frac{F_3}{F_4} \tag{5.24}$$

- The theoretical width of oval Pass 3

$$b_{3w} = \sqrt{4.35 F_3} \tag{5.25}$$

- The height of oval Pass 3 for a defined (b/h) ratio

$$h_3 = \frac{b_{3w}}{(b/h)} \tag{5.26}$$

- The width of round Pass 4 for a defined spread β_4

$$b_4 = \beta_4 h_3 \tag{5.27}$$

- Radius of curvature of oval pass

$$R_3^* = \frac{b_{3w}^2 + h_3^2}{4h_3} \tag{5.28}$$

- Mean height of the round rod produced after Pass 4

$$h_{4m} = \frac{F_4}{b_4} \tag{5.29}$$

- Theoretical diameter of roll for Pass 4

$$D_{t4} = 2\left(R_{\max,4} + \frac{G_4}{2}\right) \tag{5.30}$$

- Effective diameter of roll for Pass 4

$$D_{w4} = D_{t4} - h_{4m} \tag{5.31}$$

- Entry speed of material for Pass 4

$$v_4 = \frac{w_4}{\lambda_4} \tag{5.32}$$

- Exit speed for material for Pass 3

$$w_3 = v_4 \tag{5.33}$$

- Expression for ovality

$$\begin{aligned} O_{v4} = {} & 8.6153G_4 + 27.539b_{3w} - 0.0009N_4 + 0.0001h_3T_4 - 0.0023h_3N_4 \\ & - 0.0041G_4T_4 - 0.0269G_4N_4 - 0.0216b_{3w}T_4 - 0.0026b_{3w}N_4 \end{aligned} \tag{5.34}$$

- Throughput for Pass 4

$$T_{p4} = F_4 \times w_4 \tag{5.35}$$

- Rolling load in Pass 4

$$P_4 = Q_s F_p(2k) \tag{5.36}$$

(7) Variability in system variables
The system variables and their ranges are provided in Table 5.3.

Table 5.3 System variables and ranges for cDSP

Sr. No	Variables	Ranges
1	X_1, diameter of rod after Pass (h_4)	0.025–0.03 m
2	X_2, the coefficient of elongation for Pass 4 (λ_4)	1–3
3	X_3, the spread occurring in Pass 4 (β_4)	1–2
4	X_4, the exit velocity for Pass 4 (w_4)	0.5–3 m/s
5	X_5, the maximum radius of roll in Pass 4 ($R_{max,4}$)	0.155–0.2 m
6	X_6, the roll rpm in Pass 4 (N_4)	10–20 rpm
7	X_7, the temperature during rolling (T_4)	1270–1280 K
8	X_8, the roll gap (G_4)	0.0035–0.0055 m
9	X_9, the coefficient of friction (μ_4)	0.3–0.45

Find:
System Variables

X_1 diameter of rod after Pass 4 (h_4)*
X_2 the coefficient of elongation for Pass 4 (λ_4)
X_3 the spread occurring in Pass 4 (β_4)
X_4 the exit velocity for Pass 4 (w_4)
X_5 the maximum radius of roll in Pass 4 ($R_{\text{max},4}$)
X_6 the roll rpm in Pass 4 (N_4)
X_7 the temperature during rolling (T_4)
X_8 the roll gap (G_4)
X_9 the coefficient of friction (μ_4)

Deviation Variables

$$d_i^-, d_i^+, i = 1, 2, 3$$

Satisfy:
System Constraints

- Minimum coefficient of elongation constraint

$$\lambda_4(X_2) - 1.2 \geq 0 \tag{5.37}$$

- Maximum coefficient of elongation constraint

$$2 - \lambda_4(X_2) \geq 0 \tag{5.38}$$

- Minimum spread constraint

$$\beta_4(X_3) - 1.1 \geq 0 \tag{5.39}$$

- Maximum spread constraint

$$1.7 - \beta_4(X_3) \geq 0 \tag{5.40}$$

- Exit speed constraint

$$w_4 - v_r(X_i) \geq 0 \tag{5.41}$$

- Minimum load constraint

$$P(X_i) - P_{\min} \geq 0 \tag{5.42}$$

- Maximum load constraint

$$P_{\max} - P(X_i) \geq 0 \tag{5.43}$$

- Maximum wear constraint

$$0.0001 - \Delta R(X_i) \geq 0 \tag{5.44}$$

System Goals
Goal 1:

- Minimize Ovality

$$\frac{O_{v,\text{Target}}}{O_v(X_i)} - d_1^- + d_1^+ = 1 \tag{5.45}$$

Goal 2:

- Maximize Throughput

$$\frac{T_p(X_i)}{T_{p,\text{Target}}} + d_2^- - d_2^+ = 1 \tag{5.46}$$

Goal 3:

- Minimize Rolling Load

$$\frac{P_{\text{Target}}}{P(X_i)} - d_3^- + d_3^+ = 1 \tag{5.47}$$

Variable Bounds
Defined in Table 5.3
Bounds on deviation variables

$$d_i^-, d_i^+ \geq 0 \text{ and } d_i^- * d_i^+ = 0, \ \mathrm{i} = 1, 2, 3 \tag{5.48}$$

Minimize:

The aim for the designer using the cDSP is to minimize the over or under achievement of a goal from the target specified value. In the cDSP the objective function is represented as a weighted sum of the deviation variables and is known as the deviation function (Z). We minimize the deviation function

$$Z = \sum_{i=1}^{3} W_i\left(d_i^- + d_i^+\right); \sum_{i=1}^{3} W_i = 1 \tag{5.49}$$

The objective for us through the cDSP formulation is to minimize these deviation variables and achieve the target values of the goals as close as possible.

In the next section, we exercise the cDSP formulated for different design scenarios by changing the weights associated with the deviation variables of each goal. The results for each of these scenarios are used to construct ternary plots to help a designer visualize and explore the solution space and identify design and operating setpoints for the rolling passes to meet the identified end requirements of the process. A similar cDSP for Pass 2 is formulated with only two goals, that is, minimizing rolling load and achieving target throughput. The cDSP for Pass 2 is shown in Appendix 1.

5.6 Exploration of Solution Space

We have exercised 19 different scenarios for Pass 4. Different weights are assigned to each goal in these scenarios. Details of the scenarios are provided in Table 5.4.

Scenarios 1–3 are for a situation where the designer wants to achieve the target of one of the goals, minimizing ovality, maximizing throughput, or minimizing rolling load. For example, in scenario 1, the preference is only for achieving the ovality goal. Scenarios 4–6 are for a situation where equal preference is given to two of the goals while the third goal is not considered/relevant. Scenarios 7–12 are for situations where greater preference is given to one goal, a lower preference to the second goal while the third goal is assigned zero preference. Scenario 13 represents a situation where all three goals are given equal preferences. Scenarios 14–19 are for situations where two goals have equal preference compared to the third goal with all being non-zero.

On exercising the cDSP for these different scenarios, we obtain the design and operating setpoints for the process and the achieved values of each of the goals. Ternary plots are constructed. A ternary plot is a diagram used to plot three (input or state) variables which sum to a constant, and to show a relationship between

Table 5.4 Scenarios with weights for goals

Scenarios	W_1	W_2	W_3
1	1	0	0
2	0	1	0
3	0	0	1
4	0.5	0.5	0
5	0.5	0	0.5
6	0	0.5	0.5
7	0.25	0.75	0
8	0.25	0	0.75
9	0.75	0	0.25
10	0.75	0.25	0
11	0	0.75	0.25
12	0	0.25	0.75
13	0.33	0.34	0.33
14	0.2	0.2	0.6
15	0.4	0.2	0.4
16	0.2	0.4	0.4
17	0.6	0.2	0.2
18	0.4	0.4	0.2
19	0.2	0.6	0.2

those variables (Sabeghi et al. 2015). In our context, the axes of the ternary plots represent the assigned weights (W_1, W_2, W_3) for each of the goals and the interior color contours represent the achieved value of the particular goal for which the ternary plot is created. The achieved value is normalized to lie between 0 and 1, with 0 representing the minimum and 1 representing the maximum achieved value. These values are indicated next to the color bar for the plots. These ternary plots are used to visualize and explore the solution space and identify a feasible solution space satisfying all requirements in the best possible manner. If the designer is unsure about the region of interest in terms of weights assigned, then the ternary plots are effective tools for identifying those regions that satisfy the requirements and thus choosing a good combination of goal weights. For further information about constructing ternary plots, see Sabeghi et al. (2015). Next, we use these ternary plots to determine the weights for the goals and predict the required design setpoints.

For goal 1, a process designer is interested in identifying regions to minimize ovality to a value of nearly 0.001 m. This is an important goal and must be achieved as closely as possible since rods with ovality lead to a huge loss to the manufacturers. Here we assume that an ovality of a maximum to 0.002 m is acceptable. On analyzing Fig. 5.7, in the region identified by the orange dashed line is an ovality value very close to the specified target value is achievable. Also, higher weights are assigned

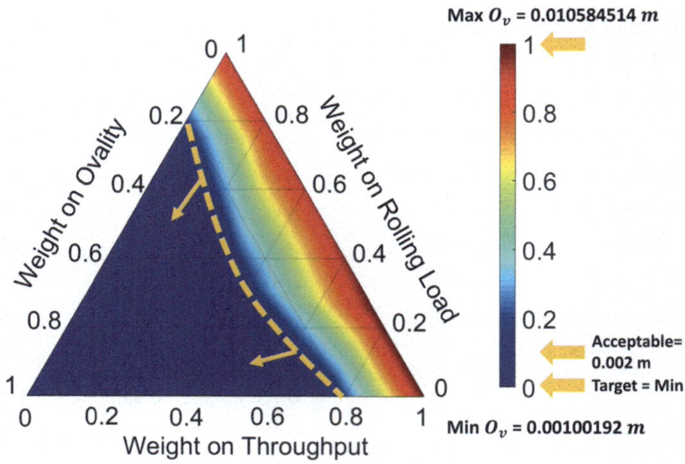

Fig. 5.7 Ternary plot for Goal 1—ovality

to the ovality goal, that is as the weight tends to 1, we approach the target value as closely as possible.

For goal 2, the process designer is interested in maximizing throughput and the target value identified is 0.0006 m^3/s. In Fig. 5.8, we see that the values in the region demarcated by the blue dashed line achieves the target.

For goal 3, the interest of the process designer is to achieve the minimum rolling load within the defined limits. The target value for this goal is 28 t. On analyzing Fig. 5.9, we see that the dark blue contour within the red dashed lines predicts the value of the goal close to the target.

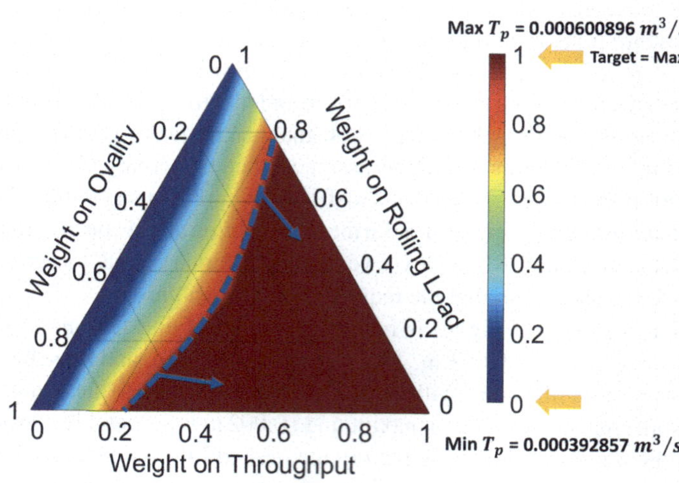

Fig. 5.8 Ternary plot for Goal 2—throughput

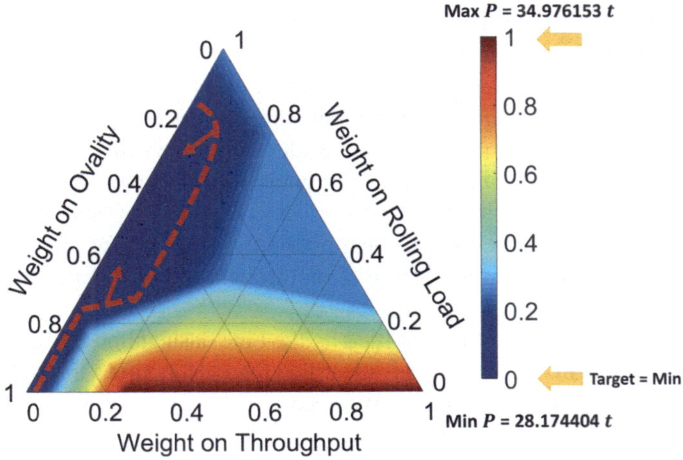

Fig. 5.9 Ternary plot for Goal 3—rolling load

Now since the designer is interested in identifying regions that satisfy all the three goals mentioned above, there is a need to visualize these design spaces together in a single ternary plot. Therefore, we superimpose plots. The superimposed plot of the regions of interest in a ternary space is shown in Fig. 5.10. In a superimposed plot, all the identified regions of interest for the three goals are merged in order to identify a single region that is common for all the goals, if it exists. If not, the designer needs to make tradeoffs among the goals. The region marked in light green satisfies the requirements for ovality and throughput, while the blue region satisfies the requirements of rolling load and ovality. There is no common region that satisfies

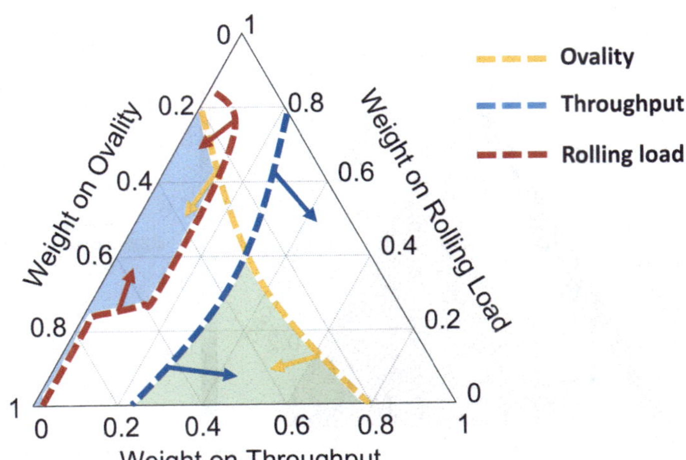

Fig. 5.10 Superimposed ternary space for all goals

all three goals simultaneously. The designer can either choose solutions from the regions identified or reformulate the constraints/goals to identify feasible spaces.

In this section, we illustrate the utility of ternary plots to reformulate a problem according to new requirements and carry out solution space exploration to support decision-making. For the problem under consideration, the ovality goal is an important goal and cannot be relaxed at all. The goals on throughput and rolling load, however, can be relaxed. This is because of the fact that we view the quality of the end product as a greater concern than productivity given that the performance criteria are met. Hence, we relax the goal on throughput even if its level drops to $0.0005\,\mathrm{m}^3/\mathrm{s}$.

This new region of interest is identified by the blue dashed line in Fig. 5.11. Any combination of weights on goals in this identified region supports a throughput value greater than or equal to $0.0005\,\mathrm{m}^3/\mathrm{s}$. We need to achieve a minimum rolling load within the lower and upper bounds defined. Since the goal of achieving a minimum of 28 t is not possible unless compromises are made on other goals, we are relaxing the rolling load value to 32 t which is within the identified bounds. The acceptable new region in the ternary plot is identified by the dashed red line in Fig. 5.12. Any combination of weights of goals in this identified region supports a rolling load value that is less than or equal to 32 t. We superimpose the new regions along with the region identified for minimizing ovality to see if there is a common region that satisfies all three goals for the new design preferences, Fig. 5.13.

In the superimposed plot for the newly identified goals, the yellow region with multiple solutions within it denoted by the letters A to G satisfies all the newly identified goals. After exploring and analyzing each solution, the designer can choose combinations from this region that meets requirements. Scenario 13 in Table 5.3 for which we have equal priority to the three goals ($W_1 = 0.33$, $W_2 = 0.34$, and $W_3 = 0.33$; point G in Fig. 5.13) satisfies the three goals as closely as possible compared

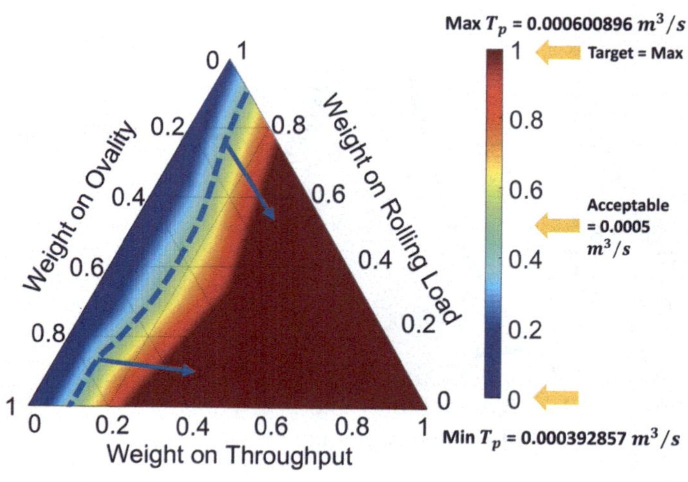

Fig. 5.11 Ternary plot for Goal 2—throughput with relaxed requirements

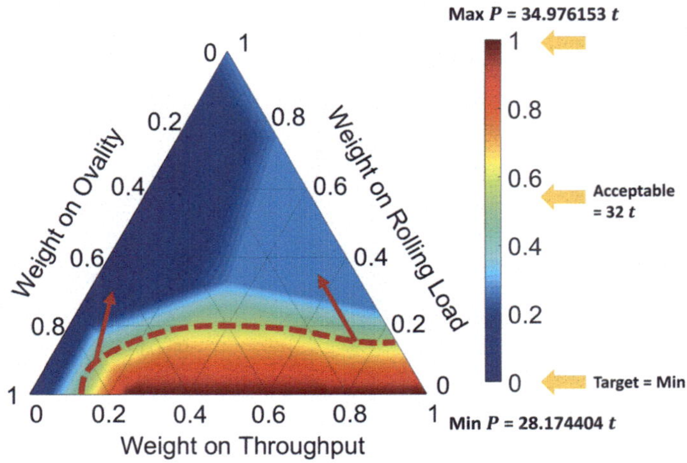

Fig. 5.12 Ternary plot for Goal 3—rolling load with relaxed requirements

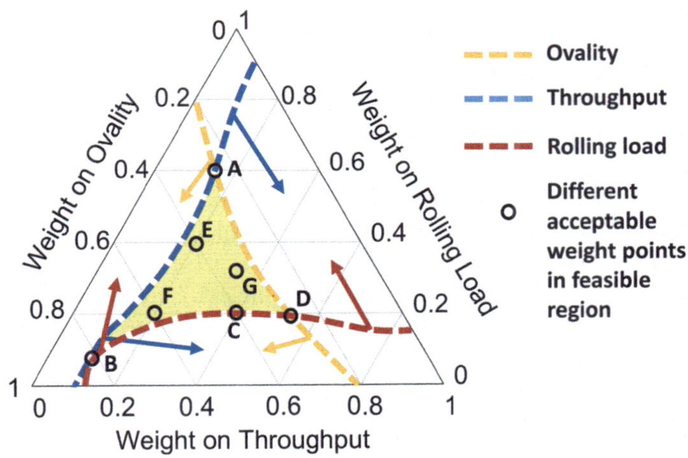

Fig. 5.13 Superimposed ternary space for all goals after changes in design preferences

to the other solutions within the region; therefore, this scenario and the weights associated with it is the best combination. Thus, a designer is able to identify those weight combinations that when used in the cDSP formulation helps in predicting the design setpoints that satisfice the conflicting goals identified. The ternary plots thus are effective tools empowering the designer to make changes in design preferences according to the demands of the problem. The designer can then analyze and explore the new scenarios in order to make effective design decisions by identifying multiple possible solutions.

Table 5.5 cDSP results for Pass 4

System variables for Pass 4 cDSP	Values obtained from running cDSP for Pass 4 (S13)
h_4	0.0260326 m
λ_4	1.3
β_4	1.15
w_4	1.12723 m/s
$R_{max,4}$	0.155012 m
N_4	17.4642 rpm
T_4	1270 K
G_4	0.004 m
μ_4	0.3

Next, we identify the system variable values for Scenario 13 obtained by solving the cDSP. These system variable values are presented in Table 5.5. We use these system variable values to design Pass 4 followed by Pass 3 by using the process design scheme described in Sect. 5.4 and illustrated in Fig. 5.2. This is followed by formulating and solving the cDSP for Pass 2. The system variable values obtained by solving the cDSP and carrying out solution space exploration for Pass 2 are presented in Table 5.6. The design of Passes 1, 2 and 3 is carried out using the results from the Pass 2 cDSP. The calculation involved in the designing of passes is provided in Appendix 2.

The design of all the passes by following the process design scheme is shown in Appendix 2. The results of the roll pass design calculations are summarized in Table 5.7 and the pass dimensions are shown in Figs. 5.14, 5.15, 5.16 and 5.17.

We discuss the design results summarized in Table 5.7 briefly here. We achieve a round rod of diameter 26 mm at the end of Pass 4 with ovality of 0.001004 m, throughput of almost 0.0006 m^3/s, and a rolling load value of almost 30 t. This is achieved with a coefficient of elongation of 1.3 and a spread of 1.15 occurring while the material is rolled in Pass 4. The entry speed of the material for Pass 4 is 0.866 m/s

Table 5.6 cDSP results for Pass 2

System variables for Pass 2 cDSP	Values obtained from running cDSP for Pass 2
h_2	0.031 m
λ_2	1.3
β_2	1.2
w_2	0.79431 m/s
$R_{max,2}$	0.155 m
G_2	0.004 m
μ_2	0.3

Table 5.7 Summary of key design results for all passes

Pass no.	Roll stand No.	Dimensions mm	Cross-section F mm^2	Coefficient		Entry speed vm/sec	Exit speed w m/s	Effective diameter D_w mm	Goals achieved		
				λ	β				Ovality O_v m	Throughput $T_p\, m^3/s$	Rolling load P t
0		Square 42 × 42	1764								
1	I	Oval 22 × 65.3	981.59	1.797	1.4	0.3401	0.611	333.3	NA	0.0006	NA
2	II	Round Ø31	755.07	1.3	1.2	0.611	0.79431	285.1	NA	0.0005997	40.82
3	III	Oval 18.3 × 55	691.93	1.0912	1.5	0.79431	0.86678	296.35	NA	0.0005999	NA
4	IV	Round Ø26	532.26	1.3	1.15	0.86678	1.1272	288.7	0.001004	0.0005999	30.002007

NA: Not applicable for the formulated problem under study

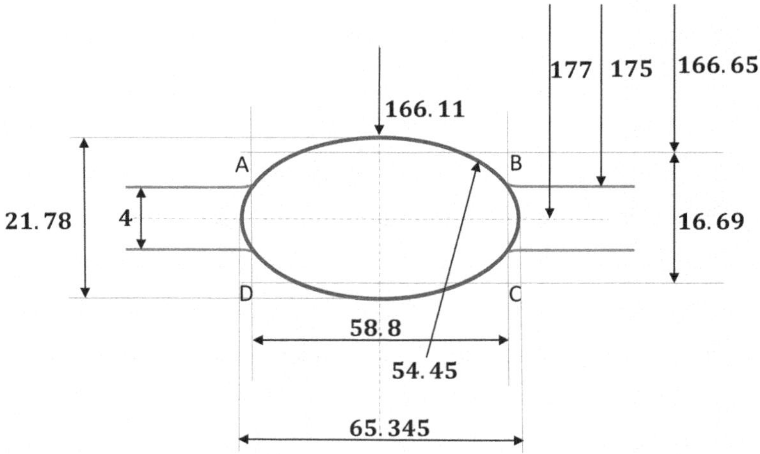

Fig. 5.14 Pass 1 dimensions designed

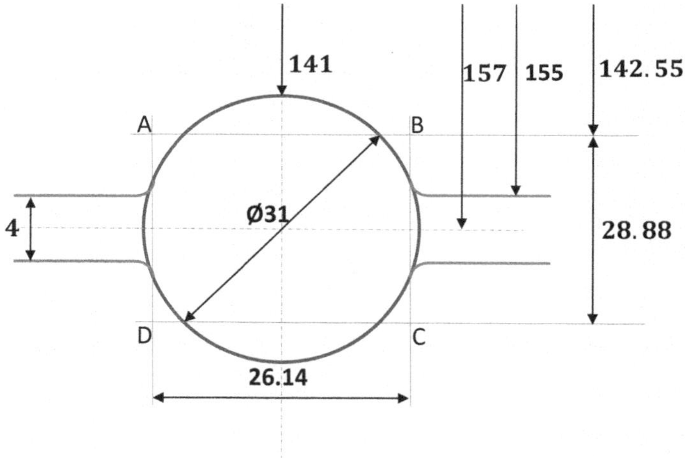

Fig. 5.15 Pass 2 dimensions designed

and exit speed is 1.127 m/s. The effective roll diameter is obtained as 288.7 mm for this pass.

The design of Pass 3 results in an oval stock of dimensions 18.3 × 55 mm. To design Pass 3 the spread value is assumed to be 1.5 and the coefficient of elongation is 1.0912. The entry speed of stock is 0.7943 m/s and the exit speed is the same as the entry speed of Pass 4. The maximum roll radius is assumed to be the same as Pass 4 and an effective roll diameter of 296.3 mm for Pass 3 is based on this assumption. The design is able to achieve/maintain a throughput of almost 0.0006 m^3/s for Pass 3.

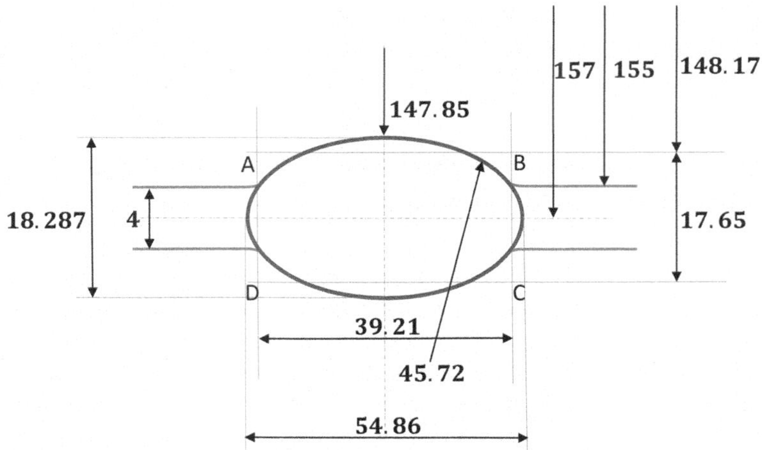

Fig. 5.16 Pass 3 dimensions designed

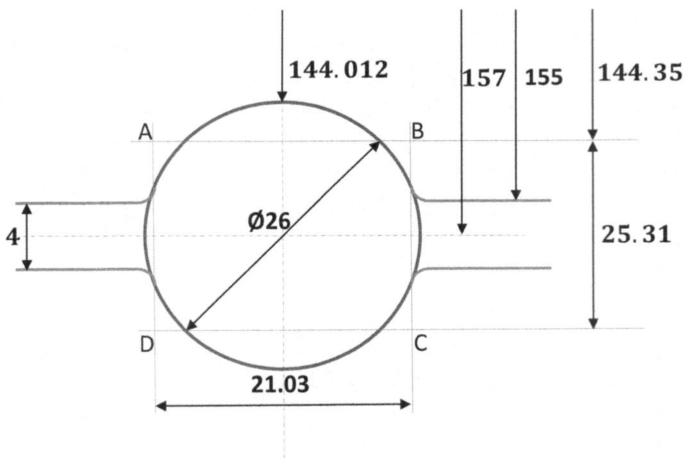

Fig. 5.17 Pass 4 dimensions designed

The design of Pass 2 results in a round stock with a diameter of 31 mm. The coefficient of elongation and spread for this Pass are 1.3 and 1.2, respectively. The entry is 0.611 m/s and the exit speed is the same as the entry speed of Pass 3. The effective diameter obtained for this pass is 285 mm. The target rolling load value of 40 *t* for Pass 2 is achieved and the throughput is maintained at 0.0006 m^3/s.

The design of Pass 1 results in an oval stock of dimensions 22 × 65.3 mm. The coefficient of elongation for this pass is 1.797. The spread value is assumed to be 1.4. The entry speed for this pass is 0.3401 m/s. The exit speed is the same as the entry speed of Pass 2. The maximum roll radius is assumed to be the same as Pass 2 and Pass 4 and the effective roll diameter is 333 mm based on this assumption. The throughput value of 0.0006 m^3/s is achieved with this configuration.

5.7 On Verification and Validation—Empirical Structural and Performance Validation

Empirical structural validation involves accepting the appropriateness of the example problems used to verify the performance of the framework and the method. The CEF and the inverse design method is first tested using the example problem presented in this chapter. In the example problem, the horizontal integration of rod rolling is considered. Using the framework and method, the integrated design of rolling passes and the final rod product is carried out in an inverse manner starting from the end goals. The example thus is appropriate to demonstrate the utility of the framework and method as it involves complex information flow across manufacturing stages that needs systematic problem formulation and exploration across stages to design the entire system. In this example, only macrostructural effects associated with hot rod rolling is considered. We direct interested readers to the following articles to gain more information on the example discussed, see (Nellippallil et al. 2016, 2017b, c, d). The example is further improved and expanded to the comprehensive problem on vertical and horizontal integration of the hot rolling process chain in Chap. 6, also see (Nellippallil et al. 2017b, c, d, 2018a, b, 2019, 2020). Using the comprehensive example problem in Chap. 6, the utility of the framework and the method is tested for the integrated design exploration of materials, products, and manufacturing processes.

5.8 Role of this Chapter in This Monograph

In this chapter, a method is proposed based on well-established empirical models and response surface models developed through simulation experiments along with the compromise decision support problem (cDSP) construct to support integrated information flow through different stages of a multistage hot rod rolling system (horizontal integration). The method presented is a specific form of the generic goal-oriented inverse design method proposed in Chap. 4. We illustrate the efficacy of the proposed goal-oriented, sequential inverse design method using hot rod rolling as an example. Here the design decisions are first made at the last rolling pass based on the end requirements of the process. We allow these design decisions to be passed to the preceding rolling passes by following the sequential relationships existing between the passes in an inverse manner. We carry out the design of individual passes by allowing design information to be passed back and forth between passes using the sequential relationships. The formulation of individual cDSPs for passes helps to organize the sequential information flow and provides the ability to the designer to consider specific goals associated with each rolling pass and integrating them with the end goals. The ternary analysis feature incorporated in the method provides the designer with the capability of exploring the solution space and identifying feasible regions that satisfice the different goals identified for a particular stage of the manufacturing process chain. The proposed method has the potential to be used for identifying design setpoints for a chain of unit operations that are connected in sequence. Once the information flow between operations and the empirical and the simulation/response surface models necessary to establish relationships are available, a designer will be able to use this method to achieve the integrated decision-based design of the product and the processes. In Chap. 6, we present a comprehensive example problem in which the results for the macrostructure space obtained will be studied further to design (identify design and operating setpoints) preceding manufacturing stages like rolling and cooling by following the proposed design method in order to realize the end product.

Appendix 1: cDSP Formulation for Pass 2 in This Chapter

In this section we describe the mathematical formulation of the compromise decision support problem (cDSP) for Pass 2 of rod rolling. The cDSP for Pass 2 incorporates the design information passed from Pass 3 to Pass 4. The cDSP reads as follows:

Given:

(1) Design information passed from Pass 3 to Pass 4
(2) Requirements at Pass 2

- Achieve target throughput (results obtained from Pass 4 design)
- Achieve target rolling load
- Target value for throughput, $T_{p4,\text{Target}} = 0.0006\,\text{m}^3/\text{s}$
- Target value for rolling load, $P_{\text{Target}} = 40\,\text{t}$
- Minimum value of rolling load, $P_{\min} = 35\,\text{t}$
- Maximum value of rolling load, $P_{\max} = 45\,\text{t}$
- Minimum value of throughput, $T_{\text{pmin}} = 0.0001\,\text{m}^3/\text{s}$
- Maximum value of throughput, $T_{\text{pmax}} = 0.0008\,\text{m}^3/\text{s}$

(3) Initial billet size = 42 × 42 mm
(4) Other parameter values for passes
(5) The regression equations and well-established empirical and theoretical correlations for the oval to round pass for Pass 2
(6) Variability in system variables

The ranges identified for the system variables are provided in Table 5.8.

Table 5.8 System variables and ranges for Pass 2 cDSP

S. no.	Variables	Ranges
1	X_1, diameter of rod after pass 2 (h_2)	0.03–0.04 m
2	X_2, the coefficient of elongation for Pass 2 (λ_2)	1–3
3	X_3, the spread occurring in Pass 2 (β_2)	1–2
4	X_4, the exit velocity for Pass 2 (w_2)	0.5–3 m/s
5	X_5, the maximum radius of roll in Pass 2 ($R_{max,2}$)	0.155–0.2 m
6	X_6, the roll gap (G_2)	0.0035–0.0055 m
7	X_7, the coefficient of friction (μ_2)	0.3–0.45

Find:
System Variables

X_1 diameter of rod after Pass 2 (h_2)
X_2 the coefficient of elongation for Pass 2 (λ_2)
X_3 the spread occurring in Pass 2 (β_2)
X_4 the exit velocity for Pass 2 (w_2)
X_5 the maximum radius of roll in Pass 2 ($R_{\max,2}$)
X_6 the roll gap (G_2)
X_7 the coefficient of friction (μ_2)

Deviation Variables

$$d_i^-, d_i^+, i = 1, 2$$

Satisfy:
System Constraints

- Minimum coefficient of elongation constraint: $\lambda_2(X_2) - 1.2 \geq 0$
- Maximum coefficient of elongation constraint: $2 - \lambda_2(X_2) \geq 0$
- Minimum spread constraint: $\beta_2(X_3) - 1.1 \geq 0$
- Maximum spread constraint: $1.7 - \beta_2(X_3) \geq 0$
- Exit speed constraint: $w_2 - v_r(X_i) \geq 0$
- Minimum load constraint: $P(X_i) - P_{\min} \geq 0$
- Maximum load constraint: $P_{\max} - P(X_i) \geq 0$
- Maximum wear constraint: $0.0001 - \Delta R(X_i) \geq 0$

System Goals
Goal 1:

- Maximize Throughput:

$$\frac{T_p(X_i)}{T_{p,\text{Target}}} + d_1^- - d_1^+ = 1$$

Goal 2:

- Minimize Rolling Load:

$$\frac{P_{\text{Target}}}{P(X_i)} - d_2^- + d_2^+ = 1$$

Variable Bounds
Defined in Table 5.8

Bounds on deviation variables

$$d_i^-, d_i^+ \geq 0 \text{ and } d_i^- * d_i^+ = 0,\ \text{i} = 1, 2$$

Minimize:
Minimize the deviation function

$$Z = \sum_{i=1}^{2} W_i \left(d_i^- + d_i^+\right); \sum_{i=1}^{2} W_i = 1$$

Appendix 2: Design Calculations (Refer to Fig. 5.2)

In this section, we describe the design calculations carried out for each pass based on the cDSP results obtained that are showcased in Tables 5.4 and 5.5. The design process is carried out following the sequential relationships that exist between passes ensuring the flow of information pattern shown in Fig. 5.3.

Step 1: Formulation of cDSP for roll Pass 4 using the information from the end product to be realized and the sequential relationship existing between roll Pass 3 and 4
The cDSP for Pass 4 is formulated in terms of the end requirements of minimizing ovality, maximizing throughput, and minimizing rolling load within the system constraints and bounds defined. The cDSP is exercised for different scenarios and ternary plots are used to identify the best region and the results are summarized in Table 5.4.

Step 2: Design of Passes 4 and 3 using the design and operating setpoints identified and the information available from end product requirements
We calculate the area of the round rod using the height value obtained for rod from cDSP results. The cross-sectional area of material after Pass 4:

$$F_4 = \frac{\pi h_4^2}{4} = 532.26\,\text{mm}^2$$

Entry speed of material for roll Pass 4:

$$v_4 = \frac{w_4}{\lambda_4} = 0.8671\,\text{m/s}$$

Throughput achieved in Pass 4:

$$T_{p4} = F_4 \times w_4 = 0.0005999\,\text{m}^3/\text{s}$$

We carry out the design calculations for Pass 3 based on the cross-sectional area of rod and elongation coefficient (cDSP result) obtained after Pass 4. We also define some requirements for Pass 3 such as meeting the throughput same as that of Pass 4.
Cross-sectional area of material after Pass 3:

$$F_3 = F_4 \times \lambda_4 = 691.93\,\text{mm}^2$$

Theoretical width of oval pass after Pass 3:

$$b_{3w} = \sqrt{4.35 \times F_3} = 54.86\,\text{mm}$$

Height of material after Pass 3 (assuming b/h ratio = 3):

$$h_3 = \frac{b_{3w}}{(b/h)} = 18.28\,\text{mm}$$

Radius of curvature of oval Pass 3:

$$R_3^* = \frac{b_{3w}^2 + h_3^2}{4h_3} = 45.72\,\text{mm}$$

Exit speed of material for roll Pass 3:

$$w_3 = v_4 = 0.8671\,\text{m/s}$$

Throughput to be maintained in Pass 3 (Given):

$$T_{p3} = T_{p4} = 0.0005999\,\text{m}^3/\text{s}$$

We carry out design calculations for Pass 4 now with the new information generated for Pass 3.
Width of round profile (approximated rectangle) after Pass 4:

$$b_4 = \beta_4 \times h_3 = 21.03\,\text{mm}$$

Mean height after Pass 4:

$$h_{4m} = \frac{F_4}{b_4} = 25.31\,\text{mm}$$

Theoretical diameter of roll for Pass 4:

$$D_{t4} = 2\left(R_{\max,4} + \frac{G_4}{2}\right) = 314\,\text{mm}$$

Effective diameter of roll for Pass 4:

$$D_{w4} = D_{t4} - h_{4m} = 288.7\,\text{mm}$$

Step 3: Formulation of cDSP for roll Pass 2 using the design information generated for Passes 3 and 4; and the sequential information existing between Passes 1 and 2; along with information on input material (billet)
The designer formulates the cDSP for Pass 2 after finding the results from Passes 3 to 4. For example, the range of the height of rod for Pass 2 is identified based on the dimensions achieved in Passes 3 and 4. Another example is the rolling load target value. Since there is a chance of having a higher rolling load during Pass 2 due to larger stock that is being rolled than Pass 4, the target, minimum and maximum values for Pass 2 are fixed after looking at the rolling load value obtained in Pass 4. The designer also fixes the target throughput value for Pass 2 after analyzing the throughput achieved in Passes 3 and 4. Thus the designer makes judgments based on the information obtained from the information as it develops.

Step 4: Design of roll Passes 2 and 1 using the design and operating setpoints identified; the information available from input material and the information from Passes 3 to 4
The cDSP results for Pass 2 presented in Table 5.5 are used to design Pass 2.

Cross-sectional area of material after Pass 2:

$$F_2 = \frac{\pi h_2^2}{4} = 755.07\,\text{mm}^2$$

Entry speed of material for roll Pass 2:

$$v_2 = \frac{w_2}{\lambda_2} = 0.611\,\text{m/s}$$

Throughput achieved in Pass 2:

$$T_{p2} = F_2 \times w_2 = 0.0005997\,\text{m}^3/\text{s}$$

Next, the design calculations for Pass 1 is carried out using Pass 2 design results and initial billet information from caster.
Cross-sectional area of material after Pass 1:

$$F_1 = F_2 \times \lambda_2 = 981.59\,\mathrm{mm}^2$$

Theoretical width of oval pass after Pass 1:

$$b_{1w} = \sqrt{4.35 \times F_1} = 65.345\,\mathrm{mm}$$

Height of material after Pass 1 (assuming $b/h = 3$):

$$h_1 = \frac{b_{1w}}{(b/h)} = 21.78\,\mathrm{mm}$$

Radius of curvature of oval Pass 1:

$$R_1^* = \frac{b_{1w}^2 + h_1^2}{4h_1} = 54.45\,\mathrm{mm}$$

Exit speed of material for roll Pass 1:

$$w_1 = v_2 = 0.611\,\mathrm{m/s}$$

Given initial billet size from caster:

$$h_0 \times b_0 = 42 \times 42\ (\mathrm{mm})$$

Cross-sectional area of initial billet:

$$F_0 = 42 \times 42 = 1764\,\mathrm{mm}^2$$

Coefficient of elongation for Pass 1:

$$\lambda_1 = \frac{F_0}{F_1} = 1.797$$

Width of oval profile (approximated rectangle) after Pass 1 (assuming $\beta_1 = 1.4$):

$$b_1 = \beta_1 \times b_0 = 58.8\,\mathrm{mm}$$

Mean height of material after Pass 1:

$$h_{1m} = \frac{F_1}{b_1} = 16.69\,\mathrm{mm}$$

Effective diameter of roll for Pass 1 (assuming a theoretical diameter for rolls in Pass 1, $D_{t1} = 350\,\text{mm}$):

$$D_{w1} = D_{t1} - h_{1m} = 333.3\,\text{mm}$$

Entry speed of material for roll Pass 1:

$$v_1 = \frac{w_1}{\lambda_1} = 0.3401\,\text{m/s}$$

Throughput to be maintained in Pass 1:

$$T_{p1} = T_{p2} = 0.0005997\,\text{m}^3/\text{s}$$

The design calculations for Pass 2 are carried out next using Pass 1 information generated followed by collecting all the results for Passes 1 and 2.

Width of round profile (approximated rectangle) after Pass 2:

$$b_2 = \beta_2 \times h_1 = 26.14\,\text{mm}$$

Mean height after Pass 2:

$$h_{2m} = \frac{F_2}{b_2} = 28.88\,\text{mm}$$

Theoretical diameter of roll for Pass 2:

$$D_{t2} = 2\left(R_{\max,2} + \frac{G_2}{2}\right) = 314\,\text{mm}$$

Effective diameter of roll for Pass 4:

$$D_{w2} = D_{t2} - h_{2m} = 285.1\,\text{mm}$$

With the information generated for Passes 1 and 2, the design calculations for Passes 3 and 4 are carried out completing design results for Passes 1, 2, 3, and 4.

Coefficient of elongation for Pass 3:

$$\lambda_3 = \frac{F_2}{F_3} = 1.091$$

Width of oval profile (approximated rectangle) after Pass 3 (assuming $\beta_3 = 1.5$):

$$b_3 = \beta_3 \times b_2 = 39.2\,\text{mm}$$

Mean height of material after Pass 3:

$$h_{3m} = \frac{F_3}{b_3} = 17.65\,\text{mm}$$

Effective diameter of roll for Pass 3 (assuming a theoretical diameter for rolls in Pass 1, $D_{t3} = 314\,\text{mm}$):

$$D_{w3} = D_{t3} - h_{3m} = 296.35\,\text{mm}$$

Entry speed of material for roll Pass 3:

$$v_3 = w_2 = 0.7943\,\text{m/s}$$

Exit speed of material for roll Pass 3:

$$w_3 = v_3 \times \lambda_3 = 0.867\,\text{m/s}$$

This completes the design of the rolling passes with the determination of all the key dimensions presented in Figs. 5.4a, b.

References

Allen, J. K., Seepersad, C. C., Mistree, F., & Savannah, G. (2006). A survey of robust design with applications to multidisciplinary and multiscale systems. *Journal of Mechanical Design, 128*(4), 832–843.

Bras, B., & Mistree, F. (1993). Robust design using compromise decision support problems. *Engineering Optimization, 21*(3), 213–239.

Ebro, M., & Howard, T. J. (2016). Robust design principles for reducing variation in functional performance. *Journal of Engineering Design, 27*(1–3), 75–92.

Galantucci, L., & Tricarico, L. (1999). Thermo-mechanical simulation of a rolling process with an FEM approach. *Journal of Materials Processing Technology*, *92*, 494–501.

Goh, C.-H., Ahmed, S., Dachowicz, A. P., Allen, J. K., & Mistree, F. (2014). Integrated multiscale robust design considering microstructure evolution and material properties in the hot rolling process. *ASME Design Automation Conference*, Paper Number: DETC2014-34157.

Jupp, S. P. (2001). Mathematical modelling of the microstructural evolution during the hot rolling of AA5083 aluminum Alloys. Ph.D. Dissertation, Department of Metals and Materials Engineering, University of British Columbia.

Lapovok, R., & Thompson, P. (1994). The mathematical basis of optimal roll pass design, engineering mathematics: the role of mathematics in modern engineering. In *Biennial Conference*.

Lapovok, R., & Thomson, P. (1997). An approach to the optimal design of rolling passes. *International Journal of Machine Tools and Manufacture, 37*(8), 1143–1154.

Lee, Y., Choi, S., & Hodgson, P. (2002). Analytical model of pass-by-pass strain in rod (or bar) rolling and its applications to prediction of austenite grain size. *Materials Science and Engineering A, 336*(1), 177–189.

Lee, Y., Choi, S., & Kim, Y. (2000). Mathematical model and experimental validation of surface profile of a workpiece in round–oval–round pass sequence. *Journal of Materials Processing Technology, 108*(1), 87–96.

McDowell, D. L., Panchal, J., Choi, H.-J., Seepersad, C., Allen, J., & Mistree, F. (2009). *Integrated design of multiscale, multifunctional materials and products*, Butterworth-Heinemann.

Michalewicz, Z. (1995). A survey of constraint handling techniques in evolutionary computation methods. *Evolutionary Programming, 4*, 135–155.

Mistree, F., Hughes, O. F., & Bras, B. (1993). Compromise decision support problem and the adaptive linear programming algorithm. *Progress in Astronautics and Aeronautics, 150*, 251.

Montgomery, D. C. (2008). *Design and analysis of experiments*, Wiley.

Montgomery, D. C., & Myers, R. H. (1995). *Response surface methodology: process and product optimization using designed experiments*. Raymond H: Meyers and Douglas C. Montgomery. A Wiley-Interscience Publications.

Nellippallil, A. B., Song, K. N., Goh, C.-H., Zagade, P., Gautham, B., Allen, J. K., & Mistree, F. (2016). A goal oriented, sequential process design of a multi-stage hot rod rolling system. *ASME Design Automation Conference*, Paper Number: DETC2016-59402.

Nellippallil, A. B., De, P. S., Gupta, A., Goyal, S., & Singh, A. K. (2017a). Hot rolling of a non-heat treatable aluminum alloy: thermo-mechanical and microstructure evolution model. *Transactions of the Indian Institute of Metals, 70*(5), 1387–1398.

Nellippallil, A. B., Rangaraj, V., Allen, J. K., Mistree, F., Gautham, B., & Singh, A. K. (2017b). A decision-based design method to explore the solution space for microstructure after cooling stage to realize the end mechanical properties of hot rolled product. In *Proceedings of the 4th World Congress on Integrated Computational Materials Engineering (ICME 2017)*, Springer.

Nellippallil, A. B., Rangaraj, V., Gautham, B., Singh, A. K., Allen, J. K., & Mistree, F. (2017c). A goal-oriented, inverse decision-based design method to achieve the vertical and horizontal integration of models in a hot rod rolling process chain. *ASME Design Automation Conference*, Paper Number: DETC2017-67570.

Nellippallil, A. B., Song, K. N., Goh, C.-H., Zagade, P., Gautham, B., Allen, J. K., & Mistree, F. (2017d). A goal-oriented, sequential, inverse design method for the horizontal integration of a multistage hot rod rolling system. *Journal of Mechanical Design, 139*(3), 031403.

Nellippallil, A. B., Mohan, P., Allen, J. K., & Mistree, F. (2018a). Robust concept exploration of materials, products and associated manufacturing processes. *ASME Design Automation Conference*, Paper Number: DETC2018-85913.

Nellippallil, A. B., Rangaraj, V., Gautham, B., Singh, A. K., Allen, J. K., & Mistree, F. (2018b). An inverse, decision-based design method for integrated design exploration of materials, products, and manufacturing processes. *Journal of Mechanical Design, 140*(11), 111403-111403-17.

Nellippallil, A. B., Mohan, P., Allen, J. K., & Mistree, F. (2019). Inverse thermo-mechanical processing (ITMP) design of a steel rod during hot rolling process. *ASME Design Automation Conference 2019*, Paper Number: IDETC2019-97390.

Nellippallil, A. B., Mohan, P., Allen, J. K., & Mistree, F. (2020). An inverse, decision-based design method for robust concept exploration. *Journal of Mechanical Design*, https://doi.org/10.1115/1.4045877.

Oduguwa, V., & Roy, R. (2001). Qualitative and quantitative knowledge in engineering system design. In *17th National Conference on Manufacturing Research*, Cardiff University, UK.

Oduguwa, V., & Roy, R. (2002a). Bi-level optimisation using genetic algorithm. In *2002 IEEE International Conference on Artificial Intelligence Systems (ICAIS 2002)*, IEEE.

Oduguwa, V., & Roy, R. (2002b). Multi-objective optimisation of rolling rod product design using meta-modelling approach. *GECCO*.

Oduguwa, V., & Roy, R. (2006). A review of rolling system design optimisation. *International Journal of Machine Tools and Manufacture, 46*(7), 912–928.

Oduguwa, V., Tiwari, A., & Roy, R. (2004). *Sequential process optimisation using genetic algorithms*. Parallel Problem Solving from Nature-PPSN VIII: Springer.

Phaniraj, M. P., Behera, B. B., & Lahiri, A. K. (2005). Thermo-mechanical modeling of two phase rolling and microstructure evolution in the hot strip mill: Part I. Prediction of rolling loads and finish rolling temperature. *Journal of Materials Processing Technology, 170*(1), 323–335.

Pietrzyk, M., Cser, L., & Lenard, J. (1999). *Mathematical and physical simulation of the properties of hot rolled products*, Elsevier.

Reddy, R., Smith, W., Mistree, F., Bras, B., Chen, W., Malhotra, A., Badhrinath, K., Lautenschlager, U., Pakala, R., & Vadde, S. (1992). DSIDES user manual. In *Systems design laboratory, department of mechanical engineering*, University of Houston, Houston, Texas.

Roberts, W. L. (1983). *Hot rolling of steel*, CRC Press.

Roy, R., Tiwari, A., Munaux, O., & Jared, G. (2000). Real-life engineering design optimisation: Features and techniques. In *CDROM Proceedings of the Fifth Online World Conference on Soft Computing in Industrial Applications (WSC-5).* Finland.

Sabeghi, M., Smith, W., Allen, J. K., & Mistree, F. (2015). Solution space exploration in model-based realization of engineered systems. *ASME Design Automation Conference*, Paper Number: DETC2015-46457.

Said, A., Lenard, J., Ragab, A., & Elkhier, M. A. (1999). The temperature, roll force and roll torque during hot bar rolling. *Journal of Materials Processing Technology, 88*(1), 147–153.

Shin, W. (1995). *Development of techniques for pass design and optimization in the rolling of shapes*. Ph.D. Dissertation, The Ohio State University.

Shinokura, T., & Takai, K. (1982). A new method for calculating spread in rod rolling. *Journal of Applied Metalworking, 2*(2), 94–99.

Shukla, R., Goyal, S., Singh, A. K., Panchal, J. H., Allen, J. K., & Mistree, F. (2014). An approach to robust process design for continuous casting of slab. *ASME Design Automation Conference*, Paper Number: DETC2014-34208.

Shukla, R., Goyal, S., Singh, A. K., Panchal, J. H., Allen, J. K., & Mistree, F. (2015). Design exploration for determining the set points of continuous casting operation: An industrial application. *Journal of Manufacturing Science and Engineering, 137*(3), 034503.

Simpson, T. W., Poplinski, J., Koch, P. N., & Allen, J. K. (2001). Metamodels for computer-based engineering design: survey and recommendations. *Engineering with Computers, 17*(2), 129–150.

Sims, R. (1954). The calculation of roll force and torque in hot rolling mills. *Proceedings of the Institution of Mechanical Engineers, 168*(1), 191–200.

Taguchi, G. (1986). Introduction to quality engineering, asian productivity organization, American Supplier Institute, Inc., Dearborn, MI.

Tennyson, G., Shukla, R., Mangal, S., Sachi, S., & Singh, A. K. (2015). ICME for process scale-up: Importance of vertical and horizontal integration of models. In *Proceedings of the 3rd World Congress on Integrated Computational Materials Engineering (ICME 2015)*, Wiley.

Tiwari, A., Oduguwa, V., & Roy, R. (2008). Rolling system design using evolutionary sequential process optimization. *IEEE Transactions on Evolutionary Computation, 12*(2), 196–202.

Wusatowski, Z. (2013). *Fundamentals of rolling*, Elsevier.

Chapter 6
Integrated Design Exploration of Materials, Products, and Manufacturing Processes Using Goal-Oriented, Inverse Design Method

6.1 Frame of Reference—Establishing Context from Chap. 1

Steel manufacturers focus on developing new grades of steel with improved properties and performance. The careful managing of material processing during steel manufacturing will lead to the development of steels with a range of mechanical properties resulting in the improved performance of products. A round rod is produced after passing the raw steel through several manufacturing processes such as casting, reheating, rolling, and cooling. This round rod forms the input material for gear production. The chemical composition of the steel including the segregation of alloying elements, the deformation history during rolling, the cooling after rolling, and the microstructure generated define the end properties of the rolled product. The presence of large numbers of design variables, constraints and bounds, conflicting goals, and sequential information/material flow during material processing makes the steel rod making process chain highly complex. Many plant trials are therefore required to produce a new steel grade with desired properties and performance. These trials are usually expensive and time-consuming. An alternative is to carry out simulation-based, integrated design exploration of the different manufacturing processes involved in exploiting the advances in computational modeling and identifying a ranged set of solutions that satisfy the requirements, both of the steel manufacturing process and the end rod product.

In practice, design is involved with the selection of a suitable material for a given application. The classical material selection approaches are being replaced by a materials design revolution that is underway in the recent past where the focus is to design the material microstructure or mesostructure to achieve certain performance requirements such as density, strength, ductility, toughness, and hardness. The demands on the microstructure placed by these multiple performance requirements are often in conflict.

A. B. Nellippallil et al., *Architecting Robust Co-Design of Materials, Products, and Manufacturing Processes*, https://doi.org/10.1007/978-3-030-45324-4_6

Our interest lies in formulating and solving the inverse problem: given the required end properties/performance, what should be the input parameters in terms of material microstructure and processing paths for the model-based realization of the material, product, and the manufacturing processes?

From a systems design perspective, we view design as a *top-down, simulation-supported, integrated, decision-based process to satisfy a ranged set of product-level performance requirements* (McDowell et al. 2009; Allen et al. 2015). Keeping with this and the discussions by Olson on Materials-by-Design (Olson 1997), we view the integrated design of materials, products, and processes as *fundamentally an inverse, goal-oriented synthesis activity in which the designer (decision-maker) aims at identifying material structures and processing paths that achieve/satisfy certain required product and process-level properties and performances.* From the standpoint of design community, design process is always the inverse process of identifying design variables to realize desired properties or performances. However, the word "inverse" is used here from the perspective of the materials design community and will be explained in the sections that follow.

The philosophical underpinning of the goal-oriented approach to materials design has been provided by Olson (1997) and reiterated by many others (McDowell et al. 2009; Horstemeyer 2012, 2018; McDowell 2018). The conventional way of modeling hierarchical processes and systems is a "bottom-up", cause and effect (deductive) approach of modeling the material's processing paths, microstructures, resulting properties, and then mapping the property relations to performance functions, as shown in Fig. 6.1. Over the years the focus in materials design has turned to provide high-throughput decision support and develop inverse methods for materials design exploration as discussed by McDowell and Kalidindi (2016). There are several works in this vein. Adams et al. (2013) present a framework that utilizes highly efficient spectral representations to arrive at invertible linkages between material structure, its properties, and the processing paths used to alter the material structure. The

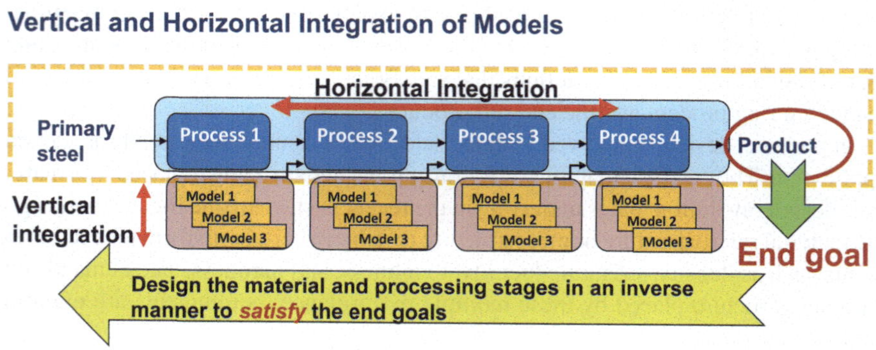

Fig. 6.1 Vertical and horizontal integration and systems-based materials and product design

materials knowledge systems approach by Kalidindi et al. (2010, 2011) showcases advances in rapid inverse design to estimate local responses. However, all these approaches including the strategy proposed by McDowell and Olson (2008) fall to specific classes of materials design problems and demand considerable knowledge and insight in mechanisms, material hierarchy, and information flow. Thus, these classes of inverse design approaches are mostly suited for detailed design and not for "design exploration" (McDowell 2018).

In our work, we seek "top-down", goals/means, inductive, or inverse methods especially at early stages of design to explore the design space of processing paths and resulting microstructures of a material *satisfying* a set of specified performance requirements (see Fig. 6.1). Approaches to pursue top-down design exploration by employing multiscale modeling and systems-based design especially at the early stages of design are addressed in limited literature. Choi et al. (2008a, b) propose the inductive design exploration method (IDEM)—a multilevel, robust design method that makes it possible to consider propagation of all three types of uncertainty (Nellippallil et al. 2018), such as that arising in hierarchical materials design problems incorporates process–structure–property relations. The two major design objectives using the IDEM for material and product design are (McDowell and Olson 2008): (i) to guide bottom-up modeling so as to conduct top-down, goal-oriented design exploration and (ii) to manage the uncertainty in chains of process–structure–property relations. Kern et al. (2017) propose pyDEM, a generalized implementation of the IDEM, as an open-source tool in the Python environment. The top-down, goal-oriented approach of materials design comes with several challenges as highlighted by McDowell et al. (2007). In this chapter, we address the challenge of incorporating the design of the material as part of a larger overall systems design process embodying the hierarchy of process–structure–property–performance set forth by Olson (1997) with consideration on supporting coordination of information and human decision-making.

To carry out design space exploration across the material processing–structure–property–performance spaces, there should be flow of information via simulation models integrated across multiple scales and across multiple manufacturing processes—defined as the vertical and horizontal integration of models. We define *vertical integration* as the integration of models and simulations for the different phenomenon that occurs at multiple length scales in a specific manufacturing process so as to generate information that can be passed to other manufacturing processes that follow. We define *horizontal integration* as the integration of different such manufacturing processes using simulation models ensuring proper flow of the information generated through vertical integration at each manufacturing process thereby establishing the processing–structure–property–performance route to realize an end product (Nellippallil et al. (2017a); Tennyson et al. 2015; Shukla et al. 2015; see Fig. 6.1).

To achieve vertical and horizontal integration of models, there must be analysis models linking different manufacturing processes and phenomena which predict the material properties associated with these processes and ensure the proper forward flow of information. Once we achieve forward modeling, we carry out top-down (goal-oriented), decision-based design exploration of the material microstructure and processing paths to achieve the required product properties. The primary mathematical construct used in the method is the compromise decision support problem (cDSP) supported by the concept exploration framework (CEF) to generate *satisficing* design solutions (Mistree et al. 1993). Our intention in solving the compromise DSP is to *satisfice* a set of goals and thus we approach the problem from the school of thought of a *satisficer*; more information are available in Mistree et al. (1994). The concept exploration framework is inspired from the robust concept exploration method (RCEM) proposed by Chen et al. (1997) to systematically generate satisficing, top-level specifications.

In Sect. 6.2, we describe the vertical and horizontal integration of models from the perspective of the steel manufacturing process chain problem focused on hot rod rolling (HRR) process that we are addressing. In Sect. 6.3, we apply the proposed goal-oriented, inverse method for the problem described. The empirical models and the response surface models for computational analysis of the problem are presented in Sect. 6.4. The mathematical formulation of the hot rod rolling process chain is provided in Sect. 6.5 and the ternary analysis for visualizing and exploring the solution space is covered in Sect. 6.6 with closing remarks in Sect. 6.7. In Table 6.1, we summarize the testing of the architecture constituents carried out in this chapter.

6.2 Integrated Design of Materials, Products, and Processes—The Steel Manufacturing Process Chain Problem (Comprehensive Example)

Many algorithms for establishing forward relationships have been developed. These models are used to predict the behavior of materials during complex manufacturing processes as the final properties of the end steel product depend on its processing route (Nellippallil et al. (2017b); Hodgson and Gibbs 1992; Majta et al. 1996; Kuziak et al. 1997; Phadke et al. 2004; Jägle 2007). It is beneficial for steel manufacturers to develop computer algorithms/tools that provide the capability to establish inverse relationships; that is, relate the end properties of the steel product as a function of process variables. These computer algorithms/tools need to be developed using mathematical models that predict the microstructure and mechanical properties of the material as a function of the manufacturing process conditions. The challenge here is in considering all the different phenomenon that happens during the processing of the material and establishing the processing–structure–property–performance relationship in an inverse manner using models. In this problem, we are interested in developing an integrated method that is generic and has the ability to relate the end mechanical properties of the material with good accuracy to the different processing and microstructure routes available for the material. The efficacy of the method is illustrated for the specific steel manufacturing process chain problem addressed in

Table 6.1 Summary of the architecture testing carried out in this chapter

Summary of the architecture testing carried out in this chapter	
Elements	*What?*
Processing elements	The functionalities of CEF and GoID—carry out systematic goal-oriented, inverse design exploration to achieve the integrated (vertical and horizontal) design exploration of materials, products, and manufacturing processes
Data elements	Vertical and horizontal integration of hot rolling process chain problem
Connecting elements	Function-based design; concept exploration framework; goal-oriented, inverse design method
Form	*How?*
Weights	Testing the utility of CEF and GoID using two comprehensive test examples
Properties	Focus on the microstructure-mediated design of hot rolling and cooling process chain for the realization of rod product
Relationship	Theoretical and empirical performance validation using the comprehensive test examples
Rationale	*Why?*
Motivation	To validate the utility of CEF and GoID functionalities in providing design decision support in the vertical and horizontal integration of process chains
Believes	• Design is a goal-oriented synthesis activity and essentially a decision-making process—the material information flow across the manufacturing process chain is therefore represented in terms of a horizontal decision workflow that involves integrated information flow from multiple scales (vertical information flow) • Decisions can be formulated mathematically, and analysis models are available for design decision-making • Design process starts with an end goal in mind—could be an end product that needs to be supplied to customers • There exist models or computational tools or data that can establish the forward information workflow for the process/product that is being considered • Analysis models for problems are typically incomplete, inaccurate, and not of equal fidelity • Designer's interest is in identifying satisficing robust solutions given the uncertainty in the problem and not single-point optimum solutions • Design is a goal-oriented activity with design requirements subject to change at any time for the problems considered
Interpretation	Understanding the efficacy of the design constructs and broader impacts

this chapter. The industry-inspired problem is contributed by Tata Consultancy Services Research and Tata Steel in India; the focus being to integrate the design of steel (material), manufacturing processes, and automotive gears (end product) (Shukla et al. 2015).

A difficulty during steelmaking is the distortion that happens in gear blanks during forging and heat treatment requiring more machining in the later stages of the manufacturing process. This distortion is mainly due to the banded microstructure that forms due to the presence of segregates. The segregation of alloying elements like manganese (Mn) occurs during casting solidification and this impacts the entire downstream processing affecting the end product mechanical properties. These segregates form due to the limited solubility of alloying elements in the melt during casting. These microsegregation patterns usually remain in the material at the later stages as complete removal of these patterns through processes like reheating is not feasible from a manufacturing standpoint as it demands large reheating time leading to increased manufacturing costs. During the hot rolling process, the deformation of these structures takes place resulting in a change in the concentration profile. The regions are flattened with alternate layers of high solute and low solute forming in the material microstructure. During the following cooling process, phase transformation occurs; also austenite to ferrite phase transformation occurs.

If the steel has hypo eutectoid composition the ferrite phase forms in regions with low austenite stabilizing solute and the remainder transforms to pearlite. Due to the alternate layers of low and high solute regions induced during hot rolling, a banded microstructure having ferrite and pearlite forms with that, finally leading to distortion in gear blanks. To manage the effects of distortion at the end of forging, these segregates must be tracked in the previous manufacturing stages and the factors must be managed effectively. These factors could be the operating setpoints needed for rolling and cooling to produce a specific microstructure. Managing these factors will affect the final mechanical properties of the product. Thus, to predict the mechanical properties of the product as a function of the composition, rolling, and cooling factors, there must be an integration of models.

6.2.1 Systematic Approach of Modeling the Hot Rolling Process Chain

The first step is the multilevel function structure creation through functional analysis, abstraction, and synthesis (Fig. 6.2). This is carried out based on the clarified problem defined in Sect. 6.2.

In terms of the steel manufacturing process chain problem clarified in Sect. 6.2, the system-level functionalities that material–product–manufacturing process system should fulfill are to ***transform*** a square billet into a rod; ***transform*** austenite phase of steel to different phases of steel like ferrite and pearlite; ***transform*** by breaking down the initial austenite grain size to smaller equiaxed grains of austenite and further to ferrite; ***manage*** banded microstructure of ferrite and pearlite; ***manage/control*** microsegregation effects caused by banded microstructure by controlling the manganese content; ***improve*** yield strength; ***improve*** tensile strength; ***improve*** hardness; and ***improve*** toughness.

The focus in this section is on the systematic mapping of phenomena to the core functions identified (Fig. 6.3). The phenomena and the input and output for the phe-

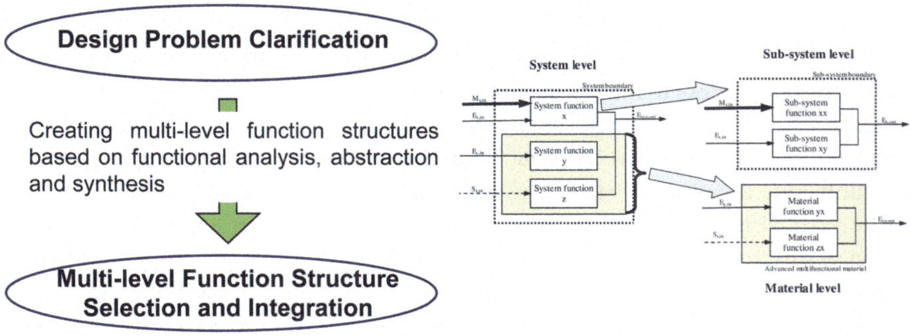

Fig. 6.2 Step 1 of systematic model integration—creation of multilevel function structure

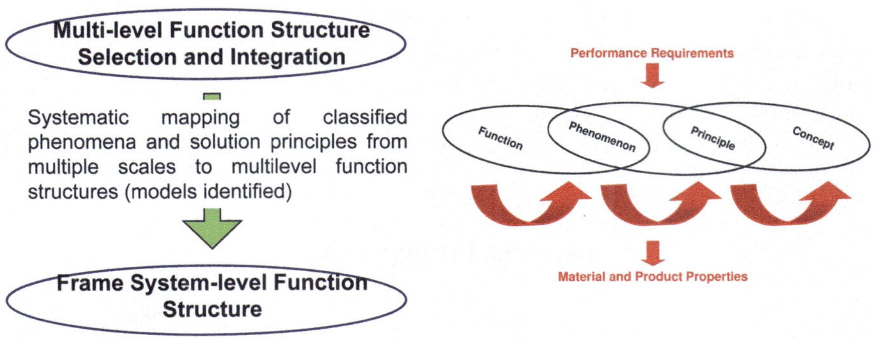

Fig. 6.3 Mapping phenomena to core functions

nomena to establish the information flow chain across the system are identified. The forward material workflow starts with the hot rolling process, which includes a hot deformation module, recrystallization module, flow stress module, and a grain growth module. The input is the chemical composition, initial austenite grain size after reheating, and the rolling schedule (strain, strain rate, interpass time, and number of passes). These are used to predict the temperature evolution, flow stress, and to estimate the austenite grain size after rolling. The output after the vertical integration of these modules is passed to the cooling process. In the vertical integration of cooling process, time–temperature transformations and simultaneous transformations must be considered for the transformations from austenite to different steel phases. This will provide a way to model the banding that occurs during cooling. Here, we consider austenite transformations to ferrite and pearlite phases only. The input to this module is the chemical composition, final austenite grain size after rolling, and the cooling conditions (cooling rate). After the vertical integration of these modules, the output is the phase fractions (final microstructure after cooling), pearlite interlamellar spacing, and the ferrite grain size. This output and the chemical composition serve as the input for the property module to predict the mechanical properties: the yield strength, tensile strength, hardness, and toughness. Through these models and simulation integrations, specific problem-dependent information is passed from one manufacturing process to the other, thereby developing a link between the manufacturing processes. This is the horizontal integration of the manufacturing processes to realize the end product (rod produced after rolling and cooling for this problem) by establishing the process–structure–property–performance relationships.

The vertical integration of models at multiple length scales and horizontal integration of different processes ensures proper flow of information across processing-microstructure-final mechanical property/performance spaces (see Fig. 6.4).

Design catalogs classifying solution principles (models for our case) associated with the phenomenon identified are developed next. The most promising and suiting solution principles for addressing the characteristics of the problem are selected and evaluated further. The design catalogs in our case include models identified to establish the relationships associated with the problem. An extensive literature search is carried out to identify the possible models for the problem addressed in this monograph. Some examples of design catalogs in terms of the models identified are shown below. The design catalog for ultimate tensile strength, yield strength, and ferrite grain in terms of different models identified is shown in Tables 6.2, 6.3, and

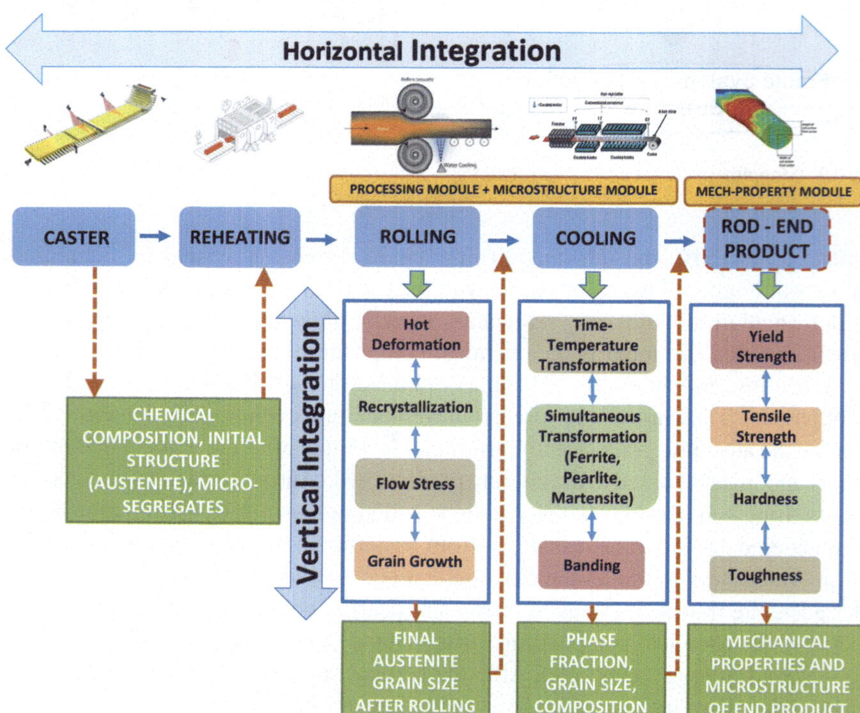

Fig. 6.4 The vertical and horizontal integration of models with information flow for the hot rod rolling process chain

Table 6.2 Design catalog for tensile strength models

Model source	Equation
Gladman et al. (1972)	$\sigma_u = X_f^{1/3}\left(246 + 1143[N]^{0.5} + 18.1(0.001D_\alpha)^{-0.5}\right) + 97[Si] + \left(1 - X_f^{1/3}\right)\left(719 + 3.5S_0^{-0.5}\right)$
Choquet et al. (1990)	$\sigma_u = 237 + 29[Mn] + 79[Si] + 5369[N] + 700[P] + 7.24X_f(0.001D_\alpha)^{-0.5} + 500(1 - X_f)$
Hodgson and Gibbs (1992)	$\sigma_u = 164.9 + 634.7[C] + 53.6[Mn] + 99.7[Si] + 651.9[P] + 472.6[Ni] + 3339[N] + 11(0.001D_\alpha)^{-0.5}$
Kuziak et al. (1997)	$\sigma_u = X_f\left(20 + 2440[N]^{0.5} + 18.5(0.001D_\alpha)^{-0.5}\right) + 750(1 - X_f) + 3\left(1 - X_f^{0.5}\right)S_0^{-0.5} + 92.5[Si]$

Table 6.3 Design catalog for yield strength models

Model source	Equation
Gladman et al. (1972)	$\sigma_y = 63[Si] + 425[N]^{0.5} + X_f^{1/3}\left(35 + 58[Mn] + 17(0.001D_\alpha)^{-0.5}\right) + \left(1 - X_f^{1/3}\right)\left(179 + 3.9S_0^{-0.5}\right)$
Gladman et al. (1977)	$\sigma_y = 88 + 37[Mn] + 83[Si] + 2900[N]^{0.5} + 17(0.001D_\alpha)^{-0.5}$
LeBon and deSaint-Martin (1977)	$\sigma_y = 190 + 15.9(0.001D_\alpha)^{-0.5}$
Kejian and Baker (1993)	$\sigma_y = \sqrt{\{70 + 37[Mn] + 83[Si] + 1500[N] + 18.6(0.001D_\alpha)^{-0.5}\}^2 + \sigma_d^2}$
Choquet et al. (1990)	$\sigma_y = 63 + 23[Mn] + 53[Si] + 5000[N] + 700[P] + X_f\left(15.4 - 30[C] + \frac{6.094}{0.8 + [Mn]}\right)(0.001D_\alpha)^{-0.5} + (1 - X_f)\left(360 + 2600[C]^2\right)$
Hodgson and Gibbs (1992)	$\sigma_y = 62.6 + 26.1[Mn] + 60.2[Si] + 759[P] + 212.9[Cu] + 3286[N] + 19.7(0.001D_\alpha)^{-0.5}$
Majta et al. (1996)	$\sigma_y = 75.4[Si] + 478[N] + 1200[P] + X_f\left(77.7 + 59.5[Mn] + 9.1(0.001D_\alpha)^{-0.5}\right) + (1 - X_f)\left(145.5 + 2.5S_0^{-0.5}\right)$
Kuziak et al. (1997)	$\sigma_y = X_f\left(77.7 + 59.5[Mn] + 9.1(0.001D_\alpha)^{-0.5}\right) + 478[N]^{0.5} + 1200[P] + (1 - X_f)\left[145.5 + 3.5S_0^{-0.5}\right]$

6.4, respectively.

Similar design catalogs are developed for other functions, and the system-level function structure based on the phenomena and solution principles is developed. The system-level function structure after the selection and integration of models showing the information flow across the processing–structure–property–performance spaces of material, product, and manufacturing processes is shown in Fig. 6.5.

The selection/identification of the solution principles from the design catalogs for the problem discussed in this monograph is provided in Sect. 6.5 of this chapter and is not detailed here.

Table 6.4 Design catalog for ferrite grain size

Steel	Model source	Equation
C-Mn	Sellars and Beynon (1984)	$D_\alpha = \left(1 - 0.45\varepsilon_r^{1/2}\right) \times \left\{1.4 + 5C_r^{-1/2} + 22\left[1 - \exp\left(-1.5 \times 10^{-2}D\right)\right]\right\}$
C-Mn $C_{eq} < 0.35$	Hodgson and Gibbs (1992)	$D_\alpha = \left(1 - 0.45\varepsilon_r^{1/2}\right) \times \left\{\left(-0.4 + 6.37C_{eq}\right) + \left(24.2 - 59C_{eq}\right)C_r^{-0.5} + 22[1 - \exp(-0.015D)]\right\}$
C-Mn $C_{eq} > 0.25$	Hodgson and Gibbs (1992)	$D_\alpha = \left(1 - 0.45\varepsilon_r^{1/2}\right) \times \left\{\left(22.6 - 57C_{eq}\right) + 3C_r^{-0.5} + 22[1 - \exp(-0.015D)]\right\}$
C-Mn	Senuma et al. (1992)	$D_\alpha = \left[5.51 \times 10^{10} D^{1.76} \exp\left(\frac{-21430}{T_{.05f}}\right) X_f\right]^{1/3}$
C-Mn	Donnay et al. (1996)	$D_\alpha = \left\{13 - 0.73\left[1000([C] + 0.1[Mn])^{0.45}\right]\right\} D^{0.3} C_r^{-0.15}$
Nb	Sellars and Beynon (1984)	$D_\alpha = \left(1 - 0.45\varepsilon_r^{1/2}\right) \times \left\{2.5 + 3C_r^{-1/2} + 20\left[1 - \exp\left(-1.5 \times 10^{-2}D\right)\right]\right\}$
V-Ti	Roberts et al. (1983)	$D_\alpha = 3.75 + 0.18D + 1.4C_r^{-0.5}$
V-Ti	Sellars and Beynon (1984)	$D_\alpha = \left(1 - 0.45\varepsilon_r^{1/2}\right) \times \left\{3 + 1.4C_r^{-1/2} + 17\left[1 - \exp\left(-1.5 \times 10^{-2}D\right)\right]\right\}$
V-T	Kuziak et al. (1997)	$D_\alpha = \frac{D}{1+(0.036+0.0233C_r^{.5})D}$

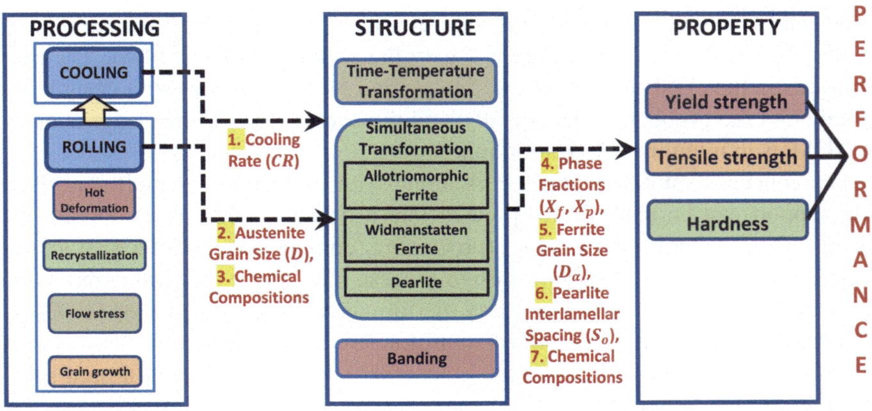

Fig. 6.5 System-level function structure and information flow chain for the problem addressed

To illustrate the goal-oriented, inverse method, we define a boundary within the problem. Here, we focus on using the proposed method to establish processing–microstructure–property relations between the rolling, cooling module (processing and microstructure), and the property module of the product that defines the end performance. In Sect. 6.3, the concept exploration framework that is used to systematically formulate the problem and identify a ranged set of satisficing solutions is discussed.

6.3 The Concept Exploration Framework—From the Standpoint of Processing–Structure–Property–Performance Relationship

The CEF is introduced in this chapter as a general framework that includes systematic steps to identify design alternatives and generate *satisficing* design solutions across the processing–microstructure–property–performance spaces. The CEF is inspired by the RCEM (Chen et al. 1997) with the addition of features (processors) to consider different material and product models and options to explore the solution space for different design scenarios. The foundational mathematical construct of the CEF is the compromise decision support problem (cDSP) (Mistree et al. 1993). The cDSP construct used here is anchored in the robust design paradigm first proposed by Taguchi. The fundamental assumption is that the models are not complete, accurate, and of equal fidelity (Taguchi 1986; Bras and Mistree 1993). The cDSP is a hybrid of mathematical programming and goal programming, and the emphasis of the designer using the cDSP is to satisfy specified target goals as closely as possible. This is achieved by seeking multiple solutions through tradeoffs among multiple conflicting goals. The solutions obtained are further evaluated by solution space exploration to identify solution regions that best satisfy the requirements identified. The four keywords in the cDSP—*Given*, *Find*, *Satisfy*, and *Minimize*—support a designer in formulating the problem systematically, and the overall goal of the designer using the cDSP is to minimize a deviation function.

The concept exploration framework along with its features of multigoal decision support can be readily incorporated into a design method that supports the design of the material and product (processing, composition and microstructure) as part of a larger overall systems design process. The framework can embody the hierarchy of process–structure–property–performance proposed by Olson (1997) by systematically accounting the information flow and mappings across these spaces and transforming overall design requirements into a set of satisficing design specifications for the material–product–manufacturing process system of interest. The designer can start with the performances/properties that he or she wants and use the CEF to identify satisficing solutions for microstructure (see Fig. 6.6). This can be further repeated to identify satisficing solutions for the processing space that meets microstructure requirements.

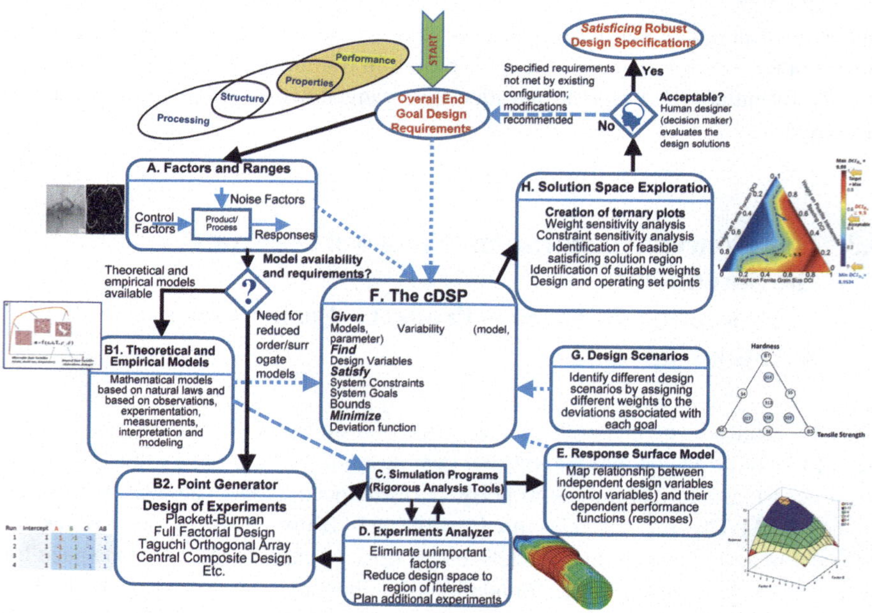

Fig. 6.6 The CEF for PSPP exploration

6.4 Application of Goal-Oriented, Inverse Design Method and CEF for the Hot Rod Rolling Process Chain Problem

In Fig. 6.7, we show the schematic application of proposed goal-oriented, inverse method to carry out the integrated design exploration of the hot rod rolling process chain problem of interest.

Step 1: Establish forward modeling and information flow across the process chain For the hot rod rolling process chain problem, the mechanical property goals and requirements for yield strength (YS), tensile strength (TS), hardness (HV), and toughness measured by impact transition temperature (ITT) are addressed in this chapter. These mechanical properties are dependent on the final microstructure after cooling, like the ferrite grain size after cooling (FGS, D_α), the phase fractions of ferrite (X_f) and pearlite ($1 - X_f$), the pearlite interlamellar spacing (S_0), and the composition variables like silicon $[Si]$, nitrogen $[N]$, phosphorous $[P]$, and manganese $[Mn]$. These microstructure factors are defined by the rate (CR) at which cooling is carried out and the final austenite grain size after rolling (AGS, D) and composition variables like carbon $[C]$ and manganese $[Mn]$. The AGS is determined by the processing carried out at rolling stage which requires the modeling of

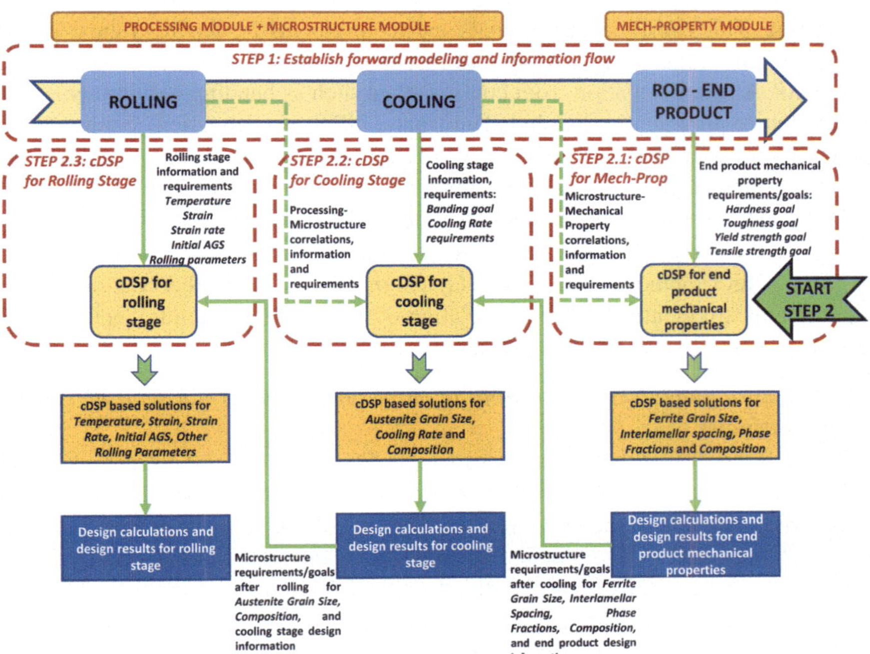

Fig. 6.7 Schematic of the proposed goal-oriented, inverse method for the hot rod rolling process chain problem

hot deformation, recrystallization, grain growth, and so on. The input to the cooling stage is D, CR, $[C]$, and $[Mn]$ from the rolling process. The outputs are D_α, X_f, and S_0, which along with the composition variables define the YS, TS, ITT, and HV of end rod produced. The models used to establish these relationships are presented in Sect. 6.5.

Step 2.1: Formulate cDSP for end mechanical properties of rod to explore the processing and microstructure space for cooling stage
In Step 2.1, the cDSP for the mechanical properties of the final end product is formulated. Information, requirements, and the correlations between mechanical properties and microstructure after cooling (ferrite grain size, pearlite interlamellar spacing, phase fractions, and composition) are communicated to this cDSP. The end mechanical property goals are requirements for yield strength, tensile strength, hardness, and toughness (impact transition temperature). On exercising the cDSP and carrying out solution space exploration of the microstructure space after cooling, the combinations for ferrite grain size (D_α), phase fractions X_f, pearlite interlamellar spacing S_0, and compositions that best satisfy the requirements for end properties are identified and are communicated to the next step. The formulation of the cDSP is provided in Sect. 6.6 and the solution space exploration is carried out in Sect. 6.7.

Step 2.2: Formulate cDSP for cooling stage to explore the processing and microstructure space of rolling
In Step 2.2, a similar process to that in Step 2.1 is carried out to formulate the cDSP for cooling. This cDSP has target goals and requirements for ferrite grain size D_α, phase fraction X_f, and compositions that are based on the solutions obtained from the first cDSP. Also, information from cooling stage such as banding requirements and cooling rate requirements are included into this cDSP. The information, requirements, and correlations of variables at the end of rolling (austenite grain size, composition) and the cooling stage parameters are communicated to the cDSP. The goals for this cDSP are target ferrite grain size and target phase fractions subject to constraints. On exercising the cDSP, the combinations for austenite grain size$D$, cooling rate CR, and composition elements like carbon (C) and manganese (Mn) that best satisfy the requirements are identified. The formulation of this cDSP is given in Sect. 6.6 and the solution space exploration is carried out in Sect. 6.7.

Step 2.3: Formulate cDSP for rolling to carry out design exploration of rolling process variables
In Step 2.3, we follow a similar procedure to Steps 2.1 and 2.2 to formulate the cDSP for rolling considering the information generated from cooling and the rolling information and requirements identified. This cDSP has a grain growth module, static, dynamic, and meta-dynamic recrystallization modules and a hot deformation module. The design problem formulation for Step 2.3 is discussed in Sect. 6.8 of this chapter as a separate comprehensive example due to the complex information flow involved in the hot rolling multistage thermo-mechanical process modeling.

6.5 Mathematical Models for Hot Rod Rolling Process Chain Design (Establishing Step 1 of Method)

The mathematical models used to formulate the hot rod rolling (HRR) process chain problem are introduced and brief descriptions are provided here. These models, the control variables, noise factors, parameters, responses, and allowable ranges are identified by carrying out Steps 1 and 2 of the CEF, as described in Sect. 4.6 in Chap. 4 and briefly addressed in Sect. 6.3. From the problem perspective, we accomplish Step 1 of the goal-oriented, inverse decision-based design method by identifying these forward models to establish the relationships described in Sect. 6.4. In Sect. 6.5.1 we describe the microstructure–mechanical property correlation models that establish relationships between the mechanical properties of the rod product and the microstructure generated after cooling stage. In Sects. 6.5.2 and 6.5.3, we describe the models for phase transformation on cooling after hot rolling.

6.5.1 Mechanical Property-Microstructure Correlation Models

The required mechanical properties for the rod are yield strength (YS), tensile strength (TS), toughness measured by impact transition temperature (ITT), and hardness (HV). Gladman et al. (1972, 1977) predict the mechanical properties of plain carbon steel products as a function of the ferrite–pearlite microstructure. Models with improved predictive power were later developed by Hodgson and Gibbs (1992), Majta et al. (1996), Kuziak et al. (1997), and Yada (1987) (Table 6.4). Details on these models and the reason for their selection are presented by Nellippallil et al. (2017c) and described next. The models identified are summarized in Table 6.5.

Model for Yield Strength (YS)

Yield strength (lower yield stress) is an important mechanical property that defines the stress at which the material begins to deform plastically. We have selected a semi-empirical model (Eq. 6.1) by Kuziak and co-authors that describes the lower yield stress YS, of carbon–manganese steels as a function of ferrite grain size after cooling D_α, cooling rate CR, ferrite fraction X_f, the pearlite interlamellar spacing S_o, and the composition elements in the steel (Kuziak et al. 1997). The model is developed following the concept of Hall–Petch relationship. The reason for the selection of this model is because of the fact that it takes into account the influence of pearlite through the factors X_f and S_o compared to other models available.

$$\begin{aligned} YS = {} & X_f\left(77.7 + 59.9 \times [Mn] + 9.1 \times (0.001 D_\alpha)^{-0.5}\right) + 478[N]^{0.5} \\ & + 1200[P] + \left(1 - X_f\right)\left[145.5 + 3.5 S_0^{-0.5}\right] \end{aligned} \tag{6.1}$$

where YS is in MPa, S_o in μm, and D_α in μm.

Table 6.5 Mechanical property models

Mechanical property	Model	References
Yield strength	$YS = X_f\left(77.7 + 59.9 \times [Mn] + 9.1 \times (0.001D_\alpha)^{-0.5}\right) + 478[N]^{0.5} + 1200[P] + (1 - X_f)\left[145.5 + 3.5S_0^{-0.5}\right]$ where YS is in MPa, S_o in μm, D_α in μm	Kuziak et al. (1997)
Tensile strength	$TS = X_f\left(20 + 2440 \times [N]^{0.5} + 18.5 \times (0.001D_\alpha)^{-0.5}\right) + 750(1 - X_f) + 3\left(1 - X_f^{0.5}\right)S_0^{-0.5} + 92.5 \times [Si]$ where TS is in MPa, S_o in μm, D_α in μm	Kuziak et al. (1997)
Hardness	$HV = X_f(361 - 0.357T_{mf} + 50[Si]) + 175(1 - X_f)$ Average austenite to ferrite transformation temperature (T_{mf}) is assumed as 700 °C	Yada (1987)
Impact transition temperature	$ITT = X_f\left(-46 - 11.5D_\alpha^{-0.5}\right) + (1 - X_f)\left(-335 + 5.6S_0^{-0.5} - 13.3p^{-0.5} + \left(3.48 \times 10^6\right)t\right) + 49[Si] + 762[N]^{0.5}$ where D_α, S_o, p and t are in mm. We have assumed the value of pearlite colony size p as 6 μm and carbide thickness t as 0.025 μm	Gladman et al. (1972)

Model for Tensile Strength (TS)

Tensile strength represents the resistance of the material to breaking when subjected to tensile loads. We have selected the model (Eq. 6.2) by Kuziak and co-authors that describes the tensile strength TS of carbon–manganese steels as a function of ferrite grain size after cooling D_α, cooling rate CR, ferrite fraction X_f, the pearlite interlamellar spacing S_o, and the composition elements in the steel (Kuziak et al. 1997). Again, we have opted for the model as it takes into account the influence of pearlite.

$$TS = X_f\left(20 + 2440 \times [N]^{0.5} + 18.5 \times (0.001D_\alpha)^{-0.5}\right) + 750\left(1 - X_f\right) + 3\left(1 - X_f^{0.5}\right)S_0^{-0.5} + 92.5 \times [Si] \tag{6.2}$$

where TS is in MPa, S_o in μm, and D_α in μm.

Model for Hardness (HV)

Hardness represents the resistance to plastic deformation usually by indentation. Hardness (HV) is represented as a function of ferrite and pearlite fractions, average austenite to ferrite transformation temperature (T_{mf}), and the weight percentage of silicon (Si) as depicted in Eq. 6.3 based on the investigation by Yada (1987).

$$HV = X_f(361 - 0.357T_{mf} + 50[Si]) + 175(1 - X_f) \tag{6.3}$$

We have assumed the value of T_{mf} as 700 °C in this work.

Model for Impact Transition Temperature (ITT)

The impact transition temperature denotes the boundary between brittle and ductile failure when subjected to impact loads and is a measure of the material's impact toughness. This temperature is extremely important during material selection. We have selected the model (Eq. 6.4) developed by Gladman et al. (1972) for impact transition temperature of high-carbon steels as a function of ferrite grain size D_α, ferrite fraction X_f, pearlite interlamellar spacing S_o, pearlite colony size p, carbide thickness t, and composition elements.

$$\begin{aligned} ITT = {} & X_f(-46 - 11.5D_\alpha^{-0.5}) + (1 - X_f) \\ & (-335 + 5.6S_0^{-0.5} - 13.3p^{-0.5} + (3.48 \times 10^6)t) + 49[Si] + 762[N]^{0.5} \end{aligned} \tag{6.4}$$

where D_α, S_o, p, and t are in mm in Eq. 6.4.

We assume the value of pearlite colony size p as 6 μm and carbide thickness t as 0.025 μm.

6.5.2 *Models for Phase Transformation on Cooling After Hot Working*

Classical nucleation and grain growth theory quantitatively describe the kinetics of decomposition of austenite. Using classical Johnson-Mehl-Avrami theory, we describe the transformation of a single phase to a product phase (Jägle 2007). The transformations that occur in steel are often simultaneous, resulting in the formation of multiple phases such as allotriomorphic ferrite, Widmanstätten ferrite, bainite, pearlite, and martensite. Therefore, one requirement for a kinetic model for the phase transformation of steel is that it must allow for simultaneous phase transformations resulting in different steel phases. Robson and Bhadeshia (1997), Jones and Bhadeshia (1997a, b) address this requirement by numerically solving all impingement equations and choosing the appropriate nucleation and grain growth equations. The simultaneous transformation of austenite into allotriomorphic ferrite, Widmanstätten ferrite, and pearlite are considered by Jones and Bhadeshia (1997a, b); details can be found in Jones and Bhadeshia (1997a, b), Jägle (2007). We have used the program STRUCTURE developed by Jones and Bhadeshia to predict the simultaneous transformation of austenite (Jones et al. 2017).

Response surface models (RSMs) are used to calculate the microstructure (allotriomorphic ferrite, Widmanstätten ferrite, and pearlite) of steel as a function of percentages of carbon $[C]$, manganese $[Mn]$, cooling rate CR, and austenite grain size $[D]$ using the STRUCTURE program. These predictors are selected because of their substantial contribution to austenite transformation and the formation of banded microstructure (Jägle 2007). Values for the other required input are based on the work of Jones and Bhadeshia (1997a, b). A fractional factorial design of experiments is carried out to develop response surface models for the transformation of austenite to ferrite and pearlite (Kuziak et al. 1997; Phadke et al. 2004) (Table 6.6). The response surface models are shown in Table 6.7. The RSMs are verified by comparing the predictions with experimental (measured) data reported by Bodnar and Hensen (1994), see Nellippallil et al. (2017d).

Table 6.6 Factors and factor levels for DoE

Level	CR (K/min)	AGS (μm)	$[C]$ (%)	$[Mn]$ (%)
1	11	30	0.18	0.7
2	55	55	0.24	1.1
3	100	100	0.3	1.5

Table 6.7 Design catalog—RSM for phase fractions

Phase fraction	Response surface model developed	R^2 value
Allotriomorphic ferrite	$X_{fa} = 1.59 - 0.26[C] - 0.00856CR - 0.0105D - 3.08[C] + 0.000826[Mn]CR + 0.0009[Mn]D + 0.7647[Mn][C] + 0.000011CR * D + 0.002CR[C] + 0.0032D[C] - 0.05058[Mn]^2 + 0.00004CR^2 + 0.000036D^2 + 2.483[C]^2$	0.98
Pearlite	$X_p = 0.206 - 0.117[Mn] - 0.0005CR - 0.00113D + 0.248[C] + 0.00032[Mn]CR + 0.000086[Mn]D + 0.9539[Mn][C] - 4.259 \times 10^{-6}CR * D + 0.00726CR[C] + 0.0023D[C] - 0.0305[Mn]^2 - 0.0000056CR^2 + 4.859 \times 10^{-6}D^2 + 0.79[C]^2$	0.99
Widmanstätten ferrite	$X_{fw} = 1 - (X_{fa} + X_p)$	–
Total ferrite	$X_f = (X_{fa} + X_{fw})$	–

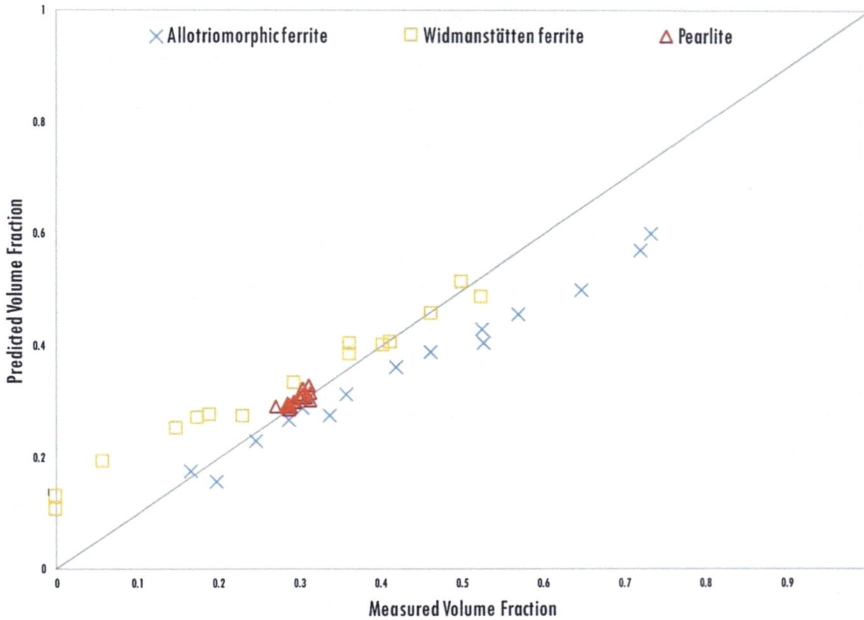

Fig. 6.8 Comparison of RSM predictions with measured volume fractions from literature (Bodnar and Hansen 1994)

Response Surface Model Verification

The developed RSMs are verified by comparing the predictions with experimental (measured) data reported by Bodnar and Hensen (1994). We observe from the comparison showcased in Fig. 6.8 that the predictions using RSM lie more or less in the vicinity of the straight line depicting the measured values. However, as we can observe from Fig. 6.8, the model is not fully accurate and the uncertainties associated with the model have to be dealt with in future analysis.

The response surface model developed by carrying out the experimental runs using the STRUCTURE program is listed in Eqs. 6.5 and 6.8 and summarized in Table 6.6. The response surface model developed for fraction of allotriomorphic ferrite X_{fa} X_{fa} with R^2 value of 0.98 is

$$\begin{aligned} X_{fa} = {} & 1.59 - 0.26[C] - 0.00856CR - 0.0105D - 3.08[C] \\ & + 0.000826[Mn]CR + 0.0009[Mn]D + 0.7647[Mn][C] \\ & + 0.000011CR * D + 0.002CR[C] + 0.0032D[C] \\ & - 0.05058[Mn]^2 + 0.00004CR^2 + 0.000036D^2 \\ & + 2.483[C]^2 \end{aligned} \tag{6.5}$$

The response surface model developed for fraction pearlite X_p with R^2 value of 0.99 is

$$X_p = 0.206 - 0.117[Mn] - 0.0005CR - 0.00113D + 0.248[C] + 0.00032[Mn]CR + 0.000086[Mn]D + 0.9539[Mn][C] - 4.259 \times 10^{-6}CR * D + 0.00726CR[C] + 0.0023D[C] - 0.0305[Mn]^2 - 0.0000056CR^2 + 4.859 \times 10^{-6}D^2 + 0.79[C]^2 \quad (6.6)$$

Widmanstätten ferrite X_{fw} is represented as the fraction that is left after transformations of austenite to allotriomorphic ferrite and pearlite.

$$X_{fw} = 1 - \left(X_{fa} + X_p\right) \quad (6.7)$$

The total ferrite is calculated as the sum of allotriomorphic ferrite and Widmanstätten ferrite.

$$X_f = \left(X_{fa} + X_{fw}\right) \quad (6.8)$$

6.5.3 Models for Ferrite Grain Size (D_α) and Pearlite Interlamellar Spacing (S_o)

As the hot worked steel cools, austenite is transformed into various phases. The most important parameters are the ferrite grain size and pearlite interlamellar spacing because they contribute to the steel's mechanical properties. The models for these parameters are summarized in Table 6.8.

Table 6.8 Design catalog—models for D_α and S_o

Parameter	Model	References
Ferrite grain size	$D_\alpha = \left(1 - 0.45\varepsilon_r^{0.5}\right) \times \{(-0.4 + 6.37C_{eq}) + \left(24.2 - 59C_{eq}\right) CR^{-0.5} + 22[1 - \exp(-0.015D)]\}$ for $C_{eq} < 0.35$ $D_\alpha = \left(1 - 0.45\varepsilon_r^{0.5}\right) \times \{(22.6 - 57C_{eq}) + 3CR^{-0.5} + 22[1 - \exp(-0.015D)]\}$ for $C_{eq} > 0.35$ where D is the final austenite grain size after rolling and ε_r is retained strain. C_{eq} is the carbon equivalent given by Eq. 6.11 $C_{eq} = (C + Mn)/6$	Hodgson and Gibbs (1992)
Pearlite interlamellar spacing	$S_o = 0.1307 + 1.027[C] - 1.993[C]^2 - 0.1108[Mn] + 0.0305CR^{-0.52}$	Kuziak et al. (1997)

Ferrite Grain Size

As the hot worked steel cools down to a lower temperature, austenite transforms into different phases. Out of the newly formed phases, the ferrite grain size is one important parameter that contributes to the strength of the compositions. The factors affecting ferrite grain size D_α are final austenite grain size after rolling D and retained strain ε_r, both related to the deformation history of the material from the rolling side, the composition, and the cooling rate which are external influences (Pietrzyk 1999). We are adopting the models by Hodgson and Gibbs (1992) (Eqs. 6.9 and 6.10) for defining ferrite grain size.

$$\begin{aligned} D_\alpha = & \left(1 - 0.45\varepsilon_r^{0.5}\right) \\ & \times \left\{\left(-0.4 + 6.37C_{eq}\right) + \left(24.2 - 59C_{eq}\right)CR^{-0.5} \right. \\ & \left. +22\left[1 - \exp(-0.015D)\right]\right\} \quad \text{for } C_{eq} > 0.35 \end{aligned} \tag{6.9}$$

$$\begin{aligned} D_\alpha = & \left(1 - 0.45\varepsilon_r^{0.5}\right) \times \{\left(22.6 - 57C_{eq}\right) + 3CR^{-0.5} \\ & + 22\left[1 - \exp(-0.015D)\right] \quad \text{for } C_{eq} > 0.35 \end{aligned} \tag{6.10}$$

where C_{eq} is the carbon equivalent given by Eq. 6.11.

$$C_{eq} = (C + Mn)/6 \tag{6.11}$$

Pearlite Interlamellar Spacing

Pearlite interlamellar spacing is a very important microstructural factor that influences the mechanical properties of steel as the steel turns more into a fully pearlitic microstructure (Vander Voort 2015). We are adopting the model (Eq. 6.12) by Kuziak et al. (1997) where pearlite interlamellar spacing S_o is defined as a function of carbon C, manganese Mn, and cooling rate CR.

$$\begin{aligned} S_o = & 0.1307 + 1.027[C] - 1.993[C]^2 - 0.1108[Mn] \\ & + 0.0305CR^{-0.52} \end{aligned} \tag{6.12}$$

In Sect. 6.6, we showcase the cDSP formulations for end product mechanical properties and cDSP for cooling stage using the models identified in Sect. 6.5.

6.6 Comprehensive Example Problem 1: Formulation of the cDSPs for Hot Rolling Process Chain Problem

In Step 2.1 of the method, we formulate the cDSP for the desired end mechanical properties of the product (Table 6.9). We then determine the end mechanical properties as a function of microstructure factors (D_α, X_f, S_0, Mn, Si, N) after cooling.

Table 6.9 The cDSP formulation for Step 2.1

Given	
(1) End requirements identified for the rod rolling process • Maximize yield strength (Goal) • Maximize tensile strength (Goal) • Maximize hardness (Goal) • Minimize ITT (Requirement) • Manage banded microstructure (Requirement) (2) Well-established empirical and theoretical correlations, RSMs, and information flow from the end of cooling to the end product mechanical properties (details provided in Sect. 6.5) (3) System variables and their ranges	
Find	
System variables	
X_1, ferrite grain size (D_α) X_2, the phase fraction of ferrite (X_f) X_3, the pearlite interlamellar spacing (S_0) X_4, manganese concentration after cooling ($[Mn]$) X_5, the composition of Si ($[Si]$) X_6, the composition of N ($[N]$)	
Deviation variables	
$d_i^-, d_i^+, i = 1, 2, 3$	
Satisfy	
System constraints	
• Minimum yield strength constraint $YS \geq 220\,\text{MPa}$	(6.13)
• Maximum yield strength constraint $YS \leq 330\,\text{MPa}$	(6.14)
• Minimum tensile strength constraint $TS \geq 450\,\text{MPa}$	(6.15)
• Maximum tensile strength constraint $TS \leq 750\,\text{MPa}$	(6.16)
• Minimum hardness constraint $HV \geq 131$	(6.17)
• Maximum hardness constraint $HV \leq 170$	(6.18)
• Minimum ITT constraint $ITT \geq -100\,°\text{C}$	(6.19)
• Maximum ITT constraint $ITT \leq 100\,°\text{C}$	(6.20)
System goals	
Goal 1:	

(continued)

Table 6.9 (continued)

• Maximize yield strength	
$\frac{YS(X_i)}{YS_{\text{Target}}} + d_1^- - d_1^+ = 1$	(6.21)
Goal 2:	
• Maximize tensile strength	
$\frac{TS(X_i)}{TS_{\text{Target}}} + d_2^- - d_2^+ = 1$	(6.22)
Goal 3:	
• Maximize hardness	
$\frac{HV(X_i)}{HV_{\text{Target}}} + d_3^- - d_3^+ = 1$	(6.23)
Variable bounds	
$8 \le X_1 \le 25$ (μm) $0.1 \le X_2 \le 0.9$ $0.15 \le X_3 \le 0.25$ (μm) $0.7 \le X_4 \le 1.5$ (%) $0.18 \le X_5 \le 0.3$ (%) $0.007 \le X_6 \le 0.009$ (%)	
Bounds on deviation variables	
$d_i^-, d_i^+ \ge 0$ and $d_i^- * d_i^+ = 0$, i = 1, 2, 3	(6.24)
Minimize	
We minimize the deviation function	
$Z = \sum_{i=1}^{3} W_i(d_i^- + d_i^+); \sum_{i=1}^{3} W_i = 1$	(6.25)

The end mechanical property goals, for example, maximizing YS, TS, and HV, are captured in the cDSP. The requirement for minimizing impact transition temperature is captured as a constraint. The possible achievement of these conflicting goals is characterized by solution space exploration. The upper and lower limits for the system variables and the maximum and minimum values for the mechanical properties are defined in the cDSP as bounds and constraints. The goal targets are $YS_{\text{Target}} = 330$ MPa, $TS_{\text{Target}} = 750$ MPa, and $HV_{\text{Target}} = 170$. The requirement for ITT is to achieve the minimum value. The requirement for managing the banded microstructure is considered during solution space exploration.

On exercising the cDSP and carrying out solution space exploration a process designer is able to solve and capture the knowledge associated with the following inverse problem: *Given the end mechanical properties of a new steel product mix, what should be the microstructure factors after the cooling stage that satisfies the requirements?*

On exercising the cDSP for different design scenarios and carrying out solution space exploration, and following the steps in concept exploration framework, we obtain the combinations for D_α, X_f, S_0, Mn, Si, N that satisfy the end mechanical properties and other requirements. The desired solutions identified for D_α, X_f, S_0

are then used as the target goals for the next cDSP (Step 2.2 of the goal-oriented, inverse method).

In Step 2.2 of the method, we formulate the cDSP for the cooling stage (Table 6.10). Using this cDSP, we relate the microstructure factors after cooling that best satisfy the first cDSP requirements as a function of the microstructure and composition factors (D, C, Mn) after the rolling and the cooling stage operating factor (CR). The target values for the goals are defined as $D_{\alpha\,\text{Target}}$, $X_{f\,\text{Target}}$, $S_{0\,\text{Target}}$ as the results from the first cDSP. On exercising this cDSP the process designer will be able to solve and capture the knowledge associated with the following inverse problem: *Given the microstructure after cooling that best satisfy the end mechanical properties of a new steel product mix, what should be the microstructure factors after rolling and the design and operating set points for cooling that satisfy the requirements identified?*

6.7 Integrated Solution Space Exploration of Hot Rod Rolling Process Chain Using the Proposed Method (Steps 2.1 and 2.2)

We have exercised 19 different scenarios for both cDSPs in Steps 2.1 and 2.2 (Table 6.11). These scenarios are selected based on judgment to effectively capture the design space for exploration in a ternary space with the different combinations of weights on goals.

We explain the significance of these scenarios using the cDSP for the end product (the cDSP in Step 2.1). For the first cDSP, Scenarios 1–3 are for a situation where the designer's interest is to achieve the target on a single goal, that is, maximizing YS, maximizing TS or maximizing X_f as closely as possible. For example, the designer's preference in Scenario 2 (for cDSP 1) is to achieve only the tensile strength goal. Scenarios 4–6 are for a situation where two goals are given equal preference, and the third goal is not assigned any preference. For example, Scenario 4 is a situation where the designer's interest is in equally maximizing YS and TS without giving any preference to the X_f goal. Scenarios 7–12 are situations where the designer gives greater preference to one goal, a lesser preference to the second goal, and zero preference to the third goal. Scenario 13 is a situation where the designer gives equal preference to all three goals. Scenarios 14–19 are situations where all the goals are assigned preferences with two of them being the same preference. The exploration of solution space is carried out by exercising the cDSPs for these scenarios and plotting the solution space obtained in a ternary space. The axes of the ternary plots are the weights assigned to each goal, and the color contour in the interior is the achieved value of the specific goal that is being addressed. From these plots, we identify feasible solution regions that satisfy our requirements and the associated weights to be assigned to each goal to achieve this solution space. To read more about the creation and interpretation of ternary plots, see Nellippallil et al. (2016).

Table 6.10 The cDSP formulation for Step 2.2

Given
(1) Target values for microstructure after cooling (the combination identified from the first cDSP as best satisfying the end goals) (2) Well-established empirical and theoretical correlations, RSMs, and complete information flow from the end of rolling to the end product mechanical properties (details provided in Sect. 6.5)
(3) System variables and their ranges
Find
System variables
X_1, Cooling rate (CR) X_2, Austenite grain size (D) X_3, the carbon concentration ($[C]$) X_4, the manganese concentration after rolling ($[Mn]$)
Deviation variables
$d_i^-, d_i^+, i = 1, 2, 3$
Satisfy
System constraints
• Minimum ferrite grain size constraint $D_\alpha \geq 8\,\mu\text{m}$ (6.26)
• Maximum ferrite grain size constraint $D_\alpha \leq 20\,\mu\text{m}$ (6.27)
• Minimum pearlite interlamellar spacing constraint $S_o \geq 0.15\,\mu\text{m}$ (6.28)
• Maximum pearlite interlamellar spacing constraint $S_o \leq 0.25\,\mu\text{m}$ (6.29)
• Minimum ferrite phase fraction constraint (manage banding) $X_f \geq 0.5$ (6.30)
• Maximum ferrite phase fraction constraint (manage banding) $X_f \leq 0.9$ (6.31)
• Maximum carbon equivalent constraint $C_{eq} \leq 0.35$ (6.32) Also included are mechanical properties constraints based on the results obtained from first cDSP solution space exploration (the acceptable ranges identified for mechanical properties)
• Minimum yield strength constraint $YS \geq YS_{\text{lower limit}}$ MPa (6.33)
• Maximum yield strength constraint $YS \leq YS_{\text{upper limit}}$ MPa (6.34)
• Minimum tensile strength constraint $TS \geq TS_{\text{lower limit}}$ MPa (6.35)
• Maximum tensile strength constraint $TS \leq TS_{\text{upper limit}}$ MPa (6.36)

(continued)

Table 6.10 (continued)

• Minimum hardness constraint $HV \geq HV_{\text{lower limit}}$ (6.37)
• Maximum hardness constraint $HV \leq HV_{\text{upper limit}}$ (6.38)
System Goals
The target values for system goals are identified from the solution space exploration carried out for the first cDSP
Goal 1: • Achieve Ferrite Grain Size Target from cDSP 1 $\frac{D_{\alpha\,\text{Target}}}{D_\alpha(X_i)} + d_1^+ - d_1^- = 1$ (6.39)
Goal 2: • Achieve Ferrite Fraction from cDSP 1 $\frac{X_f(X_i)}{X_{f\,\text{Target}}} + d_2^- - d_2^+ = 1$ (6.40)
Goal 3: • Achieve Pearlite Interlamellar Spacing Target from cDSP 1 $\frac{S_{o\,\text{Target}}}{S_o(X_i)} + d_3^+ - d_3^- = 1$ (6.41)
Variable bounds
$11 \leq X_1 \leq 100\,(\text{K/min})$ $30 \leq X_2 \leq 100\,(\mu\text{m})$ $0.18 \leq X_3 \leq 0.3\,(\%)$ $0.7 \leq X_4 \leq 1.5\,(\%)$
Bounds on deviation variables
$d_i^-, d_i^+ \geq 0 \text{and } d_i^- * d_i^+ = 0, \text{i} = 1, 2, 3$ (6.42)
Minimize
We minimize the deviation function
$Z = \sum_{i=1}^{3} W_i(d_i^- + d_i^+);\ \sum_{i=1}^{3} W_i = 1$ (6.43)

6.7.1 Solution Space Exploration of Step 2.1 cDSP

The requirement for the process designer in Step 2.1 cDSP is to achieve the goals associated with the mechanical properties of the end rod product. For Goal 1, a process designer is interested in maximizing the yield strength. The target value of 330 MPa is specified in the cDSP. On exercising the cDSP and analyzing the solution space in Fig. 6.9, we see that the red contour region identified by the blue dashed lines satisfy the requirements as closely as possible. The maximum yield strength achieved is 320 MPa and the maximum value achieved by the weight assigned to Goal 1 tends to 1. We select the region identified in Fig. 6.9 as that satisfying the requirement for YS.

Table 6.11 Scenarios with weights for goals

Scenarios	W_1	W_2	W_3
1	1	0	0
2	0	1	0
3	0	0	1
4	0.5	0.5	0
5	0.5	0	0.5
6	0	0.5	0.5
7	0.25	0.75	0
8	0.25	0	0.75
9	0.75	0	0.25
10	0.75	0.25	0
11	0	0.75	0.25
12	0	0.25	0.75
13	0.33	0.34	0.33
14	0.2	0.2	0.6
15	0.4	0.2	0.4
16	0.2	0.4	0.4
17	0.6	0.2	0.2
18	0.4	0.4	0.2
19	0.2	0.6	0.2

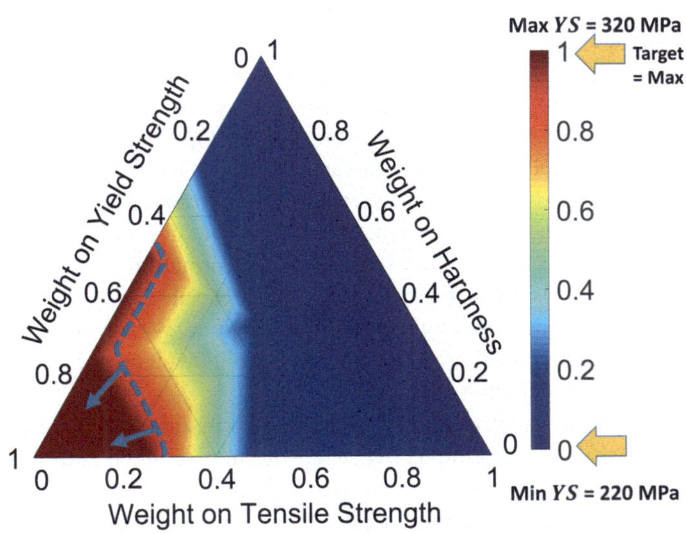

Fig. 6.9 Ternary plot for goal 1—yield strength

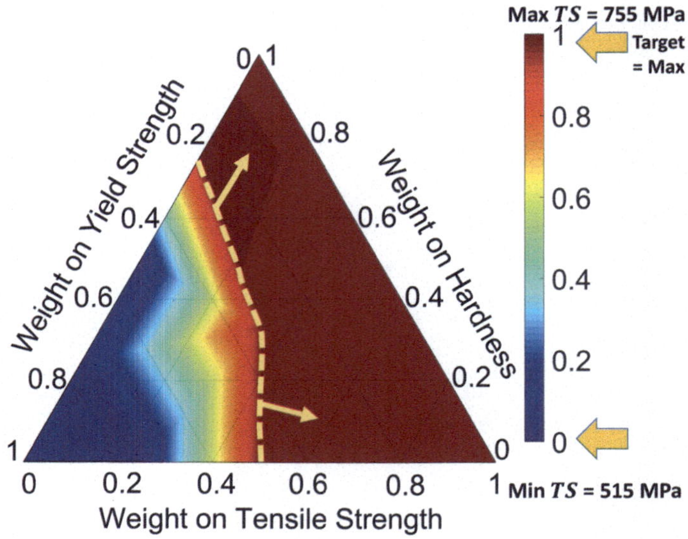

Fig. 6.10 Ternary plot for goal 2—tensile strength

For Goal 2, a process designer is interested in maximizing the tensile strength of the product. A target value of 750 MPa is specified for this goal. On analyzing Fig. 6.10, we observe that the red region marked with the light orange dashed lines satisfies this requirement. The target value of 750 MPa is achieved as we tend to the weight of 1 for the tensile strength goal. However, as the weight on the third goal (hardness) is increased, there is an increase in tensile strength as well. We achieve a value of 750 MPa for tensile strength when the weight on the hardness goal is 1. From this, we can clearly see the forward relationship that hardness and tensile strength holds with respect to the system variables identified.

For Goal 3, the process designer is interested in maximizing hardness. The hardness is a function of the ferrite fraction, silicon content, and transformational temperature of austenite to ferrite. We assumed a transformation temperature of 700 °C. From Fig. 6.11, it is clear that the hardness target value of 170 is achieved in the red contour region marked by the white dashed lines. We also observe that the requirement for hardness is achieved in regions with high weights for tensile strength confirming the relationship that we saw in Fig. 6.10.

On carrying out a parametric study with the transformation temperature value, we found that the positive relationship between tensile strength and hardness holds only at high transformation temperatures. At transformation temperatures in the range of 500–550 °C, we see that hardness tends to be greater where there is high yield strength.

Another requirement that must be strictly satisfied is the requirement for minimum impact transition temperature for the rod. From the solution space for the three goals for YS, TS, and HV, we check the region where this requirement is satisfied the best. In Fig. 6.12, the achieved values of impact transition temperature are plotted and we

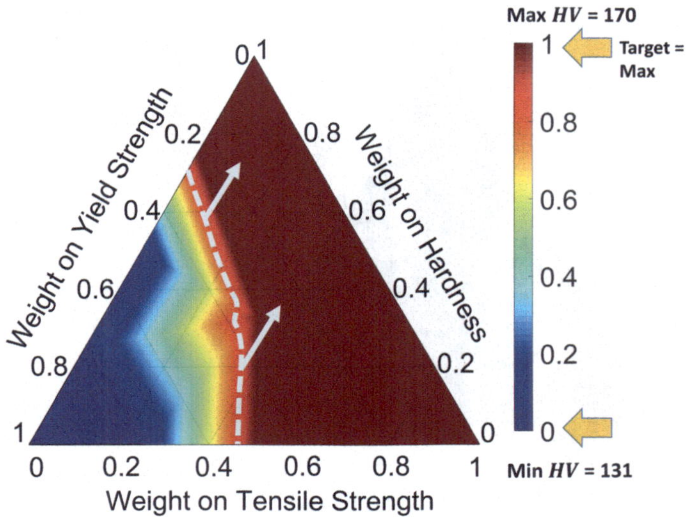

Fig. 6.11 Ternary plot for goal 3—hardness

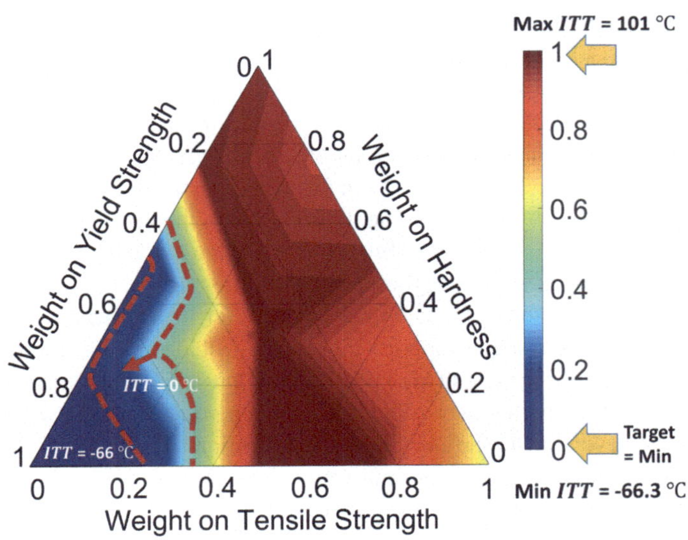

Fig. 6.12 Ternary plot—*IIT* solution space

see that the blue contour region marked by two red dashed lines is where the impact transition temperature is minimum. The first red dashed line corresponds to an ITT of 0 °C and the second dashed line closer to the blue contour region corresponds to an ITT of −66 °C. The minimum ITT is achieved in this region and corresponds to the same region where yield strength is maximized.

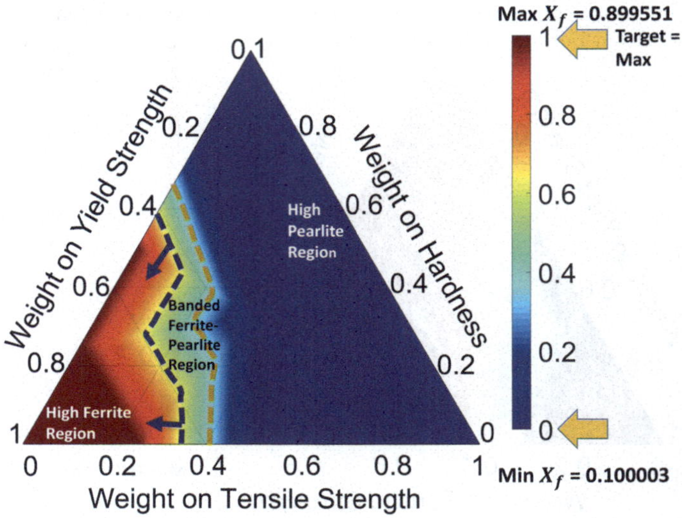

Fig. 6.13 Ternary plot—ferrite fraction solution space

On analyzing the results for the mechanical property goals and requirements, we observe that the ferrite fraction system variable plays a key role in defining the mechanical properties. A major requirement is to manage the banded microstructure. In this work, we satisfy this requirement by identifying regions with high ferrite fractions. Hence, we plot the achieved solution space for ferrite fraction with respect to the weights assigned to the three goals in Fig. 6.13. We see in Fig. 6.13 that the red contour region marked by the dark blue dashed lines is the region with the highest ferrite fraction (near 0.899). The dark blue contour region marked by the dark yellow dashed lines is the region with the highest pearlite fraction (ferrite fraction near to 0.1). The region between these two dashed lines has both ferrite and pearlite. Also, from Fig. 6.13 a high ferrite fraction supports maximizing yield strength and minimizing impact transition temperature and high pearlite fraction supports maximizing tensile strength and maximizing hardness. The banded microstructure in-between satisfies these goals, however, due to the concern about distortions in gear blanks because of these banded structures, the designer must find a region that is either highly ferritic or highly pearlitic in Fig. 6.13. To come to a decision, we superpose plots as shown in Fig. 6.14.

In the superposed plot, all the regions identified for the mechanical property goals, and the other requirements are combined to identify a single region that satisfies all the requirements, if it exists. If such a region doesn't exist, the designer must make tradeoffs among the conflicting goals. On analyzing Fig. 6.14, the requirements for maximizing tensile strength and hardness are achieved in the high pearlite fraction region while the requirements for maximizing yield strength and minimizing impact transition temperature are satisfied at the high ferrite fraction region. Hence the designer is faced with the dilemma of choosing from either the region of high ferrite or high pearlite that satisfies the goals. To make a decision, we first identify some

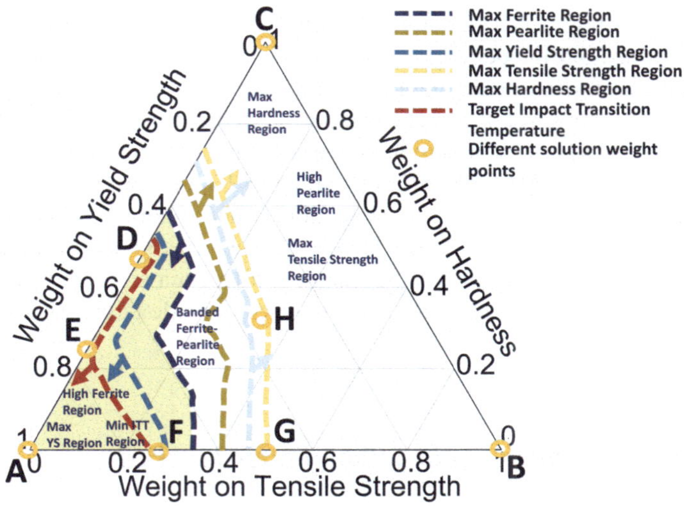

Fig. 6.14 Superposed ternary plot

solution points from the superposed plot and analyze the extent to which the goals are met. We identify eight solution points A, B, C, D, E, F, G, and H from the ternary space and the results associated with each of these solution points are summarized in Table 6.12.

From Table 6.12, we see that all the goals are satisfied by minimum values of ferrite grain size D_α and interlamellar spacing S_o. This is very important information that needs to be communicated to the preceding stages as the requirements from these stages must be to produce a material having these characteristics at the end. On analyzing the impact of ferrite fraction, we see that high yield strength and minimum ITT are satisfied when the ferrite fraction is high, while low yield strength, high ITT, high tensile strength, and high hardness occur when the ferrite fraction is low (more pearlite). As the pearlite fraction increases, the values of ITT achieved are very high (65–100 °C) which is not acceptable. Hence, we identify regions (the light

Table 6.12 Solution points selected

Sol. Pts	Microstructure Factors After Cooling				Mechanical Properties of End Rod			
	D_α μm	X_f	S_0 μm	Mn (%)	YS MPa	TS MPa	HV	ITT °C
A	8	0.9	0.15	1.49	321	516	131	-66
B	8	0.101	0.21	0.7	220	750	169.9	35
C	8	0.1	0.15	0.7	220	749	169.9	94.8
D	8	0.89	0.15	1.5	320	516	131	-66
E	8	0.89	0.15	1.49	320	516	131	-66
F	8	0.89	0.15	1.49	320	516	131	-66
G	8	0.1	0.18	1.5	228	749	169.8	65
H	8	0.113	0.15	1.49	231	749	169.4	100

yellow region in Fig. 6.14) with a high ferrite fraction, where both yield strength and impact transition temperature requirements are met while compromising on the requirements for tensile strength and hardness. From a design standpoint, the compromise does not severely affect either tensile strength and hardness. Therefore, we choose solution point A having the highest ferrite fraction. Point A achieves a YS of 321 MPa, TS of 516 MPa, HV of 131, and ITT of −66 °C.

The solutions for the microstructure space after cooling identified after exploration become the goals for the next cDSP (Step 2.2). The target goals for the cDSP for cooling, therefore, are to achieve a minimum ferrite grain size, maximum ferrite fraction, and minimum pearlite interlamellar spacing; while target values are of 8, 0.9 and 0.15 μm, respectively.

6.7.2 Solution Space Exploration of Step 2.2 cDSP

The requirement in Step 2.2 cDSP is to achieve the targets identified from the first cDSP as closely as possible. For Goal 1, the process designer is interested in minimizing ferrite grain size and the target value is 8 μm. On exercising the cDSP and analyzing the solution space for ferrite grain size in Fig. 6.15, we see that the minimum achieved value of D_α using the current configuration is 10.06 μm. Based on literature study (Kuziak et al. 1997; Pietrzyk et al. 1999), we determine that any value less than 15 μm is acceptable as the ferrite grain size after cooling. This updated requirement is met in the region identified by the red dashed lines in Fig. 6.15. As we move closer to the dark blue contour regions, the requirement for minimum D_α is closest to being satisfied.

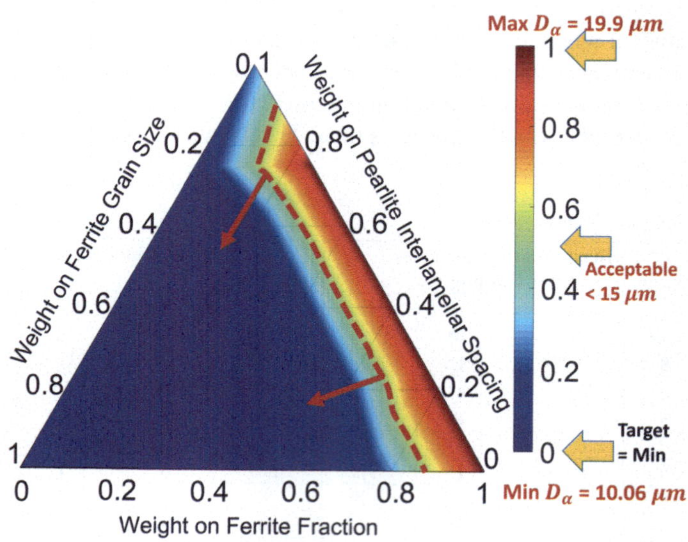

Fig. 6.15 Ternary plot—ferrite grain size

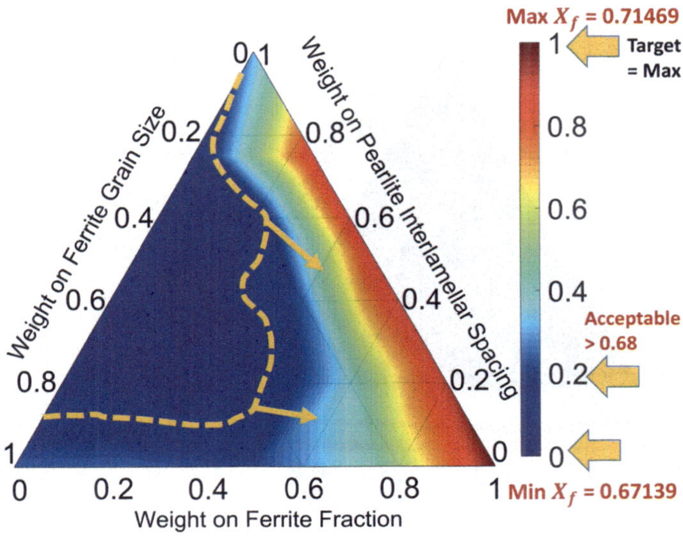

Fig. 6.16 Ternary plot—ferrite fraction

For Goal 2, the process designer must maximize ferrite fraction to the target value of 0.9. In Fig. 6.16, we see that the maximum ferrite fraction achieved is around 0.7149. Based on reported ferrite fractions after cooling from the literature (Kuziak et al. 1997), we find that any value of the ferrite fraction above 0.68 is acceptable.

The region that satisfies the requirement is marked by the dashed orange line in Fig. 6.16. As we move toward the red contour region, the maximum ferrite goal is satisfied most closely.

For Goal 3, the requirement is to minimize pearlite interlamellar spacing to a target value of 0.15 μm. On analyzing Fig. 6.17, the minimum value achieved is 0.1497 marked by the blue contour region. Based on reported values of pearlite interlamellar spacing (Kuziak et al. 1997), we define that any value less than 0.17 μm is acceptable. This region is marked by the dark blue dashed line in Fig. 6.17.

Again, to make a design decision, we superimpose all the goals in one superposed ternary plot (Fig. 6.18).

In the superposed ternary plot, the light yellow region satisfies all the requirements. To analyze further we pick six solution points, both from within the region identified and outside. Solution points C, D, and E lie within the region that satisfies all the goals in the best possible way. Solution points A, B, and F lie outside the region. The results are summarized in Table 6.13.

On analyzing the results in Table 6.13, we see that solution point A satisfies the requirement of minimizing ferrite grain size to the greatest extent and this is achieved with a high cooling rate and low value of austenite grain size. This happens because a high cooling rate results in less time for the nuclei to grow before new nuclei are formed resulting in a decrease of average grain size (Jägle 2007).

This means that there is an increased grain boundary area per volume available for nucleation resulting in more nuclei and thus smaller ferrite grain sizes. Solution point B satisfies the requirement for a high ferrite fraction and this is achieved with a

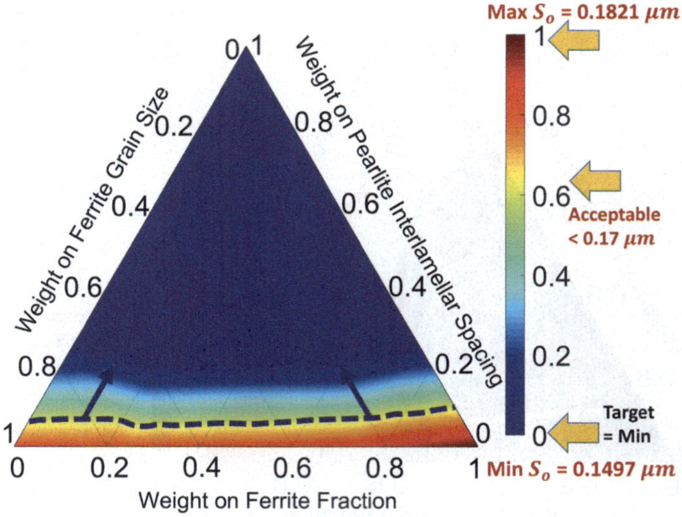

Fig. 6.17 Ternary plot—pearlite interlamellar spacing

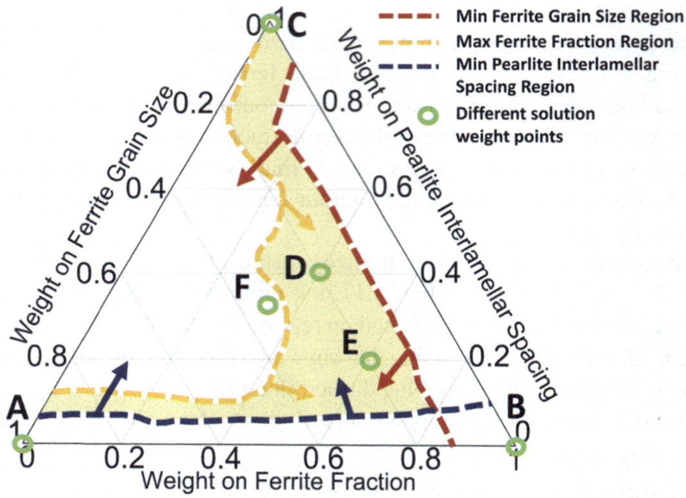

Fig. 6.18 Superposed ternary plot for all goals

low cooling rate, high austenite grain size, and low manganese. This holds true as a low cooling rate favors the growth of allotriomorphic ferrite, resulting in the overall growth of ferrite. A high austenite grain size results in an increase in Widmanstätten ferrite, while a low austenite grain size results in an increase in allotriomorphic ferrite. Both these situations need to be considered when studying the effect of austenite grain size on the ferrite fraction. Also, a low manganese content results in less banded microstructure favoring an increase in allotriomorphic ferrite. Solution

Table 6.13 Solution points selected

Sol. Pts	Processing (Cooling) and Microstructure Space after Rolling				Microstructure Space after Cooling		
	CR K/min	***D*** μm	***C*** %	***Mn*** %	D_α μm	X_f	S_0 μm
A	99.9	30	0.18	0.7	10.06	0.681	0.176
B	11	74.2	0.18	0.7	19.9	0.714	0.182
C	11	30	0.19	1.02	12.5	0.684	0.149
D	44.4	30	0.18	0.94	10.74	0.681	0.151
E	33.06	30	0.18	0.95	11.05	0.687	0.151
F	70.3	30	0.18	0.93	10.33	0.673	0.151

point C satisfies the requirement for minimum pearlite interlamellar spacing and this is achieved with both low values of cooling rate and austenite grain size. On analyzing all solutions listed in Table 6.13, we see that solution point D satisfies all the requirements to the extent possible. In Point D the values of a D_α of 10.74 μm, X_f of 0.681, and S_0 of 0.151 μm are achieved. The values for cooling rate, austenite grain size, and compositions will act as target goals for the cDSP for the last stage of rolling (cDSP in Step 2.3) following a similar format as demonstrated using cDSPs and solution space explorations in Steps 2.1 and 2.2.

6.8 Comprehensive Example Problem 2—Thermo-mechanical Processing Design of a Hot Rolled Rod Using a Goal-Oriented, Inverse Design Method

Careful controlling and managing of the thermal and deformation conditions in a metal forming process to obtain desired microstructures and mechanical properties is termed as thermo-mechanical processing (TMP) (Dieter et al. 2003). The material thermo-mechanical processing (TMP) history carried out at each process influences the final properties and performances of the end product. One major application of TMP in hot rolling is to obtain fine grains of austenite during hot deformation. The fine grains of austenite formed during hot rolling support the formation of fine grains of ferrite after phase transformation during the cooling process that follows. These refined grains of ferrite formed from fine austenite grains support improved mechanical properties of hot rolled products like strength and toughness. Our interest here lies in solving an inverse problem, that is, given the target end mechanical properties and material microstructures that the hot rolled product needs to satisfy, how to inversely design the thermo-mechanical processing during hot rolling so as to identify the austenite grain sizes and the hot rolling processing conditions that meet these end targets. We term this as the inverse thermo-mechanical processing (ITMP) design problem. The specific industry-inspired problem that we are addressing is described next.

The big picture of the industry-inspired problem that we are addressing is described in detail in Chap. 1, Sect. 6.2 of this chapter and also in Nellippallil (2018), and hence is not repeated here. The end goals are to minimize the effects of microsegregates by managing banded microstructures of ferrite and pearlite and achieve specified target values of the mechanical properties like yield strength, tensile strength, hardness, and toughness for the end rod product. Based on our inverse analysis of this problem from a goal-oriented perspective in Sect. 6.7 and Nellippallil et al. (2018a, b), we conclude that to achieve the target mechanical properties and manage banded microstructure, a specific microstructure with a high ferrite fraction, low ferrite grain size, low pearlite interlamellar spacing is preferred after cooling stage. Further, the design of microstructure after rolling is carried out and we conclude that to achieve the specific requirements for a low ferrite grain size and satisfy the mechanical property requirements, a low final austenite grain size is preferred after rolling. In this section, we address the problem dealing with hot rolling thermo-mechanical processing where the end goal is to achieve the minimum final austenite grain size (AGS) after rolling.

In order to predict the final AGS after rolling several phenomena involved in the restoration of the microstructure need to be considered. These include dynamic recovery, dynamic recrystallization, meta-dynamic recrystallization, static recovery, static recrystallization, and grain growth after recrystallization (Kuziak et al. 1997). To frame a boundary for the problem we focus on static recrystallization and grain growth that happens in a hot rolled product in the final rolling pass. We assume that the strain values are less than the critical values for initiating dynamic or meta-dynamic recrystallization in this study. Once the grains are fully recrystallized, grain growth happens and defines the final austenite grain size after rolling. Hence there is a requirement for recrystallized fraction to achieve close to 100% to initiate grain growth with a minimum requirement of at least 95% to satisfy full recrystallization.

Another important factor contributing to the microstructural phenomena during hot rolling is the grain surface to grain volume ratio (S_v). Increase in S_v tends to decrease ferrite grain size and thereby support ferrite refinement (Jägle 2007). The requirement therefore based on literature study is to achieve a target S_v value of 30 mm^{-1} with a desire to explore designs that support higher values of S_v, if possible. In Fig. 6.19, we illustrate the key aspects of the inverse thermo-mechanical processing design problem. Three aspects of the design problem are shown in the figure. In the upper portion of Fig. 6.19, we pictorially represent the hot rod rolling process. We represent the input material to the rolling unit as the billet which is typically the output of a casting unit. However, the input material for our problem could also be the thermo-mechanically processed semi-rod from a previous rolling unit if we focus on just the rolling processes. To simplify and bound our design problem, we focus on just one rolling pass with the assumption that the input is a billet with an initial austenite grain size (AGS) that is further hot rolled to a rod and sent to the cooling

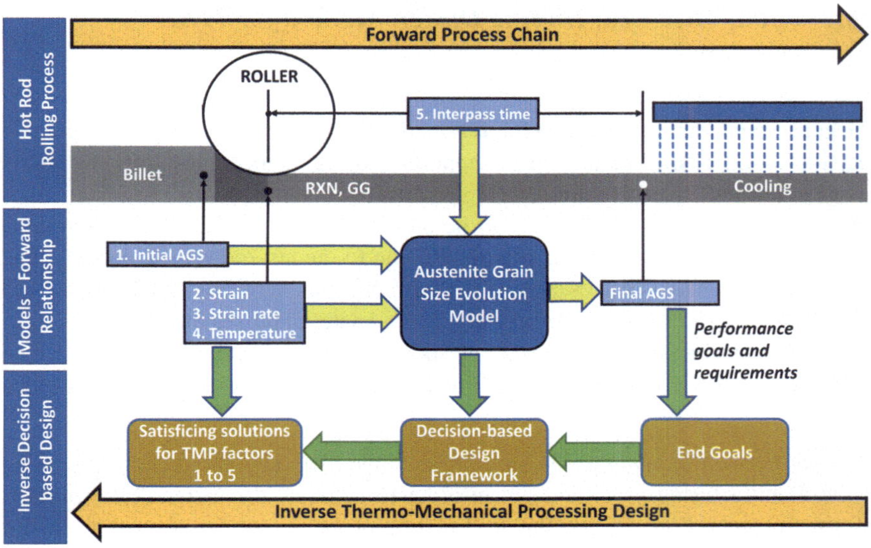

Fig. 6.19 Inverse thermo-mechanical processing (ITMP) design problem

unit where phase transformation of austenite to ferrite, pearlite, and so on occurs. The key thermo-mechanical processing factors (TMP factors) influencing AGS evolution and recrystallization are identified as strain, strain rate, initial AGS, temperature, and interpass time (Hodgson and Gibbs 1992). Mathematical models that establish relationships between the identified factors and the problem-specific requirements for final AGS, recrystallized fraction, and grain surface to volume ratio are identified. Using these models and their integration, specific problem-dependent information is generated and passed across the hot rolling process chain—defined as vertical and horizontal integration of models to establish forward material workflow (see Nellippallil et al. 2018a, b). We represent the forward model-based material workflow in the middle section of Fig. 6.19. Based on the integrated forward models and problem-specific information, we formulate the problem statement to be addressed in this paper. The *inverse thermo-mechanical processing (ITMP)* design problem statement is as follows: *Given the required microstructural targets and models to establish microstructural evolution, what should the input thermo-mechanical and microstructural parameters be for the model-based realization of the material (steel), product (rod), and the manufacturing processes (multistage hot rod rolling)?* Our goal therefore is to present a design method that supports a designer/decision-maker in inversely identifying satisficing solutions for the thermo-mechanical processing factors strain, strain rate, temperature, initial austenite grain size, and interpass time while meeting the conflicting microstructure and property/performance requirements set across each stage of the multistage hot rod rolling process chain, pictorially represented in the lower section of Fig. 6.19.

6.8.1 Goal-Oriented, Inverse Design Method for ITMP Design

The goal-oriented, inverse design (GoID) method as presented in Chap. 4 is a generic decision-based design method that supports a designer in identifying satisficing design solutions across a manufacturing process chain meeting specific design goals and requirements set at each stage of the process chain. In Fig. 6.20, we show the schematic of the GoID method with application to the hot rod rolling process. The rolling process chain is represented in the form of separate rolling stands (Stand 1 to Stand n) to produce the final rod product. As the material passes through each stand, it is thermo-mechanically processed to a specific geometry and microstructure. The GoID method applied to the hot rod rolling problem involves two phases as described below.

Phase 1: Forward modeling. The first phase in the method is the forward modeling of the manufacturing process chain in order to generate the material and product information workflow. For the hot rod rolling problem, we focus on three major aspects—hot deformation of the material during rolling, the recrystallization of austenite grains, and the grain growth that follows recrystallization (see Fig. 6.20). In Phase 1, model and simulation-based analysis is carried out to generate information workflow regarding the material and product across all the stages of rolling process (rolling stands). In the problem discussed, Phase 1 is realized by developing a finite-element-based hot rod rolling model to predict the stress, strain, strain rate, and temperature evolution as the material passes through the rolling stands. The microstructural phenomena are predicted by using an austenite grain size evolution model that predicts the recrystallized fraction, austenite grain size, and grain growth, depending on the rolling conditions and parameters. On completion of Phase 1 across the manufacturing process chain, we are able to develop the model-based information

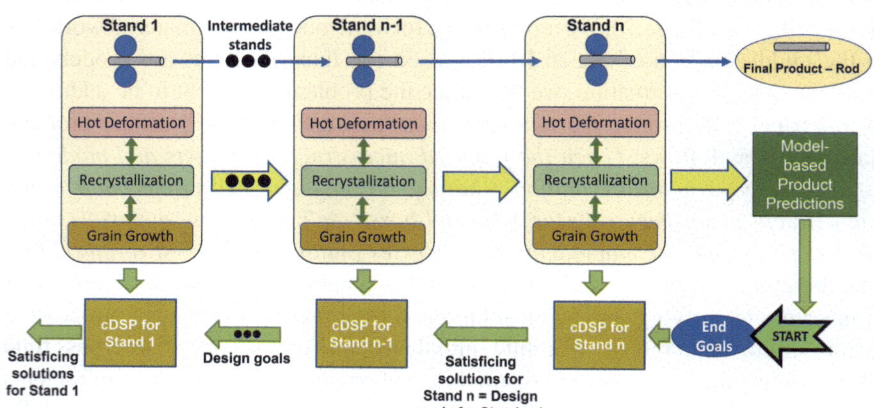

Fig. 6.20 Goal-oriented, inverse design method for hot rod rolling problem

workflow for the final product produced, illustrated in Fig. 6.20. The forward modeling phase for the hot rod rolling inverse thermo-mechanical processing problem is discussed in detail in Sect. 6.8.2.

Phase 2: Goal-Oriented Inverse Design. In this phase the design method is the process of carrying out decision-based design exploration across the manufacturing process chain starting from the end goals and requirements defined for the problem. In the lower section of Fig. 6.20, we show a schematic representation of the design exploration process. The cDSP for the rolling stand n is formulated by incorporating specific microstructural and property/performance goals for the rod produced. The cDSP is formulated and explored following the procedural steps of the CEF discussed in Chap. 4. The mathematical models identified in Phase 1 for forward modeling are used to formulate the design problem in Phase 2. For the problem discussed, we look at three end goals, namely minimize final austenite grain size, maximize static recrystallized fraction, and achieve target value for grain surface to volume ratio with a desire to explore larger values if possible. On solving the cDSP for stand n and exploring the solution space, the designer is able to identify the combinations of the input factors, such as initial austenite grain size, strain, strain rate, temperature, and interpass time, that satisfice the conflicting goals present. These combinations are then passed as goals for the next cDSP for Stand n − 1. The whole process of problem formulation and solution space exploration is repeated for rolling Stand n − 1 and satisficing solutions are identified. The process is continued until the entire problem boundary is covered depending on how this is defined with respect to the manufacturing process chain. The method is generic and can be used to design similar manufacturing process chains where there is sequential information sharing. The method supports design information coordination and goal-oriented decision support for a complex system involving sequential flow of information/energy/material.

6.8.2 Mathematical Models for Thermo-mechanical Processing

In this section, we introduce the mathematical models used to formulate the hot rod rolling thermo-mechanical processing problem. These models are used to establish the forward modeling and material information workflow for the problem. First, we discuss the finite-element model for the hot deformation during rolling process. This is followed by a discussion on the static recrystallization, grain growth, and other microstructure-specific models for the problem discussed.

Finite-Element Hot Rod Rolling Model

To predict the austenite grain size evolution, the thermo-mechanical processing information in the form of strain, strain rate, and temperature is needed. We gather this information using a finite-element (FE) hot rod rolling model; the details of this model are discussed briefly here. A four-stand coupled temperature—displacement FE hot rod rolling model is developed in ABAQUS. A simulated initial billet of

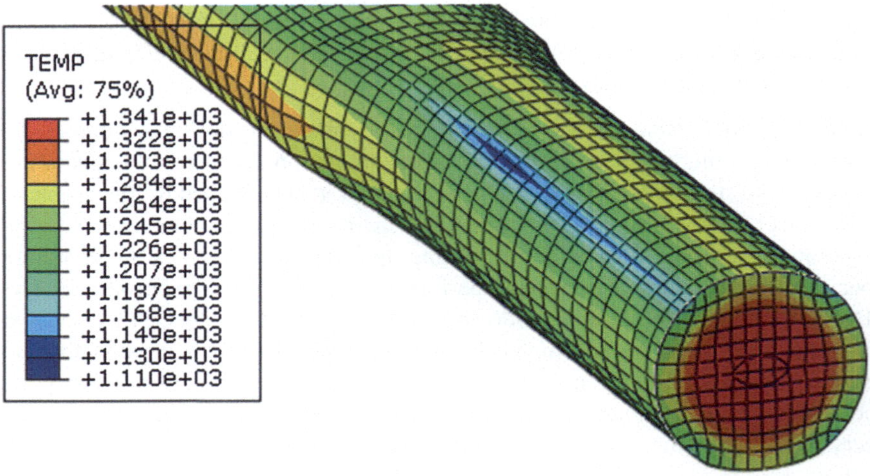

Fig. 6.21 Finite-element hot rod rolling model

square cross-section is passed through rollers and it is sequentially deformed to oval and circular cross-sections to finally produce a rod of definite diameter. The billet is modeled as a deformable body using the thermally coupled brick element C3D8RT. Material properties for steel as a function of temperature are assigned to the block. Yield stress values for steel at different plastic strains are used to model the material's plastic deformation behavior. Analytical models are used to model the rollers and the surface profile of roller grooves (Kwon et al. 2003). The rollers are modeled and meshed as discrete right bodies with R3D4 elements. The billet is assigned an initial temperature before rolling. Further details about the hot rod rolling model are available in (Nellippallil et al. 2017e). In this study, we focus only on the last rolling pass. In Fig. 6.21, we show the temperature contours for the rod after the last rolling pass. The strain, strain rate, and temperature information from the last rolling pass from the FE model are used for subsequent analysis in an austenite grain size (AGS) evolution model, discussed next.

Microstructure Evolution Models

During hot rolling, recrystallization occurs which is basically formation of new grains from old grains. Recrystallization is typically followed by grain growth where grains coarsen and arrange themselves in a uniform manner. Modeling recrystallization and grain growth have been studied over the years and several researchers have developed empirical relationships to capture these phenomena; see Pietrzyk et al. (1999) for a comprehensive list. In this formulation, we use the AGS evolution model proposed by Hodgson and Gibbs (1992). This model is used to consider static and meta-dynamic recrystallization, and grain growth. In the design study carried out, we assume that the strain (ε) is less than the critical strain (ε_c) for dynamic or meta-dynamic recrystallization to be initiated.

Table 6.14 Models specific to microstructural evolution for the hot rolling problem

Model	Equation	References
Critical strain	$\varepsilon_c = 5.6 * 10^{-4} * D_0^{0.3} * Z^{0.17}$	Hodgson and Gibbs (1992)
Static recrystallized fraction	$X = 1 - \exp\left[-0.693\left(\frac{t_{ip}}{t_{0.5}}\right)\right]$	
Time for 50% recrystallization	$t_{0.5} = 2.3 * 10^{15} * \varepsilon^{-2.5} * D_0^2 * \exp\left(\frac{230{,}000}{RT}\right)$ R is the universal gas constant	
Zener–Hollomon parameter	$Z = \dot{\varepsilon} * \exp\left(\frac{300{,}000}{RT}\right)$	
Static recrystallized grain size	$D_{SRX} = 343 * \varepsilon^{-0.5} * D_0^{0.4} \exp\left(-\frac{45{,}000}{RT}\right)$	
Grain growth	$D_{GG}^7 = D_{SRX}^7 + 1.5 * 10^{27}\left(t_{ip} - 4.32 * t_{0.5}\right) * \exp\left(-\frac{400{,}000}{RT}\right)$	
Grain surface to volume ratio	$S_v = \frac{24\varepsilon}{\pi D_0}\left[0.491 \exp(\varepsilon) + 0.155 \exp(-\varepsilon) + 0.1433 \exp(-3\varepsilon)\right]$	Suehiro et al. (1987)

Hence, we use only the static recrystallization and grain growth models proposed by Hodgson and Gibbs. The grain surface area to volume ratio is defined using the empirical model developed by Suehiro et al. (1987). In Table 6.14, we include the models used to predict the microstructural phenomenon during hot rod rolling process from the problem perspective. These are standard empirical models applicable to hot rolling of steels. For details on the applicability of these models, see Pietrzyk et al. (1999). The procedure for calculating the final austenite grain size value using the AGS evolution model developed by Hodgson and Gibbs is provided by Kwon et al. (2003) and is not repeated here. These microstructural models are dependent on the independent TMP variables of the design study, namely, initial austenite grain size (D_0), strain (ε), strain rate ($\dot{\varepsilon}$), temperature (T), and interpass time (t_{ip}). Further variables are defined in the cDSP. Given these models and the information flow chain, we are able to formulate the cDSPs starting from end goals following the GoID method. In Sect. 6.8.3, we discuss the cDSP formulated for the last rolling Stand n and the associated solution space exploration.

6.8.3 *cDSP Formulation for Roll Stand n*

Once the models are identified and integrated to establish the forward information workflow, the designer starts the inverse decision-based design exploration starting from end goals. The cDSPs are formulated starting from the last rolling pass. The formulation and exploration procedures are as described using CEF. In this section, we discuss the cDSP for the last rolling stand formulated to achieve the end product goals for the rod. The product goals for final austenite grain size, static recrystallized fraction, and grain surface to volume ratio are defined as a function of the thermo-mechanical processing factors: initial austenite grain size, strain, temperature, interpass time, and strain rate. The requirement for minimizing final AGS is defined as a minimization goal for the austenite grain size after grain growth. The requirement for achieving 100% static recrystallization is defined as a maximization goal with a constraint defined to ensure 95% of the grains are recrystallized to initiate grain growth. The grain growth constraint is defined to ensure that the grain size after grain growth is always greater than the grain after static recrystallization. A critical strain constraint is defined to ensure that the critical strain is always greater than the strain in the material deformed. This is defined to ensure the non-initiation of dynamic and meta-dynamic recrystallization in the material. An interpass time constraint is defined to ensure that there is sufficient time for recrystallization and grain growth to happen. The third goal for surface to volume ratio is defined in the minimization format to achieve a target value of 0.03 μm^{-1}. However, the desire is to explore larger values of this goal if such solutions exist in the solution space generated. On exercising this cDSP and carrying out solution space exploration, the designer is able to inversely identify the combination of thermo-mechanical processing (TMP) factors that best satisfices the end goals set for the material, product, and process. The TMP factors thus identified are communicated as goals and requirements to the cDSP for the rolling stand that precedes, and the process is repeated to cover the problem boundary defined using the goal-oriented, inverse design method proposed. The cDSP reads as follows.

Given:

(1) End requirements identified for the rod rolling process

- Minimize final austenite grain size (D_{GG})
- Maximize recrystallized grain fraction (X)
- Achieve target grain surface area to volume ratio (S_v)
- Target value for austenite grain size, $D_{GG} = 25\ \mu m$
- Target value for recrystallized grain fraction, $X = 1.0$
- Target value for grain surface area to volume ratio, $S_v = 0.03\ \mu m^{-1}$
- Minimum limit of recrystallized grain fraction, $X = 0.95$
- Maximum limit of recrystallized grain fraction, $X = 1.0$
- Grain surface area to volume ratio requirement, S_v: Explore whether larger values are possible

(2) The empirical correlations of microstructural evolutions

- Critical strain to decide whether SRX or MDRX will take place

$$\varepsilon_c = 5.6 * 10^{-4} * D_0^{0.3} * Z^{0.17} \tag{6.44}$$

- Zener–Hollomon parameter

$$Z = \dot{\varepsilon} * \exp\left(\frac{300{,}000}{RT}\right) \tag{6.45}$$

- Recrystallized grain fraction

$$X = 1 - \exp\left[-0.693\left(\frac{t_{ip}}{t_{0.5}}\right)\right] \tag{6.46}$$

- Time taken for half of the grain to recrystallize

$$t_{0.5} = 2.3 * 10^{15} * \varepsilon^{-2.5} * D_0^2 * \exp\left(\frac{230{,}000}{RT}\right) \tag{6.47}$$

- Grain size after recrystallization

$$D_{SRX} = 343 * \varepsilon^{-0.5} * D_0^{0.4} \exp\left(-\frac{45{,}000}{RT}\right) \tag{6.48}$$

- Grain size after grain growth

$$D_{GG}^7 = D_{SRX}^7 + 1.5 * 10^{27}\left(t_{ip} - 4.32 * t_{0.5}\right) * \exp\left(-\frac{400{,}000}{RT}\right) \tag{6.49}$$

- Grain surface area to volume ratio

$$S_v = \frac{24\varepsilon}{\pi D_0}\left[0.491\exp(\varepsilon) + 0.155\exp(-\varepsilon) + 0.1433\exp(-3\varepsilon)\right] \tag{6.50}$$

(3) Variability in system variables

The system variables are provided in the Find and Satisfy section.

Find:

System Variables

X_1 initial austenite grain size before the billet is rolled (D_0)—Range: 40–100 μm
X_2 strain when the rod passes under the roller (ε)—Range: 0.03–0.5
X_3 temperature during the rolling process (T)—Range: 1173–1473 K
X_4 interpass time (t_{ip})—Range: 3–13 s

X_5 strain rate ($\dot{\varepsilon}$)—Range: 5–10 s^{-1}

Satisfy:
System Constraints

- Critical strain constraint

$$\varepsilon_c - \varepsilon \geq 0 \tag{6.51}$$

- Grain growth constraint

$$D_{GG} - D_{REX} \geq 0 \tag{6.52}$$

- Interpass time constraint

$$t_{ip} - 4.32 * t_{0.5} \geq 0 \tag{6.53}$$

- Minimum recrystallized fraction constraint

$$X - 0.95 \geq 0 \tag{6.54}$$

- Maximum recrystallized fraction constraint

$$1.0 - X \geq 0 \tag{6.55}$$

System Goals
Goal 1:

- Minimize austenite grain size after recrystallization and grain growth

$$\frac{D_{GG,\text{Target}}}{D_{GG}(X_i)} - d_1^- + d_1^+ = 1 \tag{6.56}$$

Goal 2:

- Maximize recrystallized grain fraction

$$\frac{X(X_i)}{X_{\text{Target}}} + d_2^- - d_2^+ = 1 \tag{6.57}$$

Goal 3:

- Achieve grain surface to volume ratio target

$$\frac{S_{v,\text{Target}}}{S_v(X_i)} - d_3^- + d_3^+ = 1 \tag{6.58}$$

Variable Bounds

X_1 – 40 to 100 μm
X_2 – 0.03 to 0.5
X_3 – 1173 to 1473 K
X_4 – 3 to 13 s
X_5 – 5 to 10 s^{-1}

Bounds on deviation variables

$$d_i^-, d_i^+ \geq 0 \text{ and } d_i^- * d_i^+ = 0, i = 1, 2, 3 \tag{6.59}$$

Minimize:
We minimize the deviation function. Represented as DF in this formulation.

$$DF = \sum_{i=1}^{3} W_i(d_i^- + d_i^+); \sum_{i=1}^{3} W_i = 1 \tag{6.60}$$

From the cDSP, we explore the solution space by exercising the cDSP for different design scenarios, discussed next in Sect. 6.8.4.

6.8.4 *Solution Space Exploration for Thermo-mechanical Processing Using GoID*

We select 19 different design scenarios (DS) based on judgement to capture and visualize the solution space for exploration, same as shown in Table 6.11. We exercise the cDSP for each of these scenarios and develop ternary spaces for exploration. Next, we discuss the solution space exploration for each goal defined for the hot rod rolling problem. The ternary space for Goal 1 is shown in Fig. 6.22. For Goal 1, the interest of the process designer is to achieve a minimum final austenite grain size after rolling. A minimum austenite grain size supports the formation of fine ferrite grains that further improves the mechanical properties of the rod after cooling. A target value of 25 μm is defined for this goal. Based on our previous study in Nellippallil et al. (2018a, b), we have identified a value of 30 μm to be acceptable to satisfy the requirements from cooling stage after rolling. The solution region with final AGS value less than 30 μm is identified using the red dashes lines in Fig. 6.22. We see in Fig. 6.22 that the solution space thus identified lie mostly in the region with higher weight preferences assigned to Goal 1, weights between 0.5 and 1. Any combination of weights on goals from this region satisfies the AGS goal.

For Goal 2, see Fig. 6.23, the interest of the process designer is to maximize the static recrystallized fraction to 100%. A minimum recrystallized fraction of 95% is

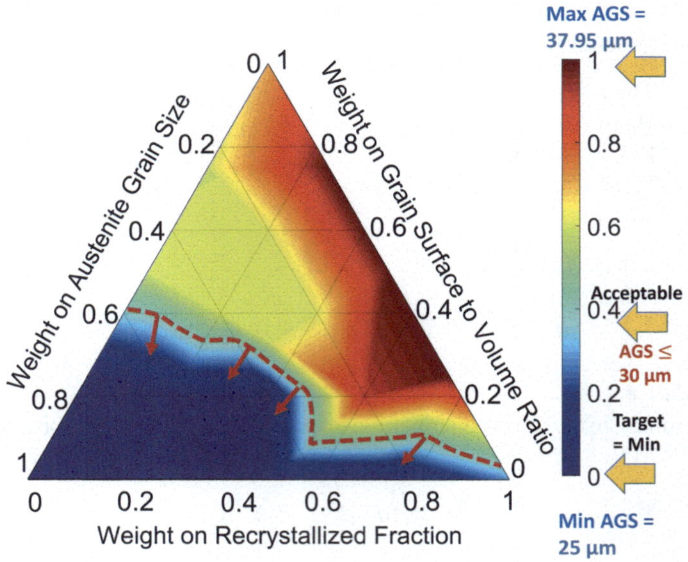

Fig. 6.22 Ternary plot for goal 1—minimize austenite grain size

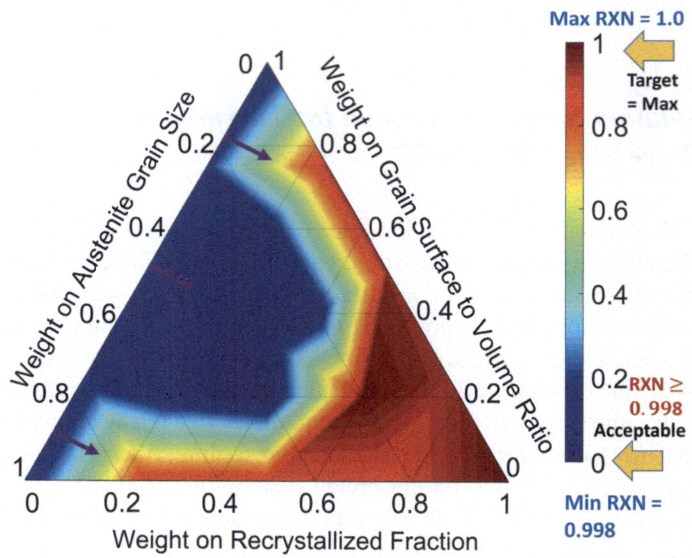

Fig. 6.23 Ternary plot for goal 2—maximize recrystallized fraction

required and this requirement is defined as a constraint in the cDSP. On analyzing the ternary space shown in Fig. 6.23, we observe that the red contour region satisfies the requirement for 100% recrystallized fraction. All the other solution regions satisfy a recrystallized fraction of 99.8% and above. Since these high values of the

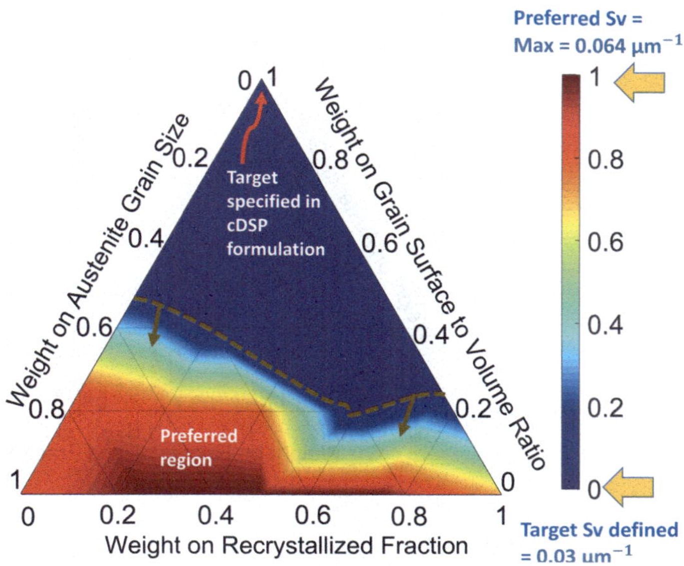

Fig. 6.24 Ternary plot for goal 3—achieve grain surface to volume ratio target

recrystallized fraction are satisfied, all the solution points are acceptable to us. This also ensures that grain growth phenomena will happen following recrystallization and therefore Goal 1 is valid from the problem perspective. We, therefore, accept all the solution points as shown using the violet arrows on the color contour in Fig. 6.23.

For Goal 3, see Fig. 6.24, the interest of the process designer is to achieve a target value of 0.03 μm^{-1} for the grain surface to volume ratio goal. On analyzing Fig. 6.24 we see that higher values for this goal are achieved in regions where the austenite grain size is minimized (red color region in Fig. 6.24). Achieving higher values of S_v is preferred as it supports the minimization of ferrite grain size and its refinement. Also, the need to minimize grain coarsening is supported by larger values of S_v. For these reasons the regions with larger values of S_v is preferred and is identified using the dashed brown lines in Fig. 6.24.

To make a decision and identify a satisficing region for all the three goals, we superpose the ternary regions (Fig. 6.25). The colored region satisfices all three goals in the best possible manner. The design scenarios (DS) listed in Table 6.11 that lie in this superposed region are in Table 6.15.

On evaluating the design solutions in Table 6.15, we observe that all the solutions are relatively close and produce responses that are relatively insensitive to changes. Based on the analysis of the results, we observe the following. The final austenite grain size tends to decrease when the initial austenite grain size is low. Thus, one requirement that can be placed on previous rolling stages is to minimize the austenite grain sizes of the rod at each rolling pass. An increase in grain surface to volume ratio is observed at those regions with low initial austenite grain size. An increased grain surface to volume ratio increases the number of nucleation sites resulting in

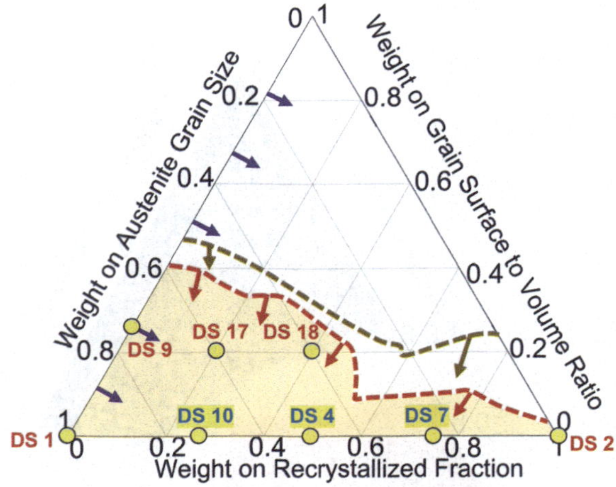

Fig. 6.25 Superposed ternary plot

Table 6.15 Evaluation of solutions

DS	Thermo-Mechanical Processing (TMP) Factors					Goals Defined for Hot Rolled Rod		
	D_0 μm	ε	T K	t_{ip} s	$\dot{\varepsilon}$ s^{-1}	D_{GG} μm	X	S_v μm^{-1}
1	40	0.353	1173	3.824	10	25	0.998	0.057
2	50	0.4	1173	8	10	28.57	1.0	0.053
4	40	0.382	1173	3.792	10	25	0.999	0.064
7	40	0.382	1173	3.792	10	25	0.999	0.064
9	40	0.352	1173	3.862	10	25	0.998	0.057
10	40	0.382	1173	3.792	10	25	0.999	0.064
17	40	0.352	1173	3.826	10	25	0.998	0.057
18	40	0.347	1173	3.948	10	25.16	0.997	0.056

more nuclei and thereby decreasing the ferrite grain sizes. This in turn improves the final mechanical properties of the rod produced after rolling and subsequent cooling processes.

From the solution points, we pick solution points from design scenarios 4, 7, and 10 that best satisfice the goals defined. The associated thermo-mechanical processing (TMP) factor values are identified and selected to achieve the system-level goals set for the problem. These include an initial austenite grain size value of 40 μm strain of 0.382, temperature of 1173 K, interpass time of 3.792 s, and strain rate of 10 s^{-1}. The generic method proposed can be used to design the previous rolling stands to meet the requirements set by the cDSP explored and the identified TMP factors.

6.8.5 Remarks Based on the Testing Carried Out Using TMP Example

Using the second comprehensive example, we present the utility of the goal-oriented, inverse design (GoID) method to achieve the integrated design exploration of materials, products, and associated manufacturing processes. Given the forward material (information) workflow using models, the proposed method can be used to support the goal-oriented inverse design exploration of material microstructures and processing paths to meet specific end properties and performances desired for the material and product. We demonstrate the utility of the method using an industry-inspired hot rolling example problem to inversely design the thermo-mechanical processing of a steel rod. The key functionality tested using this example problem is the designer's ability to carry out a microstructure-mediated thermo-mechanical processing design satisficing specific performance goals of the material (steel) and the end product (rod). The method and associated design constructs are generic and have the potential to be applied to coordinate information flow and human decision-making across processes/levels in order to realize an end goal—a key functionality that allows the application of this method to other complex system problems involving the sequential model-based flow of information across processes/levels.

6.9 On Verification and Validation

Empirical Structural Validation

Empirical structural validation involves accepting the appropriateness of the example problems used to verify the performance of the framework and the method. Using the comprehensive example problem in this chapter, the utility of the framework and the method is tested for the integrated design exploration of materials, products, and manufacturing processes. In this chapter, the design architecture in terms of Research Questions 1 and 2 developed in the monograph are applied to a multiscale, multistage materials design problem—vertical and horizontal integration and integrated design of hot rod rolling process chain, steel, and rolled rod. In this chapter, the industry-inspired problem of focus in this monograph is addressed. The bigger picture of the problem of interest and how integrated materials, product, and process design can be applied at an industrial scale is discussed in the beginning. A discussion on the specific problem (vertical and horizontal integration of the hot rolling process chain) is carried out in detail. A literature review on the hot rod rolling process is carried out. The problem—impact of segregation along the rolling to the forging process chain is discussed in detail. The problem is to design the material microstructure and processing paths to satisfy conflicting product and process-related end performances and properties in an inverse manner. The problem is modeled as an integrated design of materials, products, and manufacturing processes. In addition to the validation of design methods, the chapter is also crucial from the standpoint of the major theme addressed in this monograph. In this chapter, we discuss the validation of the proposed systematic method of model integration, inverse design method, and concept exploration framework.

Empirical Performance Validation

Empirical performance validation consists of accepting the usefulness of the outcome with respect to the initial purpose and accepting that the achieved usefulness is related to applying the framework and method. The utility of the proposed method is demonstrated by carrying out the integrated solution space exploration of the processing and microstructure spaces of the rolling and cooling processes to identify satisficing solutions that realize the end mechanical properties of the rod product in this chapter.

In this chapter, we present a goal-oriented, inverse decision-based design method supported by the concept exploration framework (CEF) to achieve the vertical and horizontal integration of models for the hot rolling and cooling stages of the steel manufacturing process chain for the production of a rod with defined properties. The method is goal-oriented and inverse because we start with end mechanical properties of the product and inversely maps the requirements to microstructure and processing spaces of the product as well as the process to identify multiple solutions that satisfy the requirements. The utility of the proposed method is demonstrated by carrying out the integrated solution space exploration of the processing and microstructure spaces of the rolling and cooling processes to identify *satisficing solutions* that help realize the end mechanical properties of the rod product. The primary advantage of the proposed method is in empowering a process designer to rapidly explore the design space for manufacturing processes using simulation models by managing the uncertainty associated with models. We believe that the ability to predict the design and operating setpoints using models reduces the need for expensive plant trials resulting in reduced time and cost involved in the production of a new grade of steel product mix with improved properties using a new class of material. The proposed method is generic and supports the integrated decision-based design of other manufacturing stages that are connected and have a sequential flow of information by identifying the design and operating setpoints that best satisfice the requirements identified. Through the proposed method and demonstration carried out in this chapter using an industry-driven problem, we propose an approach for *microstructure-mediated design* by *integrating the design of the material, product, and associated manufacturing processes* involved.

The method and its application are characterized by a confluence of different disciplines like engineering mechanics, materials science, manufacturing, and systems engineering. The functionalities offered by the method supported by CEF as illustrated using the comprehensive example problem include (selected from Chap. 4 and proved based on the testing done using the comprehensive example problems in this chapter):

- Requirements-driven, "top-down" design of system and associated subsystems by taking a goal-oriented approach,
- Human perception of a satisficing design space across process chains,
- Augmenting the human ability to make design decisions—visualizing a solution space and making logical judgments through tradeoffs to identify satisficing solution regions of interest,
- Capability to handling "n" number of design variables,
- Propagation of end goal requirements (product performance or properties) across a process chain,
- Offers flexibility in design: The capability to define new goals and requirements at each level as the method uses individual cDSPs to facilitate information flow allowing to formulate a design space at each level,
- The capability to carry out rapid, integrated design exploration of material and products using simulation models that we accept are typically incomplete and inaccurate,
- The capability to coordinate information and human decision-making,
- The CEF offers the capability to prioritize models, input factors, output responses, and computational tools in terms of their value in design, and
- Ensuring feasible design solutions that allow investing in new complex material systems with confidence.

The proposed method and the concept exploration framework are generic and support the integrated decision-based design of similar manufacturing processes involving the material and product. Given any complex systems problem that involves the sequential flow of information across processes/levels, the proposed method has the potential to be applied to support information flow and human decision-making across the processes/levels in order to realize an end goal. The verification and validation aspects are shown in Fig. 6.26.

6.10 Role of This Chapter and Connection with Other Chapters in This Monograph

In this chapter, the systems-based design architecture developed is applied to a multiscale, multistage materials design problem—vertical and horizontal integration and integrated design of hot rod rolling process chain, steel, and rolled rod. In this chapter, the industry-inspired problem of focus in this monograph is addressed. The bigger picture of the problem of interest and how integrated materials, product, and process design can be applied at an industrial scale is discussed in the beginning. A discussion on the specific problem (vertical and horizontal integration of the hot rolling process chain) is carried out in detail. A literature review on the hot rod rolling process is carried out. The problem—impact of segregation along the rolling to the forging process chain is discussed in detail. The problem is to design the material microstructure and processing paths to satisfy conflicting product and process-related

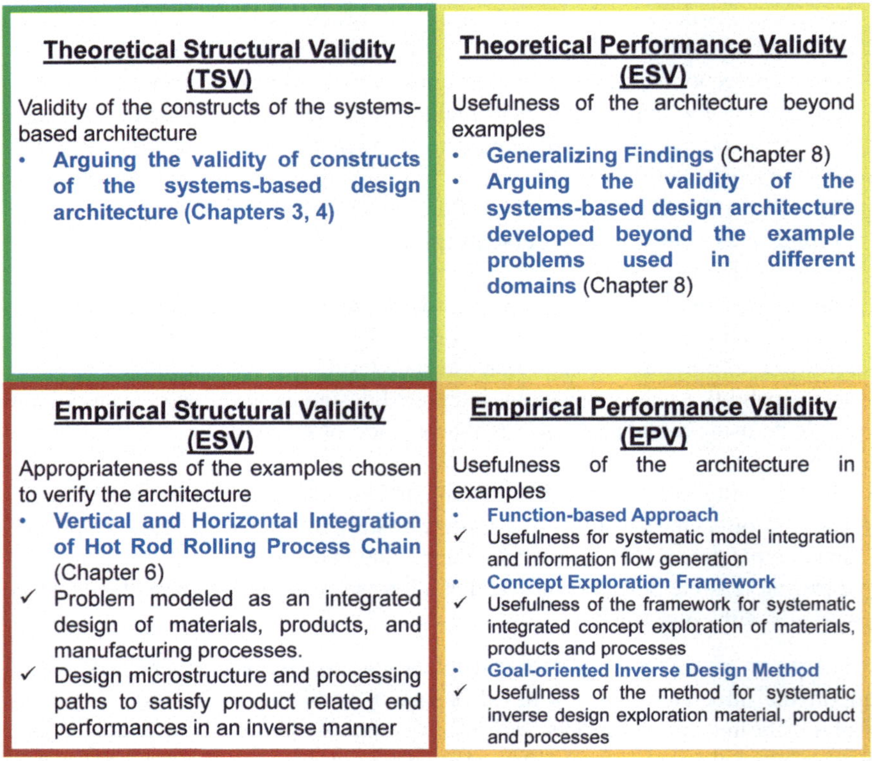

Fig. 6.26 Verification and validation aspects discussed in this chapter

end performances and properties in an inverse manner. The problem is modeled as an integrated design of materials, products, and manufacturing processes. In addition to the validation of design methods, the chapter is also crucial from the standpoint of the major theme addressed in this monograph. In this chapter, we discuss the validation of the proposed systematic method of systematic model integration (Chap. 4), concept exploration framework, and goal-oriented inverse design method (Chap. 4). Further, the problem discussed in this chapter is reformulated and utilized to demonstrate robust concept exploration of materials, products, and processes in Chap. 7. In Fig. 6.27, the validation tasks for which the example problem in this chapter is used are shown along with the connectivity with other chapters of this monograph. Interested readers are directed to the following articles for more information on the comprehensive example problems discussed in this chapter, see Nellippallil et al. (2017b, 2018a, b, 2019).

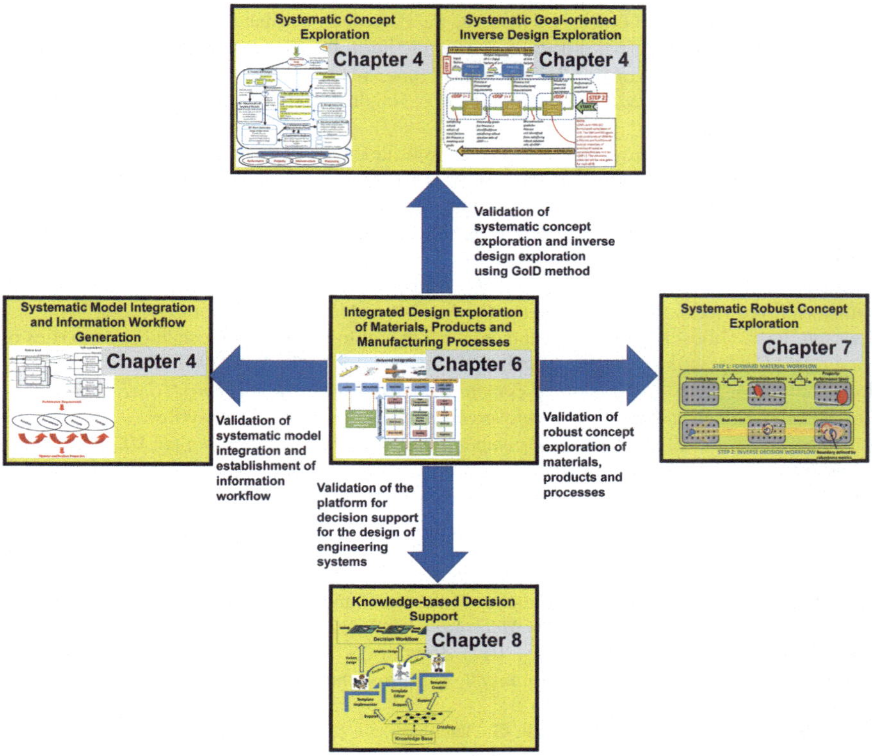

Fig. 6.27 In this chapter and connections with other chapters in this monograph

References

Adams, B. L., Kalidindi, S., & Fullwood, D. T. (2013). *Microstructure-sensitive design for performance optimization*. Butterworth-Heinemann.

Allen, J. K., Panchal, J., Mistree, F., Singh, A. K., & Gautham, B. (2015). Uncertainty management in the integrated realization of materials and components. In *Proceedings of the 3rd World Congress on Integrated Computational Materials Engineering (ICME)*. Wiley.

Bodnar, R., & Hansen, S. (1994). Effects of austenite grain size and cooling rate on Widmanstätten ferrite formation in low-alloy steels. *Metallurgical and Materials Transactions A, 25*(4), 665–675.

Bras, B., & Mistree, F. (1993). Robust design using compromise decision support problems. *Engineering Optimization, 21*(3), 213–239.

Chen, W., Allen, J. K., & Mistree, F. (1997). A robust concept exploration method for enhancing productivity in concurrent systems design. *Concurrent Engineering, 5*(3), 203–217.

Choi, H.-J., Mcdowell, D. L., Allen, J. K., & Mistree, F. (2008a). An inductive design exploration method for hierarchical systems design under uncertainty. *Engineering Optimization, 40*(4), 287–307.

Choi, H., McDowell, D. L., Allen, J. K., Rosen, D., & Mistree, F. (2008b). An inductive design exploration method for robust multiscale materials design. *Journal of Mechanical Design, 130*(3), 031402.

Choquet, P., Fabregue, P., Giusti, J., Chamont, B., Pezant, J., & Blanchet, F. (1990). Modelling of forces, structure, and final properties during the hot rolling process on the hot strip mill. In *Mathematical Modelling of Hot Rolling of Steel* (pp. 34–43).
Dieter, G. E., Kuhn, H. A., & Semiatin, S. L. (2003). Handbook of workability and process design. *ASM International.*
Donnay, B., Herman, J., Leroy, V., Lotter, U., Grossterlinden, R., & Pircher, H. (1996). Microstructure evolution of C-Mn steels in the hot-deformation process: The stripcam model. In *2nd International Conference on Modelling of Metal Rolling Processes.*
Gladman, T., McIvor, I., & Pickering, F. (1972). Some aspects of the structure-property relationships in high-C ferrite-pearlite steels. *Journal of the Iron and Steel Institute, 210*(12), 916–930.
Gladman, T., Dulieu, D., & McIvor, I. D. (1977). Structure/property relationships in high-strength micro-alloyed steels. *Proceedings of the Conference on Microalloying, 75*(1977), 32–55.
Hodgson, P., & Gibbs, R. (1992). A mathematical model to predict the mechanical properties of hot rolled C-Mn and microalloyed steels. *ISIJ International, 32*(12), 1329–1338.
Horstemeyer, M. F. (2012). *Integrated computational materials engineering (ICME) for metals: using multiscale modeling to invigorate engineering design with science*. Wiley.
Horstemeyer, M. F. (2018). *Integrated computational materials engineering (ICME) for metals: concepts and case studies*. Wiley.
Jägle, E. (2007). Modelling of microstructural banding during transformations in steel'. *Department of Materials Science & Metallurgy.*
Jones, S., & Bhadeshia, H. (1997a). Competitive formation of inter-and intragranularly nucleated ferrite. *Metallurgical and Materials Transactions A, 28*(10), 2005–2013.
Jones, S., & Bhadeshia, H. (1997b). Kinetics of the simultaneous decomposition of austenite into several transformation products. *Acta Materialia, 45*(7), 2911–2920.
Jones, S. J., Bhadeshia, H. K. D. H. (2017). Program STRUCTURE on the Materials Algorithm Project web site. Retrieved February 4, 2017, from http://www.msm.cam.ac.uk/map/steel/programs/structure.html.
Kalidindi, S. R., Niezgoda, S. R., Landi, G., Vachhani, S., & Fast, T. (2010). A novel framework for building materials knowledge systems. *Computers, Materials, & Continua, 17*(2), 103–125.
Kalidindi, S. R., Niezgoda, S. R., & Salem, A. A. (2011). Microstructure informatics using higher-order statistics and efficient data-mining protocols. *JOM Journal of the Minerals Metals and Materials Society, 63*(4), 34–41.
Kejian, H., & Baker, T. (1993). The effects of small titanium additions on the mechanical properties and the microstructures of controlled rolled niobium-bearing HSLA plate steels. *Materials Science and Engineering A, 169*(1–2), 53–65.
Kern, P. C., Priddy, M. W., Ellis, B. D., & McDowell, D. L. (2017). pyDEM: A generalized implementation of the inductive design exploration method. *Materials & Design, 134*, 293–300.
Kuziak, R., Cheng, Y.-W., Glowacki, M., & Pietrzyk, M. (1997). Modeling of the microstructure and mechanical properties of steels during thermomechanical processing. *NIST Technical Note(USA), 1393*, 72.
Kwon, H.-C., Lee, Y., Kim, S.-Y., Woo, J.-S., & Im, Y.-T. (2003). Numerical prediction of austenite grain size in round-oval-round bar rolling. *ISIJ International, 43*(5), 676–683.
LeBon, A., & deSaint-Martin, L. (1977). Using laboratory simulations to improve rolling schedules and equipment. *Proceedings of the Conference on Microalloying, 75*(1977), 90–99.
Majta, J., Kuziak, R., Pietrzyk, M., & Krzton, H. (1996). Use of the computer simulation to predict mechanical properties of C-Mn steel, after thermomechanical processing. *Journal of Materials Processing Technology, 60*(1–4), 581–588.
McDowell, D. L. (2018). Microstructure-sensitive computational structure-property relations in materials design. *Computational materials system design* (pp. 1–25). Cham: Springer.
McDowell, D. L., & Olson, G. (2008). Concurrent design of hierarchical materials and structures. *Scientific modeling and simulations* (pp. 207–240). Dordrecht: Springer.
McDowell, D. L., & Kalidindi, S. R. (2016). The materials innovation ecosystem: a key enabler for the materials genome initiative. *MRS Bulletin, 41*(4), 326–337.

McDowell, D. L., Choi, H. J., Panchal, J., Austin, R., Allen, J., & Mistree, F. (2007). Plasticity-related microstructure-property relations for materials design. In *Key engineering materials*. Trans Tech Publ.

McDowell, D. L., Panchal, J., Choi, H.-J., Seepersad, C., Allen, J., & Mistree, F. (2009). *Integrated design of multiscale, multifunctional materials and products*. Butterworth-Heinemann.

Mistree, F., Hughes, O. F., & Bras, B. (1993) Compromise decision support problem and the adaptive linear programming algorithm. *Progress in Astronautics and Aeronautics, 150*, 251.

Mistree, F., Patel, B., & Vadde, S. (1994). On modeling multiple objectives and multi-level decisions in concurrent design. *Advances in Design Automation, 69*(2), 151–161.

Nellippallil, A. B. (2018). *The integrated realization of materials, products and associated manufacturing processes*. Ph.D. Dissertation, School of Aerospace and Mechanical Engineering, University of Oklahoma, Norman.

Nellippallil, A. B., Song, K. N., Goh, C.-H., Zagade, P., Gautham, B., Allen, J. K., & Mistree, F. (2016). A goal oriented, sequential process design of a multi-stage hot rod rolling system. *ASME Design Automation Conference*, Paper Number: DETC2016-59402.

Nellippallil, A. B., De, P. S., Gupta, A., Goyal, S., & Singh, A. K. (2017a). Hot rolling of a non-heat treatable aluminum alloy: Thermo-mechanical and microstructure evolution model. *Transactions of the Indian Institute of Metals, 70*(5), 1387–1398.

Nellippallil, A. B., Song, K. N., Goh, C.-H., Zagade, P., Gautham, B., Allen, J. K., & Mistree, F. (2017b). A goal-oriented, sequential, inverse design method for the horizontal integration of a multi-stage hot rod rolling system. *Journal of Mechanical Design, 139*(3), 031403.

Nellippallil, A. B., Allen, J. K., Mistree, F., Vignesh, R., Gautham, B. P., & Singh, A. K. (2017c). A goal-oriented, inverse decision-based design method to achieve the vertical and horizontal integration of models in a hot-rod rolling process chain. *ASME Design Automation Conference*, Paper Number DETC2017-67570.

Nellippallil, A. B., Rangaraj, V., Allen, J. K., Mistree, F., Gautham, B., & Singh, A. K. (2017d). A decision-based design method to explore the solution space for microstructure after cooling stage to realize the end mechanical properties of hot rolled product. In *Proceedings of the 4th World Congress on Integrated Computational Materials Engineering (ICME 2017)*. Springer.

Nellippallil, A. B., Shukla, R., Ardham, S., Goh, C.-H., Allen, J. K., & Mistree, F. (2017e). Exploration of solution space to study thermo-mechanical behavior of AA5083 Al-alloy during hot rolling process. *ASME Design Automation Conference*, Paper Number: DETC2017-68173.

Nellippallil, A. B., Mohan, P., Allen, J. K., & Mistree, F. (2018a). Robust concept exploration of materials, products and associated manufacturing processes. *ASME Design Automation Conference*, Paper Number: DETC2018-85913.

Nellippallil, A. B., Rangaraj, V., Gautham, B., Singh, A. K., Allen, J. K., & Mistree, F. (2018b). An inverse, decision-based design method for integrated design exploration of materials, products, and manufacturing processes. *Journal of Mechanical Design, 140*(11), 111403-111403-17.

Nellippallil, A. B., Mohan, P., Allen, J. K., & Mistree, F. (2019). Inverse thermo-mechanical processing (ITMP) design of a steel rod during hot rolling process. *ASME Design Automation Conference 2019*, Paper Number: IDETC2019-97390.

Olson, G. B. (1997). Computational design of hierarchically structured materials. *Science, 277*(5330), 1237–1242.

Phadke, S., Pauskar, P., & Shivpuri, R. (2004). Computational modeling of phase transformations and mechanical properties during the cooling of hot rolled rod. *Journal of Materials Processing Technology, 150*(1), 107–115.

Pietrzyk, M., Cser, L., & Lenard, J. (1999). *Mathematical and physical simulation of the properties of hot rolled products*. Elsevier.

Roberts, W., Sandberg, A., Siwecki, T., & Werlefors, T. (1983). Prediction of microstructure development during recrystallization hot rolling of Ti–V steels. In *HSLA Steels, Technology and Applications* (pp. 67–84).

Robson, J., & Bhadeshia, H. (1997). Modelling precipitation sequences in power plant steels part 1–kinetic theory. *Materials Science and Technology, 13*(8), 631–639.

Sellars, C., & Beynon, J. (1984). *High strength low alloy steels*. Australasian Institute of Metals, Wollongong, Australia.

Senuma, T., Suehiro, M., & Yada, H. (1992). Mathematical models for predicting microstructural evolution and mechanical properties of hot strips. *ISIJ International, 32*(3), 423–432.

Shukla, R., Goyal, S., Singh, A. K., Panchal, J. H., Allen, J. K., & Mistree, F. (2015). Design exploration for determining the set points of continuous casting operation: An industrial application. *Journal of Manufacturing Science and Engineering, 137*(3), 034503.

Suehiro, M., Sato, K., Tsukano, Y., Yada, H., Senuma, T., & Matsumura, Y. (1987). Computer modeling of microstructural change and strength of low carbon steel in hot strip rolling. *Transactions of the Iron and Steel Institute of Japan, 27*(6), 439–445.

Taguchi, G. (1986). Introduction to quality engineering. Asian Productivity Organization, Distributed by the American Supplier Institute, Inc., Dearborn, MI.

Tennyson, G., Shukla, R., Mangal, S., Sachi, S., & Singh, A. K. (2015). ICME for process scale-up: Importance of vertical and horizontal integration of models. In *Proceedings of the 3rd World Congress on Integrated Computational Materials Engineering (ICME 2015)*. Wiley.

Vander Voort, G. (2015). The interlamellar spacing of pearlite. *Practical Metallography, 52*(8), 419–436.

Yada, H. (1987). Prediction of microstructural changes and mechanical properties in hot strip rolling. In *Accelerated Cooling of Rolled Steel* (pp. 105–119).

Chapter 7
Robust Concept Exploration of Materials, Products, and Associated Manufacturing Processes

7.1 Frame of Reference—Establishing Context from Chaps. 1–3

Many have highlighted the challenges associated with the top-down (goal-oriented/inverse) approach of materials design (see McDowell et al. 2009; Horstemeyer, 2012, 2018; McDowell 2018). Among these are the challenges arising due to (i) uncertain material models (that include input factors, parameters, responses, etc.) due to simplification/idealization or a lack of complete knowledge, (ii) the propagation of uncertainty from different sources as information flows from one to model to another, and (iii) the lack of domain-independent design "exploration" methods and tools for the early stages of design. An effective inverse approach for materials design must address the uncertainty of models and experiments at each scale, as well as uncertainty propagation through a chain of models and/or experiments at different levels of hierarchy with the ability to provide decision support through rapid design space exploration (McDowell 2018). We address each of these challenges next.

The sources of uncertainty could be (McDowell et al. 2009): the natural uncertainty inherent in a system, the uncertainty associated with model parameters, the uncertainty inherent in models, and the propagation of all these information flows from one model to another. Two approaches are typically followed in dealing with these sources of uncertainty—(i) mitigating uncertainty and (ii) managing uncertainty. In the first approach, the focus is to reduce/mitigate the uncertainty by seeking "perfect" models, collecting more data, developing improved methods to model, calculate, and quantify uncertainty through expensive computations. There are several recent works in this vein using Bayesian networks (see Shahan and Seepersad 2012; Li and Mahadevan 2016; Matthews et al. 2016; Mullins and Mahadevan 2016), modern data science methods, and microstructure informatics (see Panchal et al. 2013; Kalidindi 2015; McDowell and Kalidindi 2016; McDowell and LeSar 2016). However, McDowell (2018) observes that quantifying uncertainty in schemes for linking models at different length and time scale is still an immature field and formal mathematical approaches for doing this are largely undeveloped. This demands the

A. B. Nellippallil et al., *Architecting Robust Co-Design of Materials, Products, and Manufacturing Processes*, https://doi.org/10.1007/978-3-030-45324-4_7

need for the second approach of managing uncertainty by designing the system to be insensitive to the sources without reducing or eliminating them. This is done by exploring the solution space and studying the sensitivity of responses to variations in noise, control factors, and models and understanding the tradeoffs required with various compromises; defined as robust design and classified into mainly three categories (McDowell et al. 2009): (i) Type I robust design where designers seek to design insensitive to variation due to noise factors, (ii) Type II robust design where designers seek to design insensitive to variation due to control factors, and (iii) Type III robust design where designers seek to design insensitive to variation due to uncertainty inherent in models; see Chap. 3 for detailed discussion on robust design.

The focus in robust design is not on extensive optimization searches at individual levels and do not necessarily involve a large number of iterations (McDowell and Olson 2008). The practical interest here is for a ranged set of satisficing solutions that showcase good performance under variability rather than single-point solutions that are valid for a narrow range of conditions while performing poorly when the conditions are changed slightly. The term "satisficing" used in this context is coined by Simon (2013) and is foundational to our approach to decision-based design. As part of the "*satisficing*" community, our intention in this chapter from the design side is to satisfice (satisfy and suffice) a set of conflicting goals and explore robust solutions, defined as "*satisficing robust solutions*", in an inverse manner by managing uncertainty.

Given the requirement of managing uncertainty, we next address the issue of inverse design exploration using domain-independent design methods and tools. Materials design using inverse methods is gaining a lot of attention in the recent past. There are several recent efforts toward this. Adams et al. (2013) propose a framework and utilize highly efficient spectral representations to arrive at invertible linkages between material structure, its properties, and their corresponding processing paths. Kalidindi et al. (2010, 2011) propose the material knowledge systems approach and showcase the advances in rapid inverse design to estimate local responses. These inverse design approaches, however, demand considerable knowledge and insight in mechanisms, material hierarchy, and information flow and are mostly suited for detailed design and not for "design exploration" (McDowell 2018). We recognize that approaches to pursue inverse design exploration using systems-based design especially at the early stages of design are limited and need further evaluations to address hierarchical material design problems with the consideration of robustness. Choi et al. (2005) propose the compromise decision support problem formulation with error margin index (EMI) for Type I, II, and III robust design. However, their work is limited as they address only: (i) a single-objective case and not when multiple design objectives/goals are present, (ii) a single robust design formulation using EMI when all sources of uncertainty classified exist in a problem formulation, which is typically not the case when multiple goals exist. The inductive design exploration method (IDEM) proposed by Choi et al. (2008) is another multilevel, top-down, robust design exploration method that considers the propagation of model parameter and model structure uncertainty across process chains. However, there are certain limitations to IDEM. These include limitations in terms of: (i) errors due to discretization of design space resulting in the inability to capture the feasible boundary accurately that leads to loss of information, (ii) highly computationally expensive IDEM runs if accuracy is increased, (iii) the number of design variables (impossible beyond 9 variables) that can be studied thereby limiting the problem size, (iv) exploration and visualization as IDEM involves a three-dimensional visualization space using HD-EMI metric for exploration where only a maximum of three design variables can be studied at a time with the others variables taking defined values, and (v) flexibility in design as IDEM do not allow designers to incorporate new goals or requirements at different levels during the design process as the method is based on mapping to feasible spaces. Nellippallil et al. (2018b) propose an inverse, decision-based design method known as goal-oriented, inverse design (GoID) addressing some of these limitations, addressed in Chaps. 4–6. The GoID method offers: (i) an increase in problem size as there are no limitations in the number of design variables and goals to be studied, (ii) improved flexibility in the design of the various processes involved as individual design spaces are formulated at each level allowing designers to incorporate new design goals and requirements, and (iii) the capability to visualize and explore "satisficing" solutions for multiple conflicting goals. However, the idea of robustness under uncertainty is not captured in the GoID in its current form as defined in Chaps. 4–6 (Nellippallil et al. 2018b) and is, therefore, a limitation.

In this chapter, we extend the existing goal-oriented inverse design method proposed in Chap. 4 (Nellippallil et al. 2018b) to bring in robustness for multiple goals from the standpoint of Type I to III robust design across process chains. The extension embodies the introduction of specific robust design goals, constraints, and metrics to determine "*satisficing robust design*" specifications for given performance requirement ranges using the goal-oriented, inverse design method. The primary mathematical construct used in the extended inverse method is the compromise decision support problem with error margin index (Choi et al. 2005) and design capability index (Chen et al. 1999) (cDSP with EMI-DCI) supported by the concept exploration framework (CEF) to generate satisficing Type I, II, and III robust design solutions across process chains. We hypothesize that the EMIs and DCIs, when used together in search algorithms, are capable of helping the designer in designing the system robust to both model parameter and model structure uncertainty. Contributions in this chapter include the designer's ability to explore "*satisficing robust solution*" regions when multiple conflicting goals and multiple sources of uncertainty are present. The extended method and robust design constructs are suitable for problems for which clearly defined analytical models/functions are either available or can be developed to capture the problem-specific information. In this chapter, an industry-inspired example problem, namely the design of a hot rolling and cooling process chain for the production of a steel rod is used as an example to demonstrate the efficacy of the method and the associated robust design constructs for the inverse robust design under uncertainty of the material, product, and manufacturing processes. In Sect. 7.2, we describe the concept exploration framework (CEF) and the cDSP-EMI-DCI construct for robust concept exploration. In Table 7.1, we summarize the architecture constituents discussed and functionalities tested in this chapter.

Table 7.1 Summary of the elements discussed, reviewed, and tested in this chapter that form part of the architecture

Elements	*What?*
Processing elements	The functionalities of CEF and GoID—carry out robust concept exploration of process chains by managing uncertainty
Data elements	Manufacturing process chain problem with different sources of uncertainty
Connecting elements	Robust design constructs EMI and DCI integrated to the CEF and GoID
Form	*How?*
Weights	Testing the utility of CEF and GoID for robust concept exploration using a test example
Properties	Focus on robust design by managing uncertainty using the hot rolling manufacturing process chain problem
Relationship	Theoretical and empirical performance validation using the test example
Rationale	*Why?*
Motivation	To validate the utility of CEF and GoID functionalities in providing robust design decision support in process chains
Characteristics/ believes	• Analysis models are incomplete, inaccurate, and are of differing fidelity. Do not eliminate uncertainty, but manage it • Decisions can be formulated mathematically, and analysis models are available for design decision-making • Design process starts with an end goal in mind—could be an end product that needs to be supplied to customers • There exist models or computational tools or data that can establish the forward information workflow for the process/product that is being considered • Designer's interest is in identifying satisficing robust solutions given the uncertainty in the problem and not single-point optimum solutions • Design is a goal-oriented activity with design requirements subject to change at any time for the problems considered
Interpretation	Understanding the efficacy of the robust design constructs and broader impacts

7.2 The Concept Exploration Framework for Type I, II, and III Robust Design

The concept exploration framework (CEF) is a mathematical framework that includes systematic steps to generate design alternatives by exploring the solution space and identify satisficing design specifications. We recognize that a framework that supports robust concept exploration in integrated material, product, and process design should satisfy three requirements: (i) computational efficiency, (ii) generic enough to be applicable to various levels of material design hierarchy, and (iii) incorporation of Type I, II, and III robust design formulations for multiple conflicting goals. In this chapter, we update the CEF to include the compromise decision support problem with error margin index and design capability index together in a single formulation to take into account complex material and product design problems that require combinations of Type I, II, and III robust designs.

In Fig. 7.1, we show the modified CEF with the incorporation of robust design goals and constraints in the cDSP using the EMIs and DCIs. The systematic steps associated with the CEF to generate satisficing design specifications remain the same as defined in (Nellippallil et al. 2018b) and are not repeated here. The eight processors (A, B1, B2, D, E, F, G, and H) and simulation programs (C) defined in the CEF support systematic problem formulation, systematic solution space exploration, and thereby provide decision support for any complex systems problem. In this chapter,

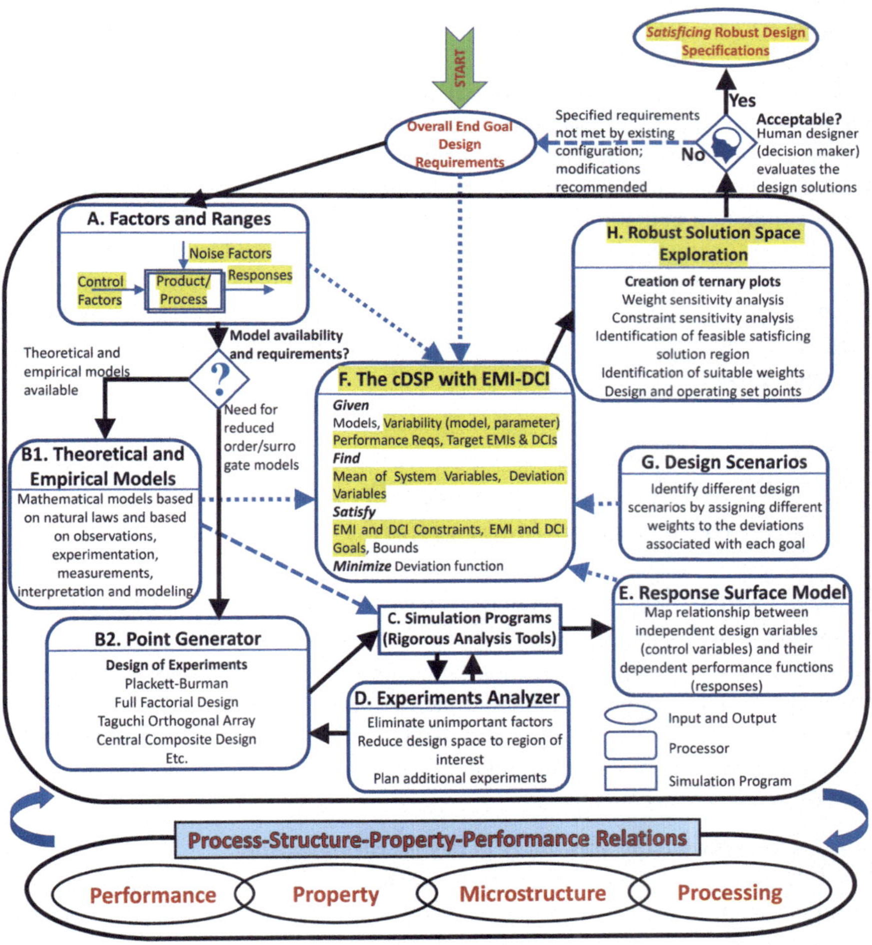

Fig. 7.1 The modified (highlighted in yellow) concept exploration framework for Type I, II, and III robust design

we address robust concept exploration for instantiating Type I, II, and III robust designs and therefore focus on processors A, F, and H of the CEF shown in Fig. 7.1. The formulation of a cDSP with EMI and DCI using the CEF involves: (a) quantification of variability and model parameter uncertainty, (b) formulation of error margin indices and design capability indices and incorporating them in the cDSP, and (c) robust decision-making by exploration of solution space by executing the cDSP with EMI-DCI. Choi et al. (2005) explain in detail on quantifying variability and model parameter uncertainty. They use a response modeling approach for quantifying response variability due to parameterizable noise factors and location and dispersion modeling approach for quantifying unparameterizable variability. However, we observe that for problems related to complex manufacturing processes involving materials and products like hot rolling and cooling, several studies are already carried out and different models defining material/process behavior are available in the literature (Gladman et al. 1972, 1977; Yada 1987; Majta et al. 1996; Kuziak et al. 1997; Pietrzyk et al. 1999; Phadke et al. 2004). These models are either based on natural laws or based on experiments/modeling. Such available theoretical and empirical models when directly used to formulate the cDSP do not require the response modeling-based approach followed by Choi et al. as the variability can be assessed directly using the functional relations. In the approach proposed using the CEF, one major assumption is that models (mean models and their variabilities) in the form of analytical functions are either available or can be developed by carefully planned design experiments. One major limitation therefore of the proposed approach is its inapplicability when models in the form of analytical equations are not available. Next, we discuss the formulation of EMI and DCI using functional relations and information available.

7.2.1 Formulation of Design Capability Indices (DCIs) and Error Margin Indices (EMIs)

DCIs and EMIs are metrics for system performance and robustness. DCIs represent the amount of safety margin against system failure due to uncertainty in the system variables while EMIs represent the margin against failure due to uncertainty in both model and design variables. Both are dimensionless. The EMIs support Type I, II, and III robust designs while DCIs support Type I and II robust designs. We briefly describe the steps involved in formulating and calculating DCIs and EMIs for two types of systems, respectively.

DCIs for systems having variability in design variables only
Step 1: Using the first-order Taylor series expansion, estimate the response variation due to variation in the design variable vector $\mathbf{x} = \{x_1, x_2, \ldots, x_n\}$. The response variation (ΔY) for small variations in design variables is

$$\Delta Y = \sum_{i=1}^{n} \left| \frac{\partial f}{\partial x_i} \right| \cdot \Delta x_i \tag{7.1}$$

Step 2: Using the mean response (μ_y) obtained from the mean response model $(f_0(x))$ and the response variation due to variation in design variables (ΔY), calculate the DCIs. For a "Larger is Better" case, the DCI is calculated as

$$\text{DCI} = \frac{\mu_y - LRL}{\Delta Y} \tag{7.2}$$

where LRL is the lower requirement limit. A DCI ≥ 1 means that the ranged set of design specifications satisfies a ranged set of design requirements and the system is robust against model parameter uncertainty. Higher the value of DCI, higher is the measure of safety against failure due to model parameter uncertainty.

EMIs for systems having variability in both models and design variables
Step 1: Assuming a system model has k uncertainty bounds, calculate the response variation (ΔY_j) for each of them for a small variation in design variables as

$$\Delta Y_j = \sum_{i=1}^{n} \left| \frac{\partial f_j}{\partial x_i} \right| \Delta x_i \tag{7.3}$$

where $j = 0, 1, 2, \ldots, k$ (number of uncertainty bounds).

In Fig. 7.2a, we show a mean response model (solid red curve) with two uncertainty bounds (the dotted curves). In the left side of Fig. 7.2a, we show the response variations of mean function and uncertainty bound functions with respect to the variations in design variables.
Step 2: After evaluating the multiple response variations of mean response function and the k uncertainty bound functions for variations in design variables, calculate the minimum and maximum responses by considering the variability in design variables and uncertainty bounds around the mean response as

$$Y_{\max} = \text{Max}\left[f_j(x) + \Delta Y_j\right] \tag{7.4}$$

and

$$Y_{\min} = \text{Min}\left[f_j(x) - \Delta Y_j\right] \tag{7.5}$$

where $j = 0, 1, 2, \ldots, k$ (number of uncertainty bounds), $f_0(x)$ is the mean response function, and $f_1(x) \ldots . f_k(x)$ are the uncertainty bound functions
Step 3: Calculate the upper and lower deviation of response at x as

$$\Delta Y_{\text{upper}} = Y_{\max} - f_o(x) \tag{7.6}$$

and

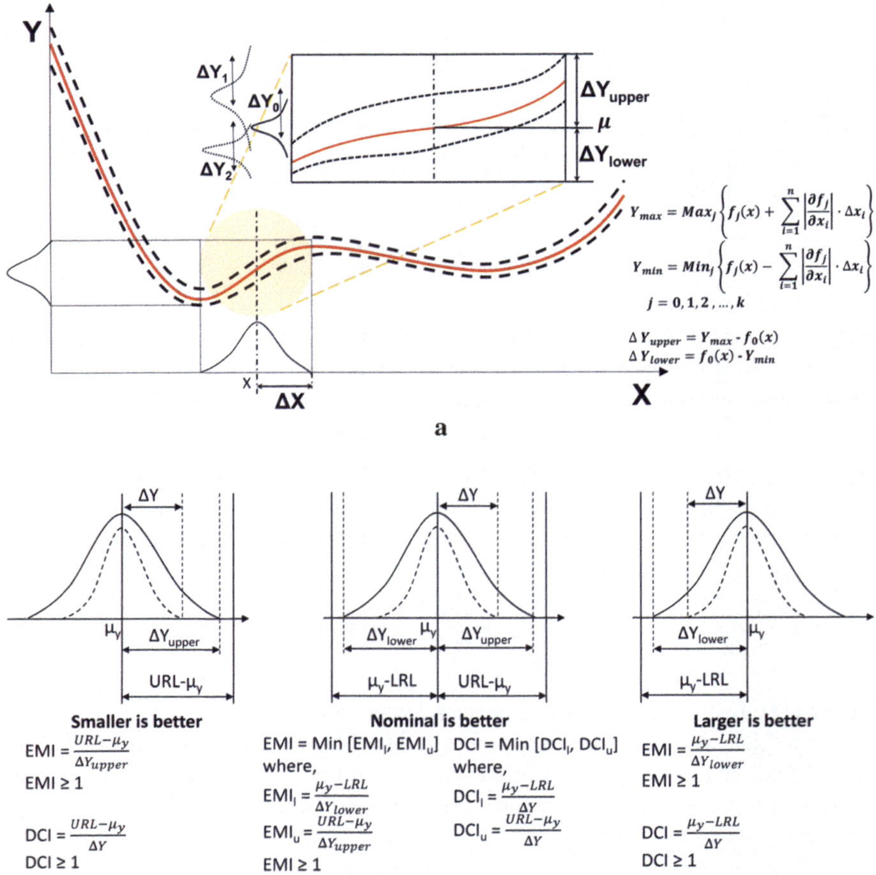

Fig. 7.2 **a** Uncertainty bound formulation for variability in design variable and model, **b** mathematical constructs of EMIs and DCIs (adapted from (Choi et al. 2005))

$$\Delta Y_{\text{lower}} = f_o(x) - Y_{\min} \tag{7.7}$$

Step 4: Using the mean response (μ_y) obtained from the mean response model $(f_0(x))$ and the upper and lower deviations $(\Delta Y_{\text{upper}}$ and $\Delta Y_{\text{lower}})$, calculate the EMIs. For a "Larger is Better" case, the EMI is calculated as

$$\text{EMI} = \frac{\mu_y - LRL}{\Delta Y_{\text{lower}}} \tag{7.8}$$

The EMI thus calculated for "Larger is Better" case will be larger when the location of μ_y is farther away from the *LRL* and/or when ΔY_{lower} gets smaller, as shown in Fig. 7.3. An EMI = 1 means that the uncertainty bound just meets the requirements

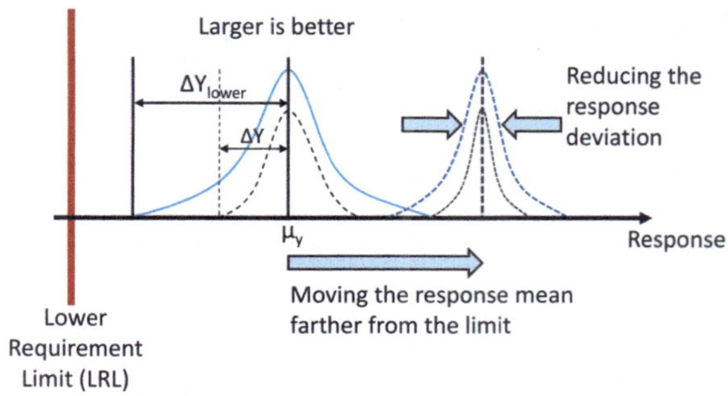

Fig. 7.3 Achieving a larger value of EMI and DCI

limit. An EMI ≤ 1 means that the requirement limit may get violated due to the uncertainty in the model and design variables. The EMI and DCI formulations for the other cases shown in Fig. 7.2b can also be derived in a similar way.

7.2.2 The cDSP with EMI-DCI for Robust Design Type I, II, and III

Core to the CEF is the foundational mathematical construct—the compromise decision support problem (cDSP). The cDSP construct is anchored in the robust design paradigm first proposed by Taguchi and Clausing (1990). The fundamental assumption here is that the models are incomplete, inaccurate, and not of equal fidelity (Mistree et al. 1993). The cDSP is a hybrid of mathematical programming and goal programming. Target values for each of the goals are defined in the cDSP and the emphasis of the designer is to satisfice these target goals as closely as possible. This is achieved by seeking multiple solutions through tradeoffs among multiple conflicting goals. The solutions obtained are further evaluated by solution space exploration to identify solution regions that best satisfice the requirements identified. There are four keywords in the cDSP—Given, Find, Satisfy and Minimize. The overall goal of the designer using the cDSP is to minimize a deviation function—a function formulated using the deviations (captured using deviation variables) that exists from the goal targets. The details regarding formulating and solving the cDSP are available in Chap. 3 as well as in (Mistree et al. 1993) and are not explained here. The mathematical formulation of the cDSP with EMI and DCI goals, constraints to achieve robust design Type I, II, and III are shown in Table 7.2.

In the cDSP formulation, mean response functions for different multiple performance goals $f_{0,i}(x)$, the upper and lower uncertainty bound functions for those goals with model uncertainty, $f_{1,i}(x)$ and $f_{2,i}(x)$ are captured. System constraints and goals in terms of EMI and DCI are formulated in the cDSP to capture the designer's requirements and the functionalities desired in the material–product system. The LRLs and URLs denote the lower and upper requirements limits for the system. The uncertain system constraints are captured as $\mathrm{EMI}_{\mathrm{constraints},i}$ (x) or $\mathrm{DCI}_{\mathrm{constraints},i}$ $(x) \geq 1$ using

Table 7.2 The mathematical form of the cDSP with EMI-DCI

cDSP with EMI-DCI for RD Type I, II, and III for multiple goals
Given
n, number of system variables
m, the total number of system goals
$m1$, number of system goals for robust design Type I, II, and III
$m2$, number of system goals for robust design Type I and II
$m = m1 + m2$
q, the number of inequality constraints
$f_{0,i}(x)$, multiple mean response functions
$f_{1,i}(x)$, multiple upper uncertainty bound functions
$f_{2,i}(x)$, multiple lower uncertainty bound functions
$g_{0,i}(x)$, multiple mean constraint functions
$g_{1,i}(x)$, multiple upper constraint bound functions
$g_{2,i}(x)$, multiple lower constraint bound functions
$\mathrm{URL_i}$ and $\mathrm{LRL_i}$, performance requirements
$\Delta \boldsymbol{x}$, deviations of system variables
$\mathrm{EMI}_{\mathrm{target},i}$, $\mathrm{EMI}_{\mathrm{targets}}$
$\mathrm{DCI}_{\mathrm{target},i}$, $\mathrm{DCI}_{\mathrm{targets}}$
Find
μ_x (mean of system variables)
d_i^+, d_i^- (deviation variables)
Satisfy
System Constraints:
$\mathrm{EMI}_{\mathrm{constraints},i}(x)$ or $\mathrm{DCI}_{\mathrm{constraints},i}(x) \geq 1\ i = 1, \ldots, q$
$\mathrm{EMI}_i(x) \geq 1\ i = 1, \ldots, m1$ **Robust solution constraints** : New constraints defined to
$\mathrm{DCI}_i(x) \geq 1\ i = 1, \ldots, m2$ ensure robust solutions under multiple conflicting goals
System Goals:
$\mathrm{EMI}_i(x)/\mathrm{EMI}_{\mathrm{target},i} + d_i^- - d_i^+ = 1 \quad i = 1, \ldots, m1$
$\mathrm{DCI}_i(x)/\mathrm{DCI}_{\mathrm{target},i} + d_i^- - d_i^+ = 1 \quad i = 1, \ldots, m2$
(Assuming there will be at least one goal for EMI and DCI)
Bounds:
$x_i^{\min} \leq x_i \leq x_i^{\max}\ \mathrm{i} = 1, \ldots, n$
$d_i^-, d_i^+ \geq 0$ and $d_i^+ \cdot d_i^- = 0 \quad i = 1, \ldots, m$
Minimize
$Z = \left[f_1\left(d_i^-, d_i^+\right), \ldots, f_k\left(d_i^-, d_i^+\right)\right]$ Preemptive
$Z = \sum W_i\left(d_i^- + d_i^+\right), \sum W_i = 1$ Archimedean

$g_i(x)$ functions depending on the type of variability. We propose new constraints, defined as ***robust solution constraints*** in our cDSP formulation to ensure the identification of robust solutions always when preferences are changed for the different goals. These constraints ensure both EMI and DCI are greater than or equal to 1 for all design scenarios during solution space exploration (in order to identify robust solutions). In this chapter, we address multiple conflicting goals and there could be situations where when achieving EMI and/or DCI greater than 1 for one goal could result in EMI and/or DCI less than 1 for the other goal. To prevent this we introduce the EMI, DCI constraints for the goal functions. This approach will result in a solution space of only robust solutions to be explored—we define it as "***robust solution space exploration***" for multiple conflicting goals. From these robust solutions, the designer chooses the range of solutions that best satisfice his/her interest. We define this as "***satisficing robust solutions***" for multiple conflicting goals under uncertainty. A designer's interest in the cDSP formulation is to minimize the deviation between the targets and what we can achieve for the multiple conflicting goals that best satisfice the individual goal requirements. The goal target values in the robust design cDSP proposed in this chapter are target values for EMI and DCI for each goal. This is defined by assigning a value greater than 1 that a designer strives to achieve for each goal but cannot fully meet. A designer can estimate the targets for the EMI and DCI goals by varying the design variables between their limits and checking EMI/DCI responses. This gives a rough estimate. A value higher than that can be defined as the target.

7.3 The Inverse Decision-Based Design Method for Robust Design Across Process Chains

We define a workflow as a sequence of computational tasks in which information flows from one process/space to another. For the integrated design of materials, products, and associated manufacturing processes, we define two types of workflow, namely, the workflow associated with simulating the behavior of the material through processing–structure–property–performance hierarchy (material workflow) and the workflow associated with the process of design in an inverse manner across process chains (inverse decision workflow). Our focus in this chapter is on the uncertainty associated with the inverse decision workflow and the analysis models embodied therein. In Fig. 7.4, we show the robust solution space exploration across the process chain considering model structure and model parameter uncertainty. The method involves two generic steps. To demonstrate the generic nature of the method we are naming the different sequential spaces as "Space X", "Space Y" and "Space Z" and the decision support constructs as "cDSP *i*" and "cDSP *i* + 1".

7.3.1 Step 1: Establish Forward Modeling and Information Flow Across Spaces X, Y, and Z

In Step 1, the designer establishes a proper forward flow of information as models are connected across different "spaces", thus generating a rough design space. These spaces X, Y, and Z could be the processing–microstructure–property–performance

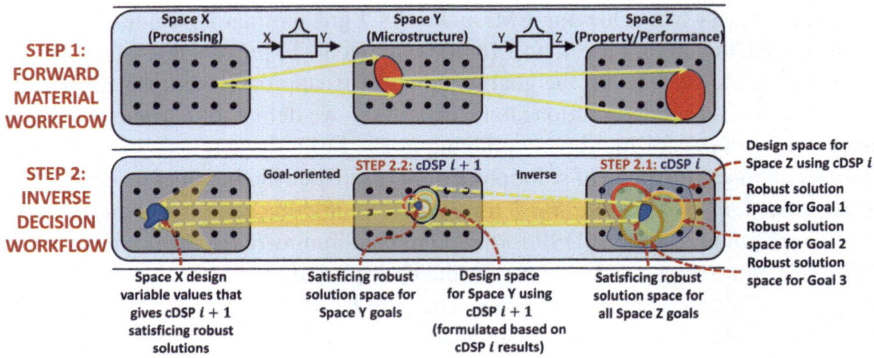

Fig. 7.4 Robust solution space exploration across the process chain in an inverse manner using the inverse design method

spaces from the materials design standpoint or any sequential model-based transformation of information across spaces/levels in a generic sense. Mathematical models are either identified or developed to establish the information flow. In Fig. 7.4, Step 1, we see that the output of one space serves as the input to the next space and this is repeated to the spaces that follow. Thus, Space X (processing space) generates output that serves as input for Space Y (the microstructure space). The output of Space Y (the microstructure identified) serves as the input for Space Z. The output of Space Z defines the property space and this directly defines the final performance characteristics of the end product. From a design standpoint, the input to a space are design variables and the output response from the space serves as input variables to the next space.

7.3.2 Step 2: Robust Solution Space Exploration Across the Process Chain in An Inverse Manner (Inverse Decision Workflow)

Step 2.1: Formulate cDSP ***i for Space Z and carry out solution space exploration***
In Step 2, we start the exploration from the rough design space for Space Z (property–performance). In the rough design space, we formulate the decision-based design space for Space Z using the cDSP construct supported by the CEF. This is done in Step 2.1 by formulating cDSP *i*, see Fig. 7.4. The decision-based design space thus formulated is represented as the light blue region in Space Z. On exercising cDSP *i* for the different conflicting goals by assigning preferences, we obtain different solution regions that satisfice individual goals identified by the three circles (red, orange, and gold color in Space Z, Fig. 7.4). If the cDSP is formulated with the robust solution constraint defined in Sect. 2.2.1 ($EMI_i\ (x) \geq 1$ and $DCI_i\ (x) \geq 1$), then the regions inside the circle denote the regions with EMI and/or DCI greater than 1, depending on

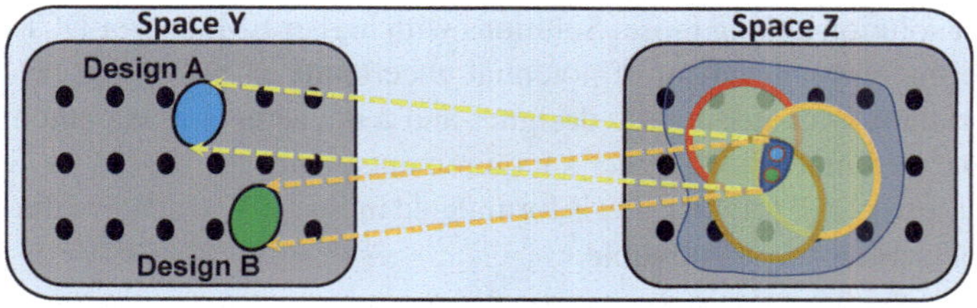

Fig. 7.5 Designs A and B produce satisficing robust solution space in Space Z

the type of goal formulation. Any region inside the circle satisfice the robust design requirements of that particular goal and there will be regions with high robustness (high EMI and/or DCI values away from 1) and low robustness (low EMI and/or DCI values that are nearer to 1) within the circle. The designer can pick solutions that achieve maximum robustness for the goal from the solutions. Since the cDSP is formulated with the defined robust solution constraints, such a solution is never reached that gives a high value of EMI and/or DCI for one goal but an EMI and/or DCI less than 1 for another goal, thus ensuring a robust solution space for all the goals. The designer can then explore the robust solution space of all the conflicting goals and identify common robust regions that best satisfice all the goals—satisficing robust solution region for all goals if it exists. This region is identified as the dark blue region inside the circles in Space Z, see Fig. 7.4. The criteria followed to select the best satisficing robust solution and how this solution is propagated are explained next using the illustration in Fig. 7.5.

While evaluating the inverse solutions, suppose there exist two design candidates, Design A and Design B, as shown in Fig. 7.5, for which the corresponding performance spaces in Space Z are either closely similar or exactly identical, then which design should the designer select? Three criteria listed below are used by the designer in this situation and a decision is made based on the evaluation of each of these criteria.

Criterion 1: Which design among Designs A and B best satisfice the designer's preferences in Space Z?
Criterion 2: Which design among Designs A and B best satisfice the designer's preferences for Space Y if such a preference exists?
Criterion 3: Which design among Designs A and B best satisfice the designer's preference for managing the potential uncertainty in the design problem?

Criterion 1 is used when both the designs result in closely similar outcomes but not exactly identical. Criterion 1 is not used when the performance space in Space Z is exactly identical for both the designs as there is no difference in the design outcomes. Criterion 2 is used when the designer has a preference among the solutions A and B as the solutions are further propagated in an inverse manner. Both Criteria 1 and 2 are evaluated based on the designer's preference for the goals, constraints, and design variable values defined for the problem. Criterion 3 is used to compare the achieved EMI and/or DCI values for both the designs A and B and then evaluate the most robust solution among those. Solutions with higher EMI and/or DCI values are better in terms of management of potential uncertainty in the design problem. All these criteria are evaluated by the designer and a decision is made among designs A and B that best meet the designer's requirements. Once the design for Space Y is identified, a new cDSP (cDSP $i + 1$) is formulated in Space Y to achieve the identified design targets as closely as possible.

Step 2.2: Formulate cDSP i + 1 for Space Y and carry out solution space exploration Once cDSP i is executed and satisficing robust solution region is identified, cDSP $i + 1$ for Space Y is formulated. This cDSP is formulated with design variable values identified from first cDSP as the Space Y goal requirements and is formulated in terms of EMI and/or DCI depending on the type of uncertainty present (the circle with the green region in Space Y represents the region identified from the previous cDSP and is the design space for the new cDSP, see Fig. 7.4). On solving the cDSP with EMI-DCI for Space Y and exploring the solution space, we obtain the robust solution regions that satisfice each goal (represented by the three circles inside the green region in Space Y, see Fig. 7.4). From these robust solutions, the designer identifies the satisficing robust region for all the goals—the blue region within the circles. The values of the Space X design variables that give this robust satisficing Space Y region are identified based on the cDSP $i + 1$ results (the blue region in the Space X, see Fig. 7.4).

Thus, using this proposed method, the designer is able to carry out decision-based robust design exploration of Spaces X, Y, and Z in an inverse manner. An application as discussed could be the identifying of material processing paths and microstructure to satisfice a ranged set of product-level performance requirements. However, the method is generic and can be applied to other complex system problems involving a sequential model-based flow of information under uncertainty. The characteristics of the problems for which the method can be applied include (i) sequential model-based information transformation from one space to another, (ii) availability of analytical models for responses as a function of control and noise factors, (iii) availability of information on model parameter and model structure uncertainty. In Sect. 7.4, we demonstrate the efficacy of the method and the associated robust design functionalities using an industry-inspired hot rolling and cooling process chain example problem for the production of a steel rod.

7.4 Test Example: Robust Concept Exploration of Material (Steel), Product (Rod), and Associated Manufacturing Processes (Hot Rolling and Cooling)

Developing new grades of steels with improved properties and performances is the focus for steel manufacturers. Developing steels with a range of mechanical properties resulting in improved performance of products is possible by carefully managing the material processing and thereby tailoring the microstructure generated. Several

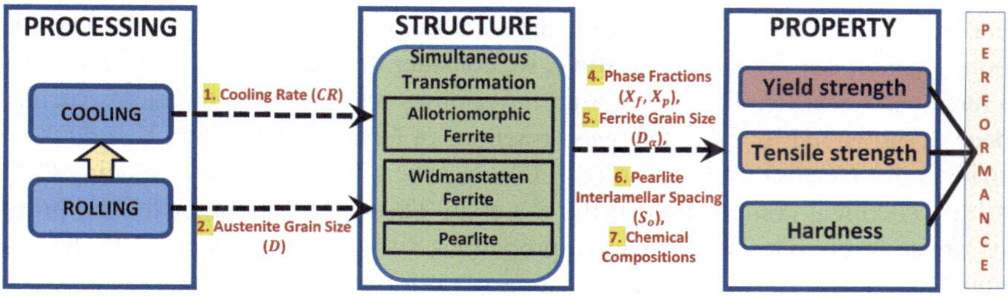

Fig. 7.6 Process–Structure–Property–Performance hierarchy for the integrated robust design of hot rolling and cooling processes to produce a steel rod—forward material workflow

manufacturing processes such as casting, reheating, rolling, and cooling are involved in the processing of a steel rod. This round rod produced is further used for gear production after forging into gear blanks. The end properties of the rolled product are influenced by the chemical composition of the steel including the segregation of alloying elements, the deformation history during rolling, the cooling after rolling, and the microstructure generated after rolling and cooling processes. The steel rod making process chain is highly complex due to large numbers of design variables, constraints and bounds, conflicting goals, and sequential information/material flow during material processing. Many plant trials that are usually expensive and time-consuming are required to produce a new steel grade with desired properties and performance. An alternative, therefore, is to carry out simulation-based, integrated design exploration of the different manufacturing processes involved by exploiting the advances in computational modeling and identifying a ranged set of robust solutions that satisfy the requirements of the processes and product.

In Fig. 7.6, we show the process–structure–property–performance hierarchy for the integrated design of hot rolling and cooling processes to produce the steel rod. Using Fig. 7.6, we capture the forward material workflow for the problem. The processing stage involves the two manufacturing processes, namely hot rolling and cooling. During hot rolling, the thermo-mechanical processing of the material happens. The modeling of the hot rolling process involves a hot deformation module, recrystallization module, grain growth module, and flow stress module (Nellippallil et al. 2017a). The inputs to the rolling process are the chemical composition, initial austenite grain size after reheating, and the rolling schedule (strain, strain rate, interpass time, and number of passes). Using these inputs we predict the temperature evolution, flow stress and calculate the final austenite grain size (AGS, D) after rolling; see (Nellippallil et al. 2017a). In our design problem, we are interested in the final AGS and it forms the input from rolling side to the microstructure space. The microstructure space is generated during the cooling process. During the cooling process, depending on the cooling rate (CR) and the final AGS from rolling and the chemical composition of the incoming steel, time–temperature transformations and simultaneous transformations take place. This results in the phase transformation of austenite to different steel phases like allotriomorphic ferrite, Widmanstätten ferrite, pearlite, and so on. Also, alternate layers of the banded microstructure of ferrite and pearlite can also form depending on the microsegregates that are present and the cooling conditions. In our design study, we consider the transformations of austenite to ferrite and pearlite. The output after the cooling process from the microstructure space as shown in Fig. 7.6 is the phase fractions of ferrite and pearlite $\left(X_f \text{ } and \text{ } X_p\right)$, ferrite grain size after transformation (FGS, D_α), pearlite interlamellar spacing (S_o), and the chemical composition of the material. These serve as input for the property space to predict the mechanical properties, such as the yield strength, tensile strength, and hardness, which are measures of performance for the final rod product. This completes the forward material workflow for the problem we are addressing and establishes the process–structure–property–performance hierarchy for the material system. In Sects. 7.4.1 and 7.4.2, we discuss the Steps 1 and 2 of the proposed method.

7.4.1 Step 1: Establish Forward Modeling and Information Flow Across the Process Chain (Material Workflow)

Identifying factors (input) and responses across process chain (see steps of CEF)
For the hot rod rolling process chain problem addressed in this chapter, the mechanical property goals and requirements are for yield strength (YS), tensile strength (TS), and hardness (HV). These mechanical properties are dependent on the final microstructure after cooling, like the ferrite grain size after cooling (FGS, D_α), the phase fractions of ferrite $\left(X_f\right)$ and pearlite $\left(1 - X_f\right)$, the pearlite interlamellar spacing (S_0), and the composition variables like silicon [Si], nitrogen [N], phosphorous [P], and manganese [Mn]. These microstructure factors are defined by the rate (CR) at which cooling is carried out and the final austenite grain size after rolling (AGS, D) and composition variables like carbon [C] and manganese [Mn].

7.4.2 Identify Models and Relationships that Map from Processing Space to Final Performance Space Across the Process Chain Taking into Account the Uncertainty in Models and Design Variables

Microstructure-Mechanical Property Correlation Models
The mechanical properties for the end rod produced are represented by yield strength (YS), tensile strength (TS), and hardness (HV). Gladman et al. (1972, 1977) were instrumental in predicting the mechanical properties of plain carbon steel products as a function of the microstructural parameters of the ferrite–pearlite microstructure. Models were later developed by Hodgson and Gibbs (Hodgson and Gibbs 1992), Majta et al. (1996), and Kuziak et al. (1997).

Models for Yield Strength and the variability associated

Over the years, different researchers have predicted yield strength as a function of different microstructural parameters. These different models when used predict values at different ranges for a given input and hence have variability associated with them in the prediction of the yield strength property. In this chapter, to demonstrate our method for inverse design and managing uncertainty, we assume the yield strength model by Gladman et al. (1972, 1977) as the mean response model $f_0(x)$ for our problem. The upper uncertainty bound function $f_1(x)$ for yield strength is the model by Hodgson and Gibbs (Hodgson and Gibbs 1992) that when used always predicts yield strength higher than the model by Gladman et al. for a given input. The lower uncertainty bound function $f_2(x)$ for yield strength is the model by Kuziak et al. (1997) that when used predicts yield strength at a lower level than the mean response model for a given input. The models thus identified for yield strength are,

Yield Strength (Mean Response Function)

$$\begin{aligned} \mathrm{YS}_{f_0(x)} &= 63[\mathrm{Si}] + 425[\mathrm{N}]^{0.5} + X_f^{1/3}\left(35 + 58[\mathrm{Mn}] + 17(0.001D_\alpha)^{-0.5}\right) \\ &\quad + \left(1 - X_f^{1/3}\right)\left(179 + 3.9S_0^{-0.5}\right) \end{aligned} \tag{7.9}$$

Yield Strength (Upper uncertainty bound function)

$$\begin{aligned} \mathrm{YS}_{f_1(x)} &= 62.6 + 26.1[\mathrm{Mn}] + 60.2[\mathrm{Si}] + 759[P] + 212.9[\mathrm{Cu}] \\ &\quad + 3286[\mathrm{N}] + 19.7(.001D_\alpha)^{-0.5} \end{aligned} \tag{7.10}$$

Yield Strength (Lower uncertainty bound function)

$$\begin{aligned} \mathrm{YS}_{f_2(x)} &= X_f\left(77.7 + 59.9 \times [\mathrm{Mn}] + 9.1 \times (0.001D_\alpha)^{-0.5}\right) + 478[\mathrm{N}]^{0.5} \\ &\quad + 1200[\mathrm{P}] + \left(1 - X_f\right)\left[145.5 + 3.5S_0^{-0.5}\right] \end{aligned} \tag{7.11}$$

The mean response function and prediction interval models are plotted in Fig. 7.7. The models are depicted as a function of the ferrite grain size (FGS, D_α) and ferrite fraction $\left(X_f\right)$ for a value of pearlite interlamellar spacing of 0.15 μm, manganese concentration of 1.5(%), nitrogen of 0.007(%), silicon of 0.36(%), phosphorous of 0.019(%), and copper of 0.08(%).

Model for Tensile Strength

We select the model by Kuziak et al., in which they describe the tensile strength *TS* of carbon–manganese steels as a function of ferrite grain size after cooling D_α, cooling rate *CR*, ferrite fraction X_f, the pearlite interlamellar spacing S_o, and the composition elements in the steel (Kuziak et al. 1997), see Eq. 7.12.

$$\begin{aligned} \mathrm{TS} &= X_f\left(20 + 2440 \times [\mathrm{N}]^{0.5} + 18.5 \times (0.001D_\alpha)^{-0.5}\right) + 750\left(1 - X_f\right) \\ &\quad + 3\left(1 - X_f^{0.5}\right)S_0^{-0.5} + 92.5 \times [\mathrm{Si}] \end{aligned} \tag{7.12}$$

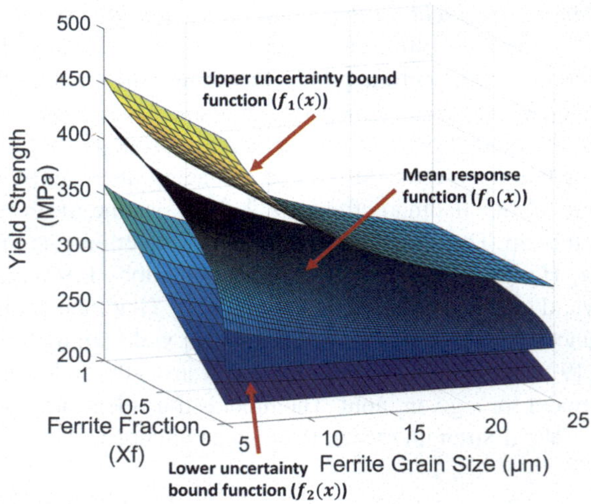

Fig. 7.7 The mean response function and the upper and lower uncertainty bound functions for yield strength

Model for Hardness

Hardness (HV) is represented as a function of ferrite and pearlite fractions, average austenite to ferrite transformation temperature (T_{mf}), and the weight percentage of silicon (Si) based on the investigation by Yada (1987), see Eq. 7.13.

$$HV = X_f(361 - 0.357T_{\mathrm{mf}} + 50[Si]) + 175\big(1 - X_f\big) \tag{7.13}$$

Processing-Microstructure Correlation Models

Model for Ferrite Fraction

We select the response surface model developed by Nellippallil et al. (2018b) for ferrite fraction, see Eq. 7.14. The model is developed by carrying out a design of experiments using the program STRUCTURE developed by Jones and Badeshia to predict the simultaneous transformation of austenite (Jones and Bhadeshia 2017). For more details on the development of the response surface model and the validation of the same, see (Montgomery and Myers 1995; Jones and Bhadeshia 1997a, b; Jägle 2007).

$$\begin{aligned} X_f = 1 - (&0.206 - 0.117[\mathrm{Mn}] \\ &- 0.0005CR - 0.00113D \\ &+ 0.248[\mathrm{C}] + 0.00032[\mathrm{Mn}]CR \\ &+ 0.000086[\mathrm{Mn}]D + 0.9539[\mathrm{Mn}][\mathrm{C}] \\ &- 4.259 \times 10^{-6}CR * D + 0.00726CR[\mathrm{C}] \\ &+ 0.0023D[\mathrm{C}] - 0.0305[\mathrm{Mn}]^2 \\ &- 0.0000056CR^2 + 4.859 \\ &\times 10^{-6}D^2 + 0.79[\mathrm{C}]^2\big) \end{aligned} \tag{7.14}$$

Model for Ferrite Grain Size
We select the model by Hodgson and Gibbs (Hodgson and Gibbs 1992) for defining ferrite grain size, see Eq. 7.15. The factors affecting ferrite grain size D_α are final austenite grain size after rolling D, retained strain ε_r, the composition of carbon and manganese measured in terms of carbon equilvalent C_{eq}, and cooling rate CR from the cooling side.

$$
\begin{aligned}
D_\alpha &= \left(1 - 0.45\varepsilon_r^{0.5}\right) \\
&\quad \times \left\{\left(-0.4 + 6.37C_{\text{eq}}\right) + \left(24.2 - 59C_{\text{eq}}\right)CR^{-0.5} + 22\left[1 - \exp(-0.015D)\right]\right\}
\end{aligned}
\tag{7.15}
$$

where $C_{\text{eq}} = (\text{C} + \text{Mn})/6$.

Model for Pearlite Interlamellar Spacing
We select the model by Kuziak et al. (1997) where pearlite interlamellar spacing S_o is defined as a function of carbon C, manganese Mn, and cooling rate CR, see Eq. 7.16.

$$S_o = 0.1307 + 1.027[\text{C}] - 1.993[\text{C}]^2 - 0.1108[\text{Mn}] + 0.0305CR^{-0.52} \tag{7.16}$$

7.4.3 Step 2: Carry Out Inverse Decision-Based Robust Design Exploration

Step 2.1: Robust Solution Space Exploration of Property–Performance Space
We start the inverse decision-based robust design exploration from property–performance space. The cDSP for the last space is formulated with EMI and DCI goals that capture both property and performance requirements for the end product. The design variables for this cDSP will be the output responses from microstructure space which forms the input for property–performance space, see Fig. 7.6. On exercising the cDSP the process designer will be able to solve and capture the knowledge associated with the following inverse problem: *Given the end mechanical properties of a new steel product mix, what should be the microstructure factors after phase transformation that satisfies the requirements identified taking into account the uncertainty associated with models and parameters associated?* The cDSP reads as follows:

cDSP for Property–Performance (Larger is Better) ***Given***
End requirements identified for the rod rolling process

- Maximize Yield Strength
- Maximize Tensile Strength
- Maximize Hardness

$f_{0,i}(x)$, multiple mean response functions
$f_{1,i}(x)$, multiple upper uncertainty bound functions
$f_{2,i}(x)$, multiple lower uncertainty bound functions
$LRL_{YS} = 200$ MPa
$LRL_{TS} = 450$ MPa
$LRL_{HV} = 130$

$EMI_{target,YS} = 3$	EMI Target for EMI goal for YS considering Type I, II and III RD
$DCI_{target,TS} = 8$	DCI Target for DCI goal for TS considering Type I and II RD
$DCI_{target,HV} = 8$	DCI Target for DCI goal for HV considering Type I and II RD

System variables, their ranges, and variability
Fixed parameters

Composition elements	Concentration (%)
C	0.18
Si	0.36
V	0.003
Cu	0.08
N	0.007
P	0.019

Find
μ_x, (Mean location of system variables)
Deviation Variables
$d_i^-, d_i^+, i = 1, 2, 3$
Satisfy
System Constraints

- Robust solution constraint for YS

  $EMI_{YS}(x) \geq 1$

- Robust solution constraint for TS

  $DCI_{TS}(x) \geq 1$

- Robust solution constraint for HV

  $DCI_{HV}(x) \geq 1$

System Goals
Goal 1:

- Maximize EMI for Yield Strength

$$\frac{\mathrm{EMI}_{\mathrm{YS}}(x)}{\mathrm{EMI}_{\mathrm{Target,\ YS}}} + d_1^- - d_1^+ = 1$$

where $\mathrm{EMI}(x) = \{f_0(x) - \mathrm{LRL}\}/\{Y_{\min} - f_0(x)\}$

where $Y_{\min} = \min\left\{\left(f_j(x) - \sum_{i=1}^{n}\left|\frac{\partial f_j}{\partial x_i}\right| \cdot \Delta x_i\right)\right\}$

Goal 2:

- Maximize DCI for Tensile Strength

$$\frac{\mathrm{DCI}_{\mathrm{TS}}(x)}{\mathrm{DCI}_{\mathrm{Target,\ TS}}} + d_2^- - d_2^+ = 1$$

Goal 3:

- Maximize DCI for Hardness

$$\frac{\mathrm{DCI}_{\mathrm{HV}}(x)}{\mathrm{DCI}_{\mathrm{Target,\ HV}}} + d_3^- - d_3^+ = 1$$

where $\mathrm{DCI}(x) = \{f_0(x) - \mathrm{LRL}\}/\Delta Y$

where $\Delta Y = \sum_{i=1}^{n}\left|\frac{\partial f_0}{\partial x_i}\right|\Delta x_i$

Variable Bounds

Defined in Table 7.3

Table 7.3 System variables, ranges, and variability

S. No	System variables (X)	Ranges	Variability (Δx)
1	X_1, ferrite grain size (D_α)	5–25 μm	[±3]
2	X_2, the phase fraction of ferrite (X_f)	0.1–1	[±0.1]
3	X_3, the pearlite interlamellar spacing (S_0)	0.15–0.25 μm	[±0.01]
4	X_4, manganese concentration after cooling ([Mn])	0.7–1.5%	[±0.1]

Bounds on deviation variables

$$d_i^-, d_i^+ \geq 0 \text{ and } d_i^- * d_i^+ = 0, \quad i = 1, 2, 3$$

Minimize

We minimize the deviation function

$$Z = \sum_{i=1}^{3} W_i\left(d_i^- + d_i^+\right); \sum_{i=1}^{3} W_i = 1$$

On exercising the cDSP for different design scenarios and carrying out robust solution space exploration, and following the steps in concept exploration framework, we obtain the combinations for D_α, X_f, S_0, and Mn that best satisfy the end mechanical properties in the presence of model structure and model parameter uncertainty. The desired solution ranges identified for D_α, X_f, S_0 are then identified as the target goals for the next cDSP (cDSP for microstructure space).

We exercise 13 different scenarios for the cDSP formulated in Sect. 7.4.2 using the computational infrastructure DSIDES (Mistree and Kamal 1985). Different weights are assigned to each goal in these scenarios. Details of the scenarios and the results obtained for the goals are provided in Table 7.4.

These scenarios are selected based on judgment to effectively capture the design space for exploration in a ternary space with the different combinations of weights on goals. Next, we explain the significance of each of these scenarios and identify robust satisficing solutions from the solution space generated in Table 7.4. We explain the significance of the scenarios using the cDSP for the property–performance space. Scenarios 1–3 are for a situation where the designer's interest is to achieve the target of one of the goals, maximizing EMI_{YS}, maximizing DCI_{TS}, or maximizing DCI_{HV} as close as possible. For example, the designer's preference in Scenario 3 is to achieve only the DCI goal for hardness. Scenarios 4–6 are for a situation where two goals are given equal preference, while the third goal is not given any preference. For example, Scenario 5 is a situation where the designer's interest is in equally

Table 7.4 Scenarios and achieved values of goals

Scenarios	w1	w2	w3	Goal 1—EMI YS	Goal 2—DCI TS	Goal 3—DCI HV
1	1	0	0	2.635684224	1.000026749	2.650971569
2	0	1	0	1.202283838	8.110528422	8.663291569
3	0	0	1	1.226883558	7.450292555	8.278751569
4	0.5	0.5	0	1.570932001	6.818501827	8.691461569
5	0.5	0	0.5	1.663214956	5.852196386	7.748441569
6	0	0.5	0.5	1.188246479	8.154843742	8.640151569
7	0.25	0.75	0	1.408711494	7.277846111	8.786341569
8	0.25	0	0.75	1.663636535	5.847812704	7.744171569
9	0.75	0	0.25	1.673002166	5.769498377	7.668461569
10	0.75	0.25	0	1.584223565	6.720117931	8.596961569
11	0	0.75	0.25	1.202283838	8.110528422	8.663291569
12	0	0.25	0.75	1.192159649	8.146318646	8.647991569
13	0.34	0.33	0.33	1.562131147	6.917735445	8.786341569

maximizing EMI_{YS} and DCI_{HV} without giving any preference to the DCI_{TS} goal. Scenarios 7–12 are situations where the designer gives greater preference to one goal, a lesser preference to the second goal and zero preference to the third goal. Scenario 13 is a situation where the designer gives equal preference to all the three goals considered. The exploration of solution space is carried out by exercising the cDSPs for these scenarios using DSIDES and plotting the solution space obtained in a ternary space. In the context of our work, the axes of the ternary plots are the weights assigned to each goal and the color contour in the interior is the achieved value of the specific goal that is being addressed. From these plots, we identify feasible solution regions that satisfice our requirements and the associated weights to be assigned to each goal to achieve this solution space. More details on the creation and interpretation of ternary plots are available in (Nellippallil et al. 2017d).

In Figs. 7.8 a–c, we show via the ternary plots the achieved values of yield strength (Goal 1), tensile strength (Goal 2), and hardness (Goal 3), respectively, for all the 13 scenarios. For Goals 1, 2, and 3, we are interested in achieving high values of EMI_{YS},

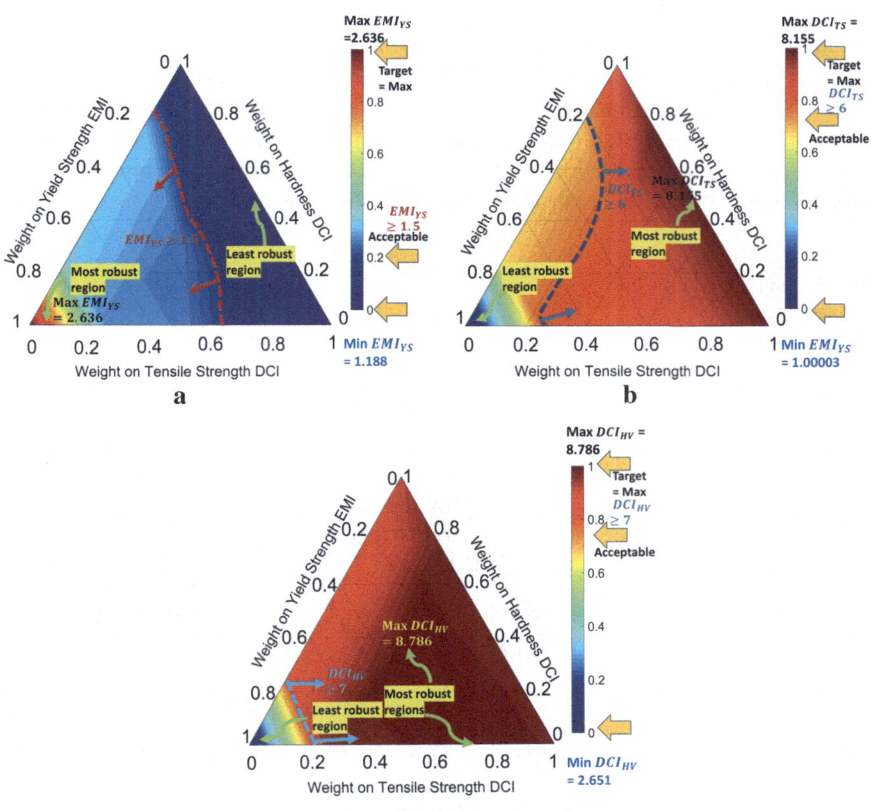

Fig. 7.8 Robust solution space for YS (9a), TS (9b), and HV (9c)

DCI_{TS}, and DCI_{HV}, respectively. We see from Figs. 7.8 a to c that the solution spaces are composed of solutions with $EMI_{YS} \geq 1$, $DCI_{TS} \geq 1$, and $DCI_{HV} \geq 1$, respectively. This ensures robust solutions under both model structure and model parameter uncertainty for Goal 1, and robust solutions under model parameter uncertainty for Goals 2 and 3. The maximum EMI_{YS}, DCI_{TS}, and DCI_{HV} are achieved in the red regions and these regions are therefore the most robust regions. The dark blue regions are the least robust in the solution spaces generated. We define acceptable compromised robust regions within the solution spaces as $EMI_{YS} \geq 1.5$, $DCI_{TS} \geq 6$, and $DCI_{HV} \geq 7$ identified by the dashed lines. Any solution points lying within these regions are acceptable for us as these points satisfice the requirement for the three goals under uncertainty.

Since we are interested in identifying *satisficing robust solution* regions for the multiple conflicting goals, we plot the superposed plot with all the robust solution spaces of interest as shown in Fig. 7.9. The light yellow region identified in Fig. 7.9, satisfice the robust design requirements identified for the conflicting mechanical property goals. In Fig. 7.9, we highlight three points A, B, and C. A is the most robust region for YS with high EMI but lowest for TS and HV with low DCIs. Similarly, B is the most robust region for TS with high DCI_{TS} but lowest for YS with low EMI_{YS}. Point C (Scenario 13 in Table 7.4) lying inside the satisficing robust solution space achieves the highest DCI_{HV} and is the most robust region for HV goal satisficing the robust design requirements of other goals. We select Point C and the solution region around it as the robust solution of interest and this information is passed to the cDSP for microstructure space (Table 7.5).

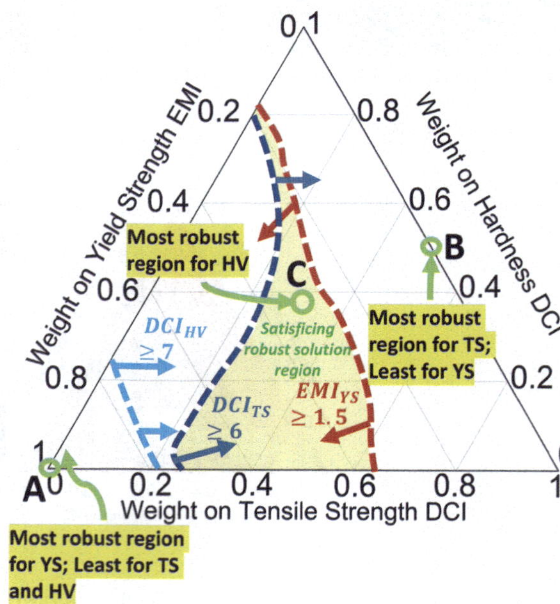

Fig. 7.9 Superposed robust solution spaces

Table 7.5 Microstructure information for next cDSP

Sol. Pt	Microstructure factors (solutions identified are passed as microstructure requirements to next cDSP)				Mechanical properties of rod (achieved robust values)		
	X_f	D_α (μm)	S_0 (μm)	Mn (%)	YS (MPa)	TS (MPa)	HV
C	0.1 [±0.1]	24.7 [±3]	0.15 [±0.01]	0.7 [±0.1]	245	747	170

Step 2.2: Carry out inverse decision-based design exploration of microstructure space

We carry out the inverse decision-based design exploration of microstructure space with information coming from the first cDSP as our requirements. The three goals for X_f, D_α, and S_O are defined with the designer's interest to achieve the values identified in Table 7.5 as close as possible. Upper requirement limits ($\mathrm{URL}_{X_f} = 0.75$, $\mathrm{URL}_{D_\alpha} = 30\,\mu\mathrm{m}$, $\mathrm{URL}_{S_O} = 0.2\,\mu\mathrm{m}$) for the three goals are defined since the individual goal requirements are to achieve as small value as possible (smaller is better case). In this formulation, we are considering only the model parameter uncertainty with an assumption that the model structure uncertainty does not exist. The cDSP for the microstructure space is thus formulated with the DCI goals that capture microstructure requirements identified under model parameter uncertainty. The design variables for this cDSP will be the output responses from processing space which forms the input for microstructure space, see Fig. 7.6. The cDSP with DCI reads as follows:

cDSP for Microstructure Space (Smaller is Better), RD I&II *Given*

End requirements identified for the rod rolling process

- Minimize Ferrite Fraction
- Minimize Ferrite Grain Size
- Minimize Pearlite Interlamellar Spacing

$f_{0,i}(x)$, multiple mean response functions
$\mathrm{URL}_{X_f} = 0.75$
$\mathrm{URL}_{D_\alpha} = 30\,\mu\mathrm{m}$
$\mathrm{URL}_{S_O} = 0.2\,\mu\mathrm{m}$
$\mathrm{DCI}_{\mathrm{target},}X_f = 10$
$\mathrm{DCI}_{\mathrm{target},}D_\alpha = 10$
$\mathrm{DCI}_{\mathrm{target},}S_O = 200$
System variables, their ranges, and variability.

Find

μ_x, (Mean location of system variables)
Deviation Variables
$d_i^-, d_i^+, i = 1, 2, 3$

Satisfy
System Constraints

- Robust solution constraint for Ferrite Fraction

$$\mathrm{DCI}_{X_f}(x) \geq 1$$

- Robust solution constraint for Ferrite Grain Size

$$\mathrm{DCI}_{D_\alpha}(x) \geq 1$$

- Robust solution constraint for Pearlite Interlamellar Sp.

$$\mathrm{DCI}_{S_O}(x) \geq 1$$

System Goals
Goal 1:

- Maximize DCI for Ferrite Fraction

$$\frac{\mathrm{DCI}_{X_f}(\mathrm{x})}{\mathrm{DCI}_{\mathrm{Target},X_f}} + d_1^- - d_1^+ = 1$$

Goal 2:

- Maximize DCI for Ferrite Grain Size

$$\frac{\mathrm{DCI}_{D_\alpha}(x)}{\mathrm{DCI}_{\mathrm{Target},D_\alpha}} + d_2^- - d_2^+ = 1$$

Goal 3:

- Maximize DCI for Pearlite Interlamellar Spacing

$$\frac{\mathrm{DCI}_{S_O}(x)}{\mathrm{DCI}_{\mathrm{Target},S_O}} + d_1^- - d_1^+ = 1$$

where $\mathrm{DCI}(x) = \{\mathrm{URL} - f_0(x)\}/\Delta Y$
where $\Delta Y = \sum_{i=1}^{n} \left|\frac{\partial f_0}{\partial x_i}\right| \Delta x_i$

Variable Bounds
Defined in Table 7.6

Table 7.6 System variables, ranges, and variability

Sr. No	System variables (X)	Ranges	Variability (Δx)
1	X_1, cooling rate (*CR*)	11–100 K/min	[±10]
2	X_2, austenite grain size (*D*)	30–100 μm	[±10]

Bounds on deviation variables

$$d_i^-, d_i^+ \geq 0 \text{ and } d_i^- * d_i^+ = 0, \quad i = 1, 2, 3$$

Minimize

We minimize the deviation function

$$Z = \sum_{i=1}^{3} W_i(d_i^- + d_i^+); \sum_{i=1}^{3} W_i = 1$$

On exercising the cDSP for different design scenarios and carrying out robust solution space exploration, and following the steps in the concept exploration framework, we obtain the combinations for *CR* and *D*, and the variables from processing space that best satisfice the microstructure requirements in the presence of model parameter uncertainty. The cDSP formulated for microstructure space is exercised for 13 different scenarios (same scenarios as in Table 7.4) by assigning weights to the goals using DSIDES.

In Fig. 7.10 a–c, we show the robust solution space for ferrite fraction, X_f (Goal 1), D_α (Goal 2), and S_O (Goal 3), respectively. Our interests in Goals 1, 2, and 3 are to achieve high DCI values for X_f, D_α, and S_O, respectively. The ternary spaces are made of $\text{DCI}_{X_f} \geq 1$, $\text{DCI}_{D_\alpha} \geq 1$, and $\text{DCI}_{S_O} \geq 1$ ensuring robust solutions under model parameter uncertainty associated with the design variables. We compromise and identify the regions with $\text{DCI}_{X_f} \geq 7$, $\text{DCI}_{D_\alpha} \geq 9.5$, and $\text{DCI}_{S_O} \geq 150$ as the robust satisficing regions of interest under uncertainty as shown in Fig. 7.10.

To identify satisficing robust solution regions for microstructure, we plot the superposed plot shown in Fig. 7.11 with all the robust solution spaces of interest. In the superposed ternary plot, we see that the light yellow region satisfices all the identified microstructure requirements under model parameter uncertainty. To analyze further we pick three solution points from the region identified. Solution points A, B, and C lie within the region that satisfies all the robust design goals in the best possible manner. The results associated with the selected points are summarized in Table 7.7.

On analyzing the results in Table 7.7, we see that the solutions identified from the satisficing robust region in the ternary space show a very small deviation in

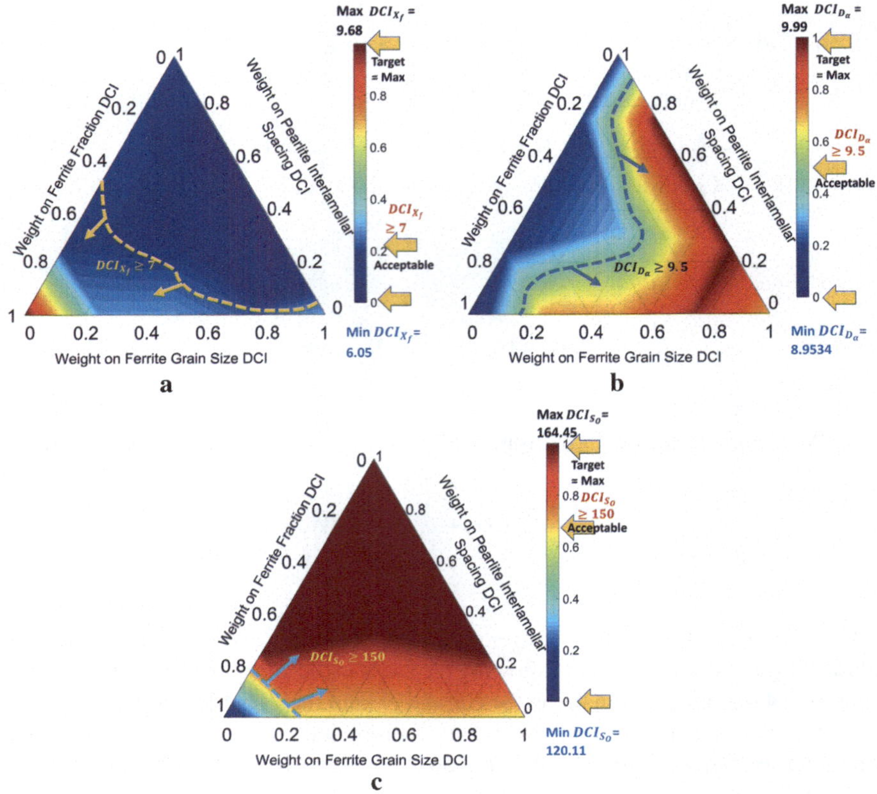

Fig. 7.10 Robust solution space for X_f (**a**) D_α (**b**) and S_O (**c**)

Fig. 7.11 Superposed robust solution space

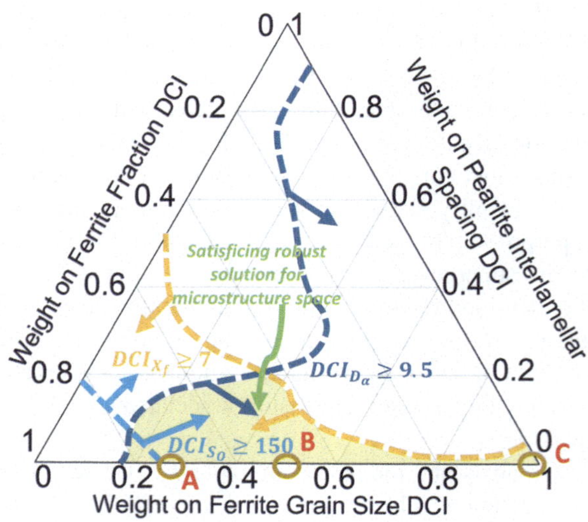

Table 7.7 Solution points selected

Sol. Pts	Factors from processing space		Microstructure space		
	CR (K/min)	D (μm)	X_f	D_α (μm)	S_0 (μm)
A	94	46	0.69147	13.1038	0.176
B	93.9	46	0.691	13.103	0.1763
C	93.7611	45.7	0.69125	13.0554	0.1763

performance from each other. The processing variable values associated with the solution points in this region result in robust solutions of microstructure under the model parameter uncertainty considered in this design problem.

Thus, using this proposed inverse method, we are able to carry out top-down driven, decision-based robust design exploration of processing paths, and material microstructure to satisfy a ranged set of product-level performance requirements. The inverse method proposed is generic and can be applied to similar problems with information flow from one process to another to design the system under different types of uncertainty classified in this chapter.

7.5 Discussion: Robustness Under Model Structure and Model Parameter Uncertainty Using EMI and DCI

In this section, based on the design study carried out, we discuss the usefulness of the robust design metrics EMI and DCI used in this chapter for designing a system under model structure and model parameter uncertainty by carrying out a comparative study. To illustrate the same, we use the yield strength model proposed by Gladman et al. (Eq. 7.9), which we used as the mean response function for YS in the cDSP formulated for property–performance space. We explore three formulations: In the first, we formulate a single-goal cDSP with EMI for the yield strength mean model with the uncertainty bounds defined by the yield strength models by Hodgson and Gibbs (Eq. 7.10) and Kuziak et al. (Eq. 7.11; the formulation is the same as in the first cDSP, except there is only one goal which is for maximizing the EMI for yield strength). In second, we formulate a single-goal cDSP with DCI goal for the yield strength mean model with consideration of only model structure uncertainty defined in the first cDSP. Third, we formulate a single-objective traditional optimization problem for maximizing the mean yield strength function.

The results associated with this comparative study are plotted in Fig. 7.12 with ferrite fraction and ferrite grain size as the input factors for the yield strength model. We see that the cDSP with EMI predicts a mean response value of 288.755 MPa. The corresponding EMI value for the solution point is 2.63568. The formulation with DCI predicts yield strength at 306.08 MPa and is higher than the EMI prediction. The DCI value at this point is 5.37195. However, the EMI value when calculated is only 1.85375. This means that the EMI is less for the solution point that is identified using

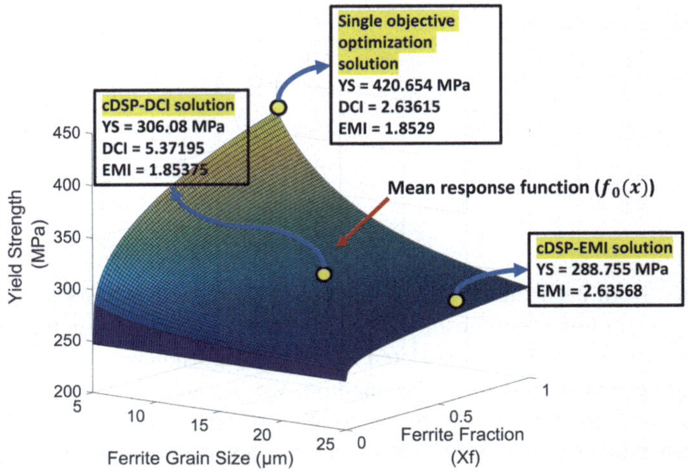

Fig. 7.12 Solutions obtained for yield strength as a single goal using different formulations—a comparative study

DCI formulation compared to the solution point identified using an EMI formulation. This is because the DCI formulation overlooks the uncertainty associated with the model and thus achieves a lower EMI value for the design solutions. Next, on analyzing the solution obtained via the single-objective optimization formulation, we see that the optimal solution predicts the highest response for yield strength (YS = 420.654 MPa). However, both the DCI and EMI values are low for the optimization solution point when calculated. This means that the optimal solution points obtained are prone to both model structure and model parameter uncertainty and are less robust compared to the solutions obtained via cDSP-DCI and cDSP-EMI.

We infer from this comparative study, the advantage of EMI and DCI formulations for complex material, product, and process systems as the design solutions are more robust against model structure uncertainty and model parameter uncertainty. The limitation here with the EMI and DCI is the inability to capture the designer's preference since the EMI and DCI are calculated as a combination of mean and response variations. This limitation can be overcome by separating the mathematical combinations of mean and performance variance and formulating them as two individual goals in the cDSP and repeating the same for multiple goals.

7.6 Role of in This Chapter in This Monograph—Closing Remarks

In this chapter, we present robust concept exploration by extending the goal-oriented, inverse design method proposed by Nellippallil et al. (2018b) to identify *satisficing robust solutions* across process chains. The extension embodies the introduction of specific robust design goals and new robust solution constraints anchored in the mathematical constructs of error margin indices (EMIs) and design capability indices (DCIs) to determine "satisficing" robust design specifications for multiple conflicting goals from the standpoint of Type I, II, and III robust design across process chains. The utility of the proposed method, robust design goals, constraints, and metrics proposed is demonstrated by carrying out the solution space exploration of the processing and microstructure spaces of the rolling and cooling processes to identify *satisficing robust solutions* to realize the end mechanical properties of the rod product. More details on the example problem used in this chapter are available in Chap. 6 and the following articles (see Nellippallil et al. 2017b, c, 2018a, b, 2019, 2020). The contributions in this chapter are the following:

(1) Previous work by Choi et al. (2005) on robust design using EMI address only a single-objective case. In this chapter, we address multiple conflicting goals and how satisficing robust solutions can be explored across a process chain using the inverse design method.
(2) We make it possible for a designer to handle multiple robust design formulations (using both EMI and DCI) in one problem formulation. This allows designers to look at robust design Type I, II for one goal (via DCI) and Type I, II, and III for another goal (via EMI) in a single problem formulation depending on the problem requirements.

Based on these contributions, the key functionalities of the GoID method with robust design metrics, goals, and constraints when compared to other domain-independent design exploration methods include:

- an increase in problem size as there are no limitations in the number of design variables and goals to be studied,
- improved flexibility in the design of the various processes involved as individual design spaces are formulated at each level allowing designers to incorporate new design goals and requirements,
- the capability to identify robust satisficing solutions for multiple conflicting goals under multiple sources of uncertainty, and
- the capability to visualize and explore solutions that are relatively insensitive to the sources of uncertainty identified.

The method and associated robust design constructs are generic and can be applied to coordinate information flow and human decision-making across processes/levels in order to realize an end goal by managing the sources of uncertainty—a key functionality that allows the application of this method to other complex system problems involving sequential model-based flow of information under uncertainty.

References

Adams, B. L., Kalidindi, S., & Fullwood, D. T. (2013). *Microstructure-sensitive design for performance optimization*. Butterworth-Heinemann.

Chen, W., Simpson, T. W., Allen, J. K., & Mistree, F. (1999). Satisfying ranged sets of design requirements using design capability indices as metrics. *Engineering Optimization, 31*(5), 615–619.

Choi, H.-J., Austin, R., Allen, J. K., McDowell, D. L., Mistree, F., & Benson, D. J. (2005). An approach for robust design of reactive power metal mixtures based on non-deterministic micro-scale shock simulation. *Journal of Computer-Aided Materials Design, 12*(1), 57–85.

Choi, H.-J., Mcdowell, D. L., Allen, J. K., & Mistree, F. (2008). An inductive design exploration method for hierarchical systems design under uncertainty. *Engineering Optimization, 40*(4), 287–307.

Gladman, T., McIvor, I., & Pickering, F. (1972). Some aspects of the structure-property relationships in high-C ferrite-pearlite steels. *The Journal of the Iron and Steel Institute, 210*(12), 916–930.

Gladman, T., Dulieu, D., & McIvor, I. D. (1977). Structure/property relationships in high-strength micro-alloyed steels. In *Proceedings of the Conference on Microalloying* (Vol. 75, No. 1977, pp. 32–55).

Hodgson, P., & Gibbs, R. (1992). A mathematical model to predict the mechanical properties of hot rolled C-Mn and microalloyed steels. *ISIJ International, 32*(12), 1329–1338.

Horstemeyer, M. F. (2012). *Integrated computational materials engineering (ICME) for metals: Using multiscale modeling to invigorate engineering design with science*. Wiley.

Horstemeyer, M. F. (2018). *Integrated computational materials engineering (ICME) for metals: Concepts and case studies*, Wiley.

Jägle, E. (2007). Modelling of microstructural banding during transformations in steel. Master of Philosophy Dissertation, Department of Materials Science & Metallurgy, University of Cambridge.

Jones, S., & Bhadeshia, H. (1997a). Competitive formation of inter-and intragranularly nucleated ferrite. *Metallurgical and Materials Transactions A, 28*(10), 2005–2013.

Jones, S., & Bhadeshia, H. (1997b). Kinetics of the simultaneous decomposition of austenite into several transformation products. *Acta Materialia, 45*(7), 2911–2920.

Jones, S. J., & Bhadeshia, H. K. D. H (2017) Program STRUCTURE on the materials algorithm project web site. http://www.msm.cam.ac.uk/map/steel/programs/structure.html.

Kalidindi, S. R. (2015). *Hierarchical materials informatics: Novel analytics for materials data*, Elsevier.

Kalidindi, S. R., Niezgoda, S. R., Landi, G., Vachhani, S., & Fast, T. (2010). A novel framework for building materials knowledge systems. *Computers, Materials, & Continua, 17*(2), 103–125.

Kalidindi, S. R., Niezgoda, S. R., & Salem, A. A. (2011). Microstructure informatics using higher-order statistics and efficient data-mining protocols. *JOM Journal of the Minerals Metals and Materials Society, 63*(4), 34–41.

Kuziak, R., Cheng, Y.-W., Glowacki, M. & Pietrzyk, M. (1997). Modeling of the microstructure and mechanical properties of steels during thermomechanical processing. *NIST Technical Note(USA), 1393*, 72.

Li, C. & Mahadevan, S. (2016). Role of calibration, validation, and relevance in multi-level uncertainty integration. *Reliability Engineering & System Safety, 148*, 32–43.

Majta, J., Kuziak, R., Pietrzyk, M., & Krzton, H. (1996). Use of the computer simulation to predict mechanical properties of C-Mn steel, after thermomechanical processing. *Journal of Materials Processing Technology, 60*(1–4), 581–588.

Matthews, J., Klatt, T., Morris, C., Seepersad, C. C., Haberman, M., & Shahan, D. (2016). Hierarchical design of negative stiffness metamaterials using a bayesian network classifier. *Journal of Mechanical Design, 138*(4), 041404.

McDowell, D. L. (2018). Microstructure-sensitive computational structure-property relations in materials design. In *Computational materials system design* (pp. 1–25), Springer.

McDowell, D. L., & Olson, G. (2008). Concurrent design of hierarchical materials and structures. In *Scientific modeling and simulations* (pp. 207–240), Springer.

McDowell, D. L., & Kalidindi, S. R. (2016). The materials innovation ecosystem: A key enabler for the materials genome initiative. *MRS Bulletin, 41*(4), 326–337.

McDowell, D. L., & LeSar, R. A. (2016). The need for microstructure informatics in process–structure–property relations. *MRS Bulletin, 41*(8), 587–593.

McDowell, D. L., Panchal, J., Choi, H.-J., Seepersad, C., Allen, J., & Mistree, F. (2009). *Integrated design of multiscale, multifunctional materials and products*, Butterworth-Heinemann.

Mistree, F., & Kamal, S. (1985). *DSIDES: Decision support in the design of engineering systems*, University of Houston.

Mistree, F., Hughes, O. F. & Bras, B. (1993). Compromise decision support problem and the adaptive linear programming algorithm. *Progress in Astronautics and Aeronautics, 150*, 251–251.

Montgomery, D. C., & Myers, R. H. (1995). *Response surface methodology: Process and product optimization using designed experiments*. Raymond H: Meyers and Douglas C. Montgomery. A Wiley-Interscience Publications.

Mullins, J., & Mahadevan, S. (2016). Bayesian uncertainty integration for model calibration, validation, and prediction. *Journal of Verification, Validation and Uncertainty Quantification, 1*(1), 011006.

Nellippallil, A. B., De, P. S., Gupta, A., Goyal, S., & Singh, A. K. (2017a) Hot rolling of a non-heat treatable aluminum Alloy: Thermo-mechanical and microstructure evolution model. *Transactions of the Indian Institute of Metals, 70*(5), 1387–1398.

Nellippallil, A. B., Rangaraj, V., Allen, J. K., Mistree, F., Gautham, B., & Singh, A. K. (2017b). A decision-based design method to explore the solution space for microstructure after cooling stage to realize the end mechanical properties of hot rolled product. In *Proceedings of the 4th World Congress on Integrated Computational Materials Engineering (ICME 2017)*, Springer.

Nellippallil, A. B., Rangaraj, V., Gautham, B., Singh, A. K., Allen, J. K., & Mistree, F. (2017c). A goal-oriented, inverse decision-based design method to achieve the vertical and horizontal integration of models in a hot rod rolling process chain. *ASME Design Automation Conference*, Paper Number: DETC2017-67570.

Nellippallil, A. B., Song, K. N., Goh, C.-H., Zagade, P., Gautham, B., Allen, J. K., et al. (2017d). A goal-oriented, sequential, inverse design method for the horizontal integration of a multistage hot rod rolling system. *Journal of Mechanical Design, 139*(3), 031403.

Nellippallil, A. B., Mohan, P., Allen, J. K., & Mistree, F. (2018a). Robust concept exploration of materials, products and associated manufacturing processes. *ASME Design Automation Conference*, Paper Number: DETC2018-85913.

Nellippallil, A. B., Rangaraj, V., Gautham, B., Singh, A. K., Allen, J. K., & Mistree, F. (2018b). An inverse, decision-based design method for integrated design exploration of materials, products, and manufacturing processes. *Journal of Mechanical Design, 140*(11), 111403.

Nellippallil, A. B., Mohan, P., Allen, J. K., & Mistree, F. (2019). Inverse thermo-mechanical processing (ITMP) design of a steel rod during hot rolling process. *ASME Design Automation Conference 2019*, Paper Number: IDETC2019-97390.

Nellippallil, A. B., Mohan, P., Allen, J. K., & Mistree, F. (2020). An inverse, decision-based design method for robust concept exploration. *Journal of Mechanical Design*, https://doi.org/10.1115/1.4045877.

Panchal, J. H., Kalidindi, S. R., & McDowell, D. L. (2013). Key computational modeling issues in integrated computational materials engineering. *Computer-Aided Design, 45*(1), 4–25.

Phadke, S., Pauskar, P., & Shivpuri, R. (2004). Computational modeling of phase transformations and mechanical properties during the cooling of hot rolled rod. *Journal of Materials Processing Technology, 150*(1), 107–115.

Pietrzyk, M., Cser, L., & Lenard, J. (1999). *Mathematical and physical simulation of the properties of hot rolled products*, Elsevier.

Shahan, D. W., & Seepersad, C. C. (2012). Bayesian network classifiers for set-based collaborative design. *Journal of Mechanical Design, 134*(7), 071001.

Simon, H. A. (2013). *Administrative behavior*, Simon and Schuster.
Taguchi, G., & Clausing, D. (1990). Robust quality. *Harvard Business Review, 68*(1), 65–75.
Yada, H. (1987). Prediction of microstructural changes and mechanical properties in hot strip rolling. *Accelerated Cooling of Rolled Steel*, 105–119.

Chapter 8
Advancing the Vision for the Systems-Based Design Architecture Via a Cloud-Based Platform for Decision Support in the Design of Engineered Systems (CB-PDSIDES)

Achieving integrated materials and product design in the digitized world necessitates facilitating a network of participants (material scientists, systems designers, software developers, end customers) to share material/product/manufacturing process/market data, information, knowledge, and resources instantly and in an integrated fashion, and thereby collaborate so as to facilitate a cost-effective co-creation of value supporting open innovation. In this monograph, we define this as the capability to carry out ***co-design***. New product development models like cloud-based design and cloud-based manufacturing that form part of the current digital manufacturing and design innovation (DMDI) paradigm supports people/stakeholders involved in the value chain to perform co-design for realizing materials, products, and manufacturing processes. In this chapter, we present the architecture and functionalities of a cloud-based computational platform anchored in the decision-based design paradigm to facilitate mass collaboration and open innovation, thereby supporting integrated materials and product realization. We begin the chapter by checking if the objectives planned for the monograph are addressed. We carry out a self-reflection of what has been achieved in past chapters and identify enabling technologies that require advancements to further develop the vision of the integrated design of materials, products, and processes as required by the field of integrated computational materials engineering (ICME) in the digital age. In that context, we recognize the value of a platform for facilitating integrated materials, products, and manufacturing process design in the digital age and present PDSIDES and cloud-based PDSIDES as examples to achieve the same. We illustrate the efficacy of the cloud-based platform using the hot rolling example problem to produce a steel rod, see Chap. 6 for details on the example problem. Using this example, we illustrate the utility of the cloud-based platform in carrying out seamless, yet controllable, information, knowledge, and resource sharing, thereby supporting the integrated design of materials, products, and manufacturing processes. In Table 8.1, we summarize the elements discussed, reviewed, and tested in this chapter.

A. B. Nellippallil et al., *Architecting Robust Co-Design of Materials, Products, and Manufacturing Processes*, https://doi.org/10.1007/978-3-030-45324-4_8

Table 8.1 Summary of the elements discussed, reviewed, and tested in Chap. 8

Summary of the elements discussed, reviewed and tested in Chap. 8 that form part of the architecture	
Elements	*What*?
Processing elements	The functionalities of PDSIDES and cloud-based PDSIDES
Data elements	Customer-product designer–materials designer connectivity via the cloud
Connecting elements	Cloud-based design and manufacturing
Form	*How*?
Weights	Competencies and core functionalities of CB-PDSIDES
Properties	Focus on emerging digital manufacturing and design innovation technologies
Relationship	Theoretical performance validation of the architecture
Rationale	*Why*?
Motivation	To present the architecture and functionalities of a cloud-based computational platform to facilitate mass collaboration and open innovation, thereby supporting integrated material and product realization in the industry
Characteristics/beliefs	• Digital manufacturing and design innovation paradigm facilitates a network of participants to share material/product/manufacturing process/market data, information, knowledge, and resources instantly and collaborate so as to facilitate a cost-effective co-creation of value supporting open innovation—the capability to carry out co-design • Design is a goal-oriented synthesis activity and essentially a decision-making process—the material information that flow across the manufacturing process chain are therefore represented in terms of a horizontal decision workflow that involves integrated information flow from multiple scales (vertical information flow) • Decisions can be formulated mathematically, and analysis models are available for design decision-making • Design process starts with an end goal in mind—could be an end product that needs to be supplied to customers • There exist models or computational tools or data that can establish the forward information workflow for the process/product that is being considered • Analysis models for problems are typically incomplete, inaccurate, and not of equal fidelity • Designer's interest is in identifying satisficing robust solutions given the uncertainty in the problem and not single-point optimum solutions • Design is a goal-oriented activity with design requirements subject to change at any time for the problems considered
Interpretation	Traditional product development methods need to be updated and bridged to the new developments happening in the globalized world like cloud-based design and manufacturing

8.1 Summarizing Chaps. 1 to 7 of the Monograph

A materials design transformation is underway in the recent past where the focus is to design (not select) the material microstructure and processing paths to achieve multiple property or performance requirements that are often in conflict. The advancements in computer simulations have resulted in the speeding up of the process of discovering new materials and has paved the way for rapid assessment of process–structure–property–performance relationships of materials, products, and processes. This has led to the simulation-based design of material microstructure (microstructure-mediated design) to satisfy multiple properties or performance goals of the product/process/system, thereby replacing the classical material design and selection approaches. The foundational premise for this monograph is that systems-based materials design techniques offer the potential for tailoring materials, their processing paths, and the end products that employ these materials in an integrated fashion for challenging applications to satisfy conflicting product- and process-level property and performance requirements. The primary goal in this monograph is to establish some of the scientific foundations and tools that are needed for the integrated realization of materials, products, and manufacturing processes using simulation models that are typically incomplete, inaccurate, and not of equal fidelity by managing the uncertainty associated. Accordingly, the interest in this monograph lies in establishing a ***systems-based design architecture*** that includes system-level synthesis methods and tools that are required for the integrated design of complex materials, products, and associated manufacturing processes starting from the end requirements. Hence the primary research question: *What are the theoretical, mathematical, and computational foundations needed for establishing a comprehensive systems-based design architecture to realize the integrated design of the product, its environment, manufacturing processes, and material as a system?* Major challenges addressed are: (a) integration of models (material, process, and product) to establish processing–structure–property–performance relationships, (b) goal-oriented inverse design of material microstructures and processing paths to meet multiple conflicting performance/property requirements, (c) robust concept exploration by managing uncertainty across process chains, and (d) systematic, domain-independent, modular, reconfigurable, reusable, computer interpretable, archivable, and multiobjective decision support in the early stages of design to different users.

Approach: In order to address these challenges, the primary hypothesis in this monograph is to establish the theoretical, mathematical, and computational foundations for: (1) forward material, product, and process workflows through systematic identification and integration of models to define the processing–structure–property–performance relationships; (2) a concept exploration framework supporting systematic formulation of design problems facilitating robust design exploration by bringing together robust design principles and multiobjective decision-making protocols; (3) a generic, goal-oriented, inverse decision-based design method that uses (1) and (2) to facilitate the systems-based inverse design of material microstructures and processing paths to meet multiple product-level performance/property requirements, thereby generating the problem-specific inverse decision workflow; and (4) integrating the workflows with a cloud-based knowledge platform anchored in modeling decision-related knowledge facilitating capture, execution, and reuse of the knowledge associated with (1), (2), and (3). This establishes a comprehensive systems-based design architecture to realize the integrated design of the product, its environment, manufacturing processes, and material as a system.

Validation: The systems-based design architecture for the integrated realization of materials, products, and associated manufacturing processes is validated using the validation-square approach that consists of theoretical and empirical validation. Empirical validation of the design architecture is carried out using an industry-driven problem, namely the "Integrated Design of Steel (Material), Manufacturing Processes (Rolling and Cooling), and Hot Rolled Rods (Product) for Automotive Gears". Specific subproblems are formulated within this problem domain to address the various research questions identified in this monograph, see Chaps. 5, 6, and 7.

Contributions: The contributions from the monograph are categorized into new knowledge in four research domains: (a) systematic model integration (vertical and horizontal) for integrated material and product workflows, (b) goal-oriented, inverse decision support, (c) robust concept exploration of process chains with multiple conflicting goals, and (d) cloud-based decision support for rapid and robust design exploration in simulation-based integrated material, product, and process design (the cloud-based decision support platform is addressed in this chapter in the sections that follow). The creation of new knowledge in this monograph is associated with the development of a systems-based design architecture involving systematic function-based approach of formulating forward material workflows, a concept exploration framework for systematic design exploration, an inverse decision-based design method, and robust design metrics, all integrated with a knowledge-based cloud platform for decision support. The theoretical, mathematical, and computational foundations for the design architecture are presented and tested using example problems in this monograph. These foundations support designers by facilitating the rapid and robust exploration of the design and solution spaces to identify material microstructures and processing paths that satisfy conflicting property and performance for complex materials, products, and processes by managing uncertainty. The key components of the systems-based design architecture presented in this monograph are shown in Fig. 8.1. The key components include: *(1) Systematic Model Integration and Information Flow, (2) Systematic Concept Exploration, (3) Systematic Inverse Design Exploration, (4) Systematic Robust Concept Exploration by Managing Uncertainty, (5) Knowledge-based Platform for Decision Support, and (6) Cloud-Based Decision Support.* We discussed key components 1–4 in Chaps. 4–7. These key components support the architecture in adding value to the emerging digital manufacturing and design innovation paradigm by means of the following:

- Systematic function structure-based identification and integration of material, process, and product models and workflows to define processing–structure–property–performance mapping and information workflow (addressed in Chaps. 4, 5, and 6),

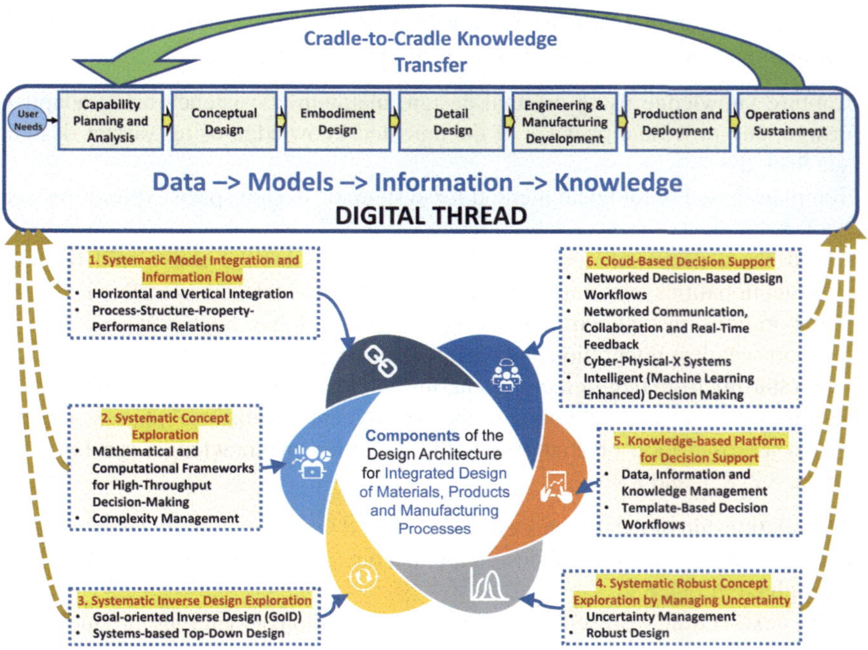

Fig. 8.1 Components of the design architecture adding value to the digital manufacturing and design innovation paradigm

- A reusable, expandable computational framework supporting vertical and horizontal integration of models to identify material structures and processing paths that satisfy ranged set of product and manufacturing process-level property and performance requirements (addressed in Chaps. 4 and 6),
- A framework supporting systematic design and solution space exploration (addressed in Chaps. 4, 5, and 6),
- A generic method for inverse design of materials and products across process chains (addressed in Chaps. 5 and 6),
- Metrics, robust design constraints, and goals for facilitating robust design by managing uncertainty across process chains for multiple conflicting goals (addressed in Chap. 7).

Given these contributions from the previous chapters, in this chapter, we present the architecture and functionalities of a cloud-based computational platform anchored in modeling decision-related knowledge facilitating capture, execution, and reuse of the knowledge associated with the above-said tools, constructs, methods, and frameworks, thereby supporting integrated material and product realization in the digital era. The cloud-based computational platform for decision support is associated with key components 5 and 6 of the systems-based design architecture (see Fig. 8.1) and will be discussed in detail in the sections that follow. Key components 5 and 6 support the architecture in adding value to the emerging digital manufacturing and design innovation paradigm by means of the following:

- Capture knowledge using original design, maintain consistency using adaptive design, and provide a package of documented knowledge using variant design, see Sect. 8.3;
- Template-based ontological method for systematic design space exploration, see Sect. 8.3;
- Cloud-based decision support platform that embodies the above said features and functionalities enabling networked decision-based design workflows, design collaborations, instant communication, and feedback, see Sect. 8.3;
- Support seamless integration of data, information, and knowledge across the value chain starting from requirements gathering and ideation to design, manufacturing, testing, supply chain, and finally to the sustainment and disposal phases—resulting in a closed-loop of cradle-to-cradle transition of knowledge, see Fig. 8.1, Sects. 8.2, 8.3, and 8.4.

The relationship of the research efforts presented in the previous chapters with the constructs of the systems-based design architecture presented in the current chapter is highlighted in Fig. 8.2.

In the next section, we discuss the integrated realization of materials, products, and manufacturing processes in the digital age and establish the need for a cloud-based decision support platform.

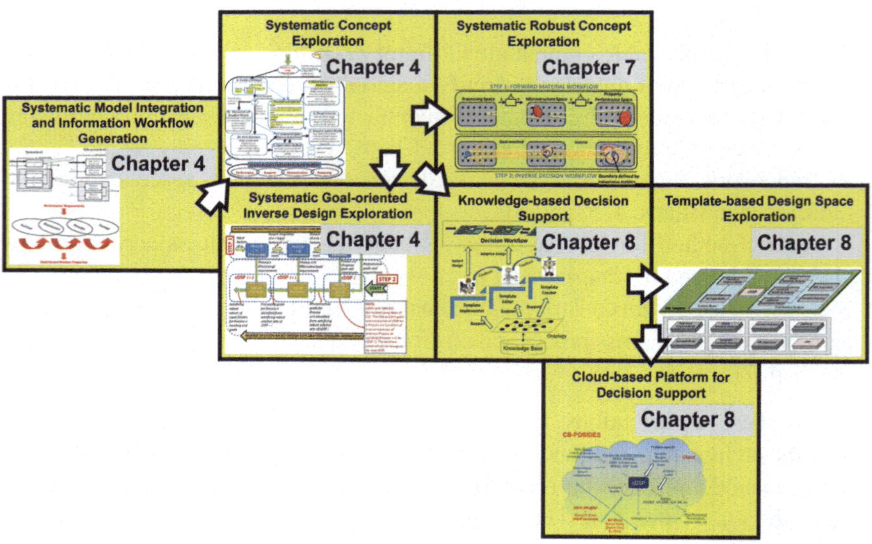

Fig. 8.2 Relationship of the research efforts presented in the previous chapters with the constructs of the systems-based design architecture presented in the current chapter

8.2 Integrated Materials, Products, and Manufacturing Process Design in the Digital Era

8.2.1 Realizing the Co-design of Materials, Products, and Manufacturing Processes

The integrated computational materials engineering (ICME) (Pollock et al. 2008; Horstemeyer 2012) community aims at exploiting the advancements in computational modeling tools and simulations supported by knowledge-based engineering tools for realizing the systems-based, integrated design exploration of materials, products, and manufacturing processes meeting end customer requirements. In order to foster integrated materials and product design as required by ICME in the current digital world and realize its full potentials, there is a need to facilitate co-design. *We define co-design as the capability of a network of participants, that includes material scientists, systems designers, software developers, end customers to come together and share material/product/manufacturing process/market data, information, knowledge, and resources instantly and in an integrated fashion, and thereby collaborate so as to facilitate a cost-effective co-creation of value supporting open innovation*. This necessitates the creation of computational platforms that enable seamless, yet controllable, information, knowledge and resource sharing supporting the integrated design of materials, products, and manufacturing processes. The domain-independent platform thus developed should support model development; model and simulation software integration; problem formulation; design and solution space exploration and visualization; data, information and knowledge management (capture, store, retrieve); and knowledge-based design guidance so as to provide design decision support for materials and product designers who might be collaborating from different parts of the world. The platform from the larger context of ICME should support designers in coming with smart and intelligent manufacturing procedures for product realization with reduced design and production time in developing robust products. Digital manufacturing and design innovation paradigm, also known as DMDI, and its new product development models/technologies like cloud-based design and cloud-based manufacturing support this need. As discussed by Thames and Schaefer (Thames and Schaefer 2017), technologies such as cloud-based design and manufacturing systems, the internet of things (IoT), the industrial internet of things, and social-product development are driven by technologies and innovations that are disruptive leading to massive creation of value to those involved in the market sectors. These disruptive transformations are a result of the convergence of industrial systems with advanced computing technologies, sensors, and ubiquitous communication systems (Thames and Schaefer 2017). In this chapter, we address how the technologies associated with DMDI, especially cloud-based design can change the way we realize materials and products and the way they are designed.

One major change that has happened with these new technologies is the power shift from the hierarchical business models that used to exist in the industry to cooperative collaboration networks with a mindset of sharing to gain. This is true in the case of design also. Using traditional product realization design paradigms, like Beitz et al. (1996) systematic design approach and Suh's axiomatic design (Suh 1990), a designer is able to describe product development as a series of core transformations where information is shared sequentially. However, these traditional design paradigms are not competent to address the changing needs and technologies associated with integrated materials and product design. "Neither Pahl and Beitz design method nor Suh's Axiomatic Design offers a framework that facilitates seamless information, knowledge and resource sharing, or aids participants of global value co-creation networks in identifying potential collaboration partners or resource providers" (Franke et al. 2006; Wu et al. 2012; Thames and Schaefer 2017).

The need, therefore, identified in this chapter from the larger context of ICME is for a platform that facilitates a network of participants involved in materials and product realization to share information, knowledge, and design/manufacturing resources so as to facilitate co-creation of value in a more cost-effective manner. Thus, traditional product development methods need to be updated and bridged to the new developments happening in the globalized world like cloud-based design and manufacturing, the foundations for which are provided in (Wu et al. 2012). In the next section, we discuss the shifts happening in design and manufacturing in the digital era.

8.2.2 Digital Manufacturing and Design Innovation Paradigm

A major transformation happening in the industry is the interconnection and interplay of computers and other technologies resulting in a shift in the way traditional products and services are realized. This has resulted in the new paradigm of digital manufacturing and design innovation (DMDI). The goal here is to use the advances in computing power, sensing technologies, and cloud-based instant communication to improve the lifecycle of a product through integrated information flow, intelligent decisions, and instant feedback. Digital manufacturing is defined by the Digital Manufacturing and Design Innovation Institute (DMDII) as *an integrated suite of tools that work with product definition data to support tool design, manufacturing process design, visualization, modeling and simulation, data analytics, and other analyses necessary to optimize the manufacturing process*. Through digital manufacturing, industries are able to quickly, securely, and authoritatively share data, information, knowledge, and resources during the design process spanning the entire lifecycle of the product. The data thus generated for a product's lifecycle is captured as a ***digital thread***. The digital thread enables users to link computer-based digital twins for the product/service that the industry is focusing on along with data across the lifecycle of the product/service. The digital twin can be viewed as a computer-based integrated system of data, models, and analyses that can be used in the design, manufacturing, support, supply chain, and disposal, essentially across the entire lifecycle of the product connected through the digital thread. The US Air Force (USAF) (2013) defines digital thread as follows: "*the creation and use of a digital surrogate of a materiel system to allow dynamic, real-time assessment of the system's current and future capabilities to inform decisions in the Capability Planning and Analysis, Preliminary Design, and Detailed Design, Manufacturing, and Sustainment acquisition phases*". According to Hehenberger and Bradley (2016), the ***digital surrogate*** is, therefore, a physics-based technical description of the system resulting from the generation, management, and application of data, models, and information from authoritative sources across the system's lifecycle.

The concept of digital thread supports the seamless integration of data, information, and knowledge across the value chain, starting from requirements gathering and ideation to design, manufacturing, testing, supply chain, and finally to the sustainment and disposal phases—resulting in a closed-loop of cradle-to-cradle transition of knowledge (see Fig. 8.3). In Chap. 2, Fig. 2.7, we show the cradle-to-grave transition of knowledge for a product across manufacturing process chains, as represented by Horstemeyer (2012). Due to the current digital transformations, the cradle-to-grave transition of knowledge, which is an open-loop knowledge transfer system, is being replaced with the cradle-to-cradle closed-loop knowledge transfer system that is supported by instant feedback, sensor-based controls, and process deviation management through intelligent decisions. In Fig. 8.3, we show how the digital thread supports seamless data, model, information, and knowledge integration across the value chain resulting in a highly integrated and intelligent digital design, manufacturing, supply chain, sustainment, and disposal enterprise. Along with the digital thread, the other components that support the digital manufacturing and design paradigm are Advanced Analysis, Intelligent Machining, Advanced Manufacturing Enterprise, and Digital Manufacturing Security. Readers interested in each of these topics are directed to the Digital Manufacturing and Design course offered by the University at Buffalo through Coursera where each of these topics is covered extensively. In the next section, we identify the core competencies needed for a successful platform in facilitating product realization in the digital industrial era.

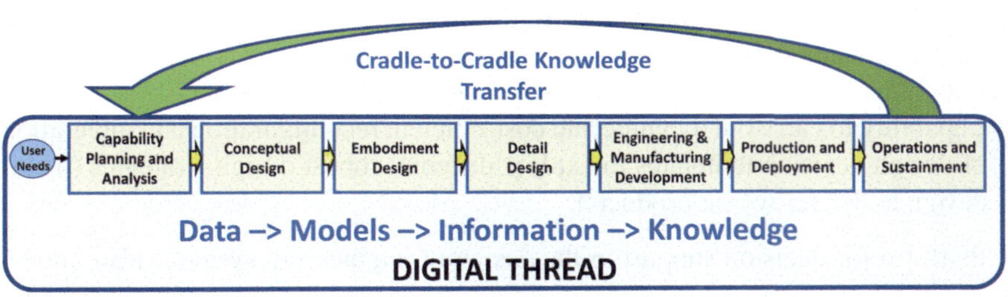

Fig. 8.3 Cradle-to-cradle knowledge transfer across the product lifecycle using digital thread. The figure is inspired by the information in the US Air Force Presentation, see Kraft (2013)

8.2.3 *Core Competencies Needed for Product Realization in the Digital Age*

Model-based realization of complex systems necessitates designers to deal with models that are typically incomplete, inaccurate, and not of equal fidelity. One focus of the integrated materials and product design research in industry is in mitigating the uncertainty in models by seeking "perfect" models, collecting more data, developing improved methods to model, calculate and quantify uncertainty through expensive computations. There are several challenges associated with this, as highlighted by McDowell (2018). The alternative to this is to focus on managing the uncertainty by designing the material/product systems to be insensitive to the sources of uncertainty without reducing or eliminating them—also defined as robust concept exploration (Nellippallil et al. 2018b), which is our focus in this monograph. Successful institutionalization of integrated materials and product design in the industry, however, requires the development of platforms that can enable human designers to carry out materials and product design using simulation models by providing design decision support.

The core competencies needed for a designer using the platform developed for implementing integrated materials and product realization in the digital industrial era are identified as follows:

- Capability to integrate models and simulation tools spanning different processes and length scales (typically defined as vertical and horizontal integration in the ICME context, see Chap. 2 for details),
- Capability to define computational workflows involving decision making, spanning multiple activities and users; define modular, reusable sub-workflows for specific processes,
- Capability to connect to external databases on materials, products, and processes,
- Capability to provide knowledge-guided assistance to different types of users in design-related decision-making,
- Capability to carry out collaborative, multidisciplinary design, and privacy control,
- Capability to manage complexity (reduced cost of computation via surrogate models/metamodels),
- Capability to explore and visualize the design and solution space,
- Capability to carry out dynamic and cost-efficient reconfiguration and integration of design decision templates to explore different robust design strategies (meta-design to deliver robust products).

Platform for decision support in the design of engineering systems, also known as PDSIDES, is a knowledge-based platform that embodies these capabilities and supports a designer/user to realize the integrated decision-based design of materials and products in this digital age. In the sections that follow we discuss PDSIDES in greater detail. The rest of the chapter is organized as follows. In Sect. 8.3, we discuss platform PDSIDES and propose the cloud-based platform CB-PDSIDES. In

Sect. 8.4, the core functionalities offered by CB-PDSIDES are discussed from the context of integrated materials and product design. The demonstration of the key core functionalities of CB-PDSIDES is carried out in Sect. 8.5. We discuss some of the broader applications of CB-PDSIDES in Sect. 8.6. We end the chapter with our closing remarks on advancing the vision of integrated materials, products, and manufacturing process design as required by ICME and identifying opportunities that are worthy of further investigation in Sect. 8.7.

8.3 Cloud-Based Platform for Decision Support in the Design of Engineering Systems (CB-PDSIDES)

8.3.1 Platform for Decision Support in the Design of Engineering Systems (PDSIDES)

Mistree and co-authors (Mistree et al. 1990) define design as the conversion of information that characterizes the needs and requirements for a product into knowledge about the product. **The underlying philosophy in the definition of design by Mistree and co-authors (Mistree et al.** 1990) **is that the designer starting with the functional requirements that are desired (the goal that designer wishes to achieve) should be able to work backward to explore effective design solutions.** This philosophy is adopted in this chapter for design—as a goal-oriented activity. As noted by Gero (1990), the designing involves transforming requirements—generally termed *functions*—into design descriptions. *Decision-based design* (Shupe 1988; Mistree et al. 1990) is a term coined to emphasize a different perspective to develop methods for design. The principal role of a human designer in decision-based design (DBD) is to make decisions given the information available. From an engineering perspective, decisions exclusively deal with the allocation of resources in some form, usually as capital expenditures. Thus, the definition of a decision here is "an irrevocable allocation of resources" (Hazelrigg 1996). We believe that there are two types of decisions that a human designer can make: selection and compromise decisions. A complex design can be represented by modeling workflow of compromise and selection decisions; achieved using PDSIDES.

PDSIDES (Ming et al. 2018) is a "knowledge-based" platform for decision support in the design of engineering systems that is anchored in modeling decision-related knowledge with templates using ontologies to facilitate execution and reuse. The two primary constructs required for the realization of decisions within PDSIDES are (Ming et al. 2018): (1) decision support problem (DSP) construct, and (2) ontology. Three types of platform users are defined according to the amount of knowledge they have for operating the decision template, namely, template creators, template editors, and template implementers. Template creators are domain experts, responsible for creating decision templates for original design, which require the greatest novelty. Template editors are senior designers who have sufficient knowledge and experience in a specific domain and are responsible for editing (or tailoring) existing decision templates in adaptive design; this requires the original templates to be adapted for new applications. Template implementers are designers who have basic knowledge and typically little knowledge or interest in the analysis embodied in the template; they are responsible for executing existing decision templates that result in variant designs that require only parametric changes in the original decision templates. Ming et al. (2017a) present the ontologies developed for selection decision, compromise decision, and hierarchical coupled decisions. The ontologies are defined using the software tool, Protégé (Musen 2015). The classes and properties of the relevant ontologies are presented by Ming and co-authors, see (Ming et al. 2017a, b). Ming et al. (2018) present the very first version of platform PDSIDES which is web-based and deployed in the local server of the Systems Realization Laboratory at the University of Oklahoma.

8.3.2 Primary Constructs Used in PDSIDES

The primary constructs used in PDSIDES include (Ming et al. 2018):

Decision Support Problem
PDSIDES is designed from a decision-based design (DBD) perspective, wherein decisions serve as markers to identify the progression of design from initiation to implementation to termination (Mistree et al. 1990). We recognize that the implementation of DBD can take many forms, such as (Hazelrigg 1998); our implementation being the decision support problem (DSP) technique (Muster and Mistree 1988). The key to the DSP technique is the notion that there are two types of decisions, namely selection and compromise, and that a complex design can be represented by modeling workflow of compromise and selection decisions. The design of complex systems may require the formulation and resolution of a series of coupled decisions, in which case the hierarchical DSP construct based on the selection decision support problem (sDSP) and the compromise decision support problem (cDSP) is used as the model to support hierarchical decision-making; for the detailed mathematical model, see (Smith 1985; Bascaran et al. 1987). A tailored computational system, known as DSIDES—decision support in the design of engineering systems (Reddy et al. 1996)—is integrated into PDSIDES to solve such problems.

Decision Template
One of our primary goals in designing PDSIDES is that designers can rapidly create decision models for the specific design problems they have by using the DSP constructs, and making decisions, and finally the produced decision knowledge can be stored and reused by other users for similar designs. To achieve this goal, the DSPs are represented as computational decision templates in PDSIDES. Decision templates, originally proposed by Panchal and co-authors (Panchal et al. 2004), make it possible to model the compromise DSP so that the template is reusable and computer

interpretable. The idea is extended to model the selection DSP and hierarchical DSP as templates (see Ming et al. 2017a, b). The key to the computational DSP templates is the modularization of the DSP constructs and the separation of declarative and procedural knowledge, which allows both to be reused across problems.

Ontology

In order to store and reuse the knowledge archived in the DSP templates in a computational environment, there needs to be a formal representation scheme. Ontologies are defined by Gruber (1993) as explicit formal specifications of terms, and relations among them are increasingly used for knowledge modeling in engineering design (see Rockwell et al. 2010; Zhan et al. 2010). Ontological frameworks are also being developed for realizing the needs of ICME in industry (see Yeddula et al. 2016; Reddy et al. 2017). In PDSIDES, ontology is used to formally represent the knowledge (including declarative and procedural knowledge) archived in the DSP templates. There are two popular paradigms for ontology formalism, namely web ontology language (OWL) and frame (Wang et al. 2006). The frame paradigm is chosen because it is based on a closed-world assumption wherein everything is prohibited until it is permitted, which is suitable for modeling the highly constrained DSPs. In frame-based ontology, terms are defined as classes, and relations are defined as slots. For detailed specification of the classes and slots, see earlier works on PDSIDES (Ming et al. 2017a, b).

The advantages of the use of ontology in PDSIDES are as follows (Ming et al. 2018):

- ***Facilitate knowledge sharing***. This is embodied in two aspects, namely knowledge sharing among different users in PDSIDES and knowledge sharing between PDSIDES and other product lifecycle management (PLM) platforms using the DSP ontologies which are essentially a common language for design decision-making in PDSIDES.
- ***Facilitate knowledge population***. The DSP ontologies are the abstractive representations of the templates, which are very convenient for instantiating different instances with specific information.
- ***Facilitate knowledge retrieval***. The DSP ontologies capture the complex semantic relationships among the modules and templates, which allow it to support semantic-based retrieval that can respond to comprehensive query needs. For details about semantic retrieval, see (Mocko et al. 2007).
- ***Facilitating consistency maintaining***. Modification of the original templates usually happens in adaptive or variant design, which may lead to inconsistency of the modified templates since the arrangement or values of the modules are changed. The DSP ontologies support rule-based reasoning and appropriately handle the inconsistency, which is discussed in Sect. 8.3.3.

8.3.3 Platform Overview

An overview of PDSIDES is illustrated in Fig. 8.4. PDSIDES is divided into three parts: knowledge, users, and decision-based design. What follows is the description of the platform from the bottom-up that includes how these three parts are connected to enable the functionalities.

At the bottom of PDSIDES, decision-related knowledge is stored in the knowledge base. The knowledge including declarative knowledge such as *problem statement, alternatives, attributes, variables, parameters,* and *constraints*, and procedural knowledge such as consistency rules and computing codes (for calculating, e.g., expected utility of a sDSP template) are organized by a holistic ontology; see (Ming et al. 2018) for details. In the middle part are the three types of users, namely the template creator, template editor, and template implementer. The three types of users embody three different levels of knowledge (represented by the stairs in Fig. 8.4). The top-level is the template creator who is responsible for creating the DSP templates, the middle level is the template editor who is responsible for editing DSP templates, and the bottom level is the template implementer who is responsible for implementing the DSP templates. The interactions among the three types of users are a closed-loop, where the template operational guidance is passed downwards

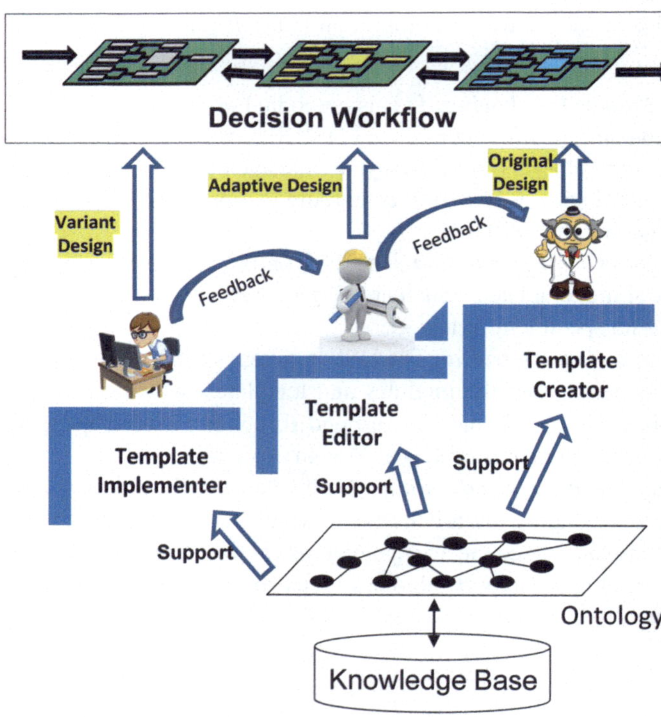

Fig. 8.4 PDSIDES overview (Ming et al. 2018)

from the creator to the editor to the implementer, and the feedback of operating the templates is sent upwards from the implementer to the editor and then to the creator. The creation, edit, and implementation of the DSP templates are all facilitated using the holistic ontology. The top part of PDSIDES is about decision-based design. In PDSIDES, design is classified into three types, namely original design, adaptive design, and variant design; all are realized from a decision-based perspective using the DSP templates. In specific design cases, the underlying decision workflow is represented by networked DSP templates that can be exercised by three types of users through creating, editing, and implementing.

8.3.4 Users and Working Scenarios

The definitions of three types of users are introduced and their associated working scenarios are described in detail in this section.

Template Creator: Template creators are domain experts and are responsible for creating DSP templates for original design that calls for new concepts.

Template Editor: Template editors are senior designers who have sufficient knowledge and experience in a specific domain and are responsible for editing (or tailoring) existing decision templates in adaptive design, and this requires the original templates to be adapted for new applications.

Template Implementer: Template implementers are designers who have basic knowledge and typically little knowledge or interest in the analysis embodied in the template; they are responsible for executing existing decision templates that result in variant designs that require only parametric changes to the original decision templates.

8.3.5 Cloud-Based PDSIDES

PDSIDES has the potential to support a human designer with the identified core competencies for product realization when integrated with the cloud. In Fig. 8.5, we illustrate the concepts underlying the foundations and principles of CB-PDSIDES as described in this chapter. To integrate PDSIDES with cloud and bring-in the concepts of cloud computing and collaboration into product design and manufacturing, we adopt the definition for cloud-based design and manufacturing as proposed by Wu and co-authors (2012): "*Cloud-Based Design and Manufacturing refers to a product realization model that enables collective open innovation and rapid product development with minimum costs through a social networking and negotiation platform between*

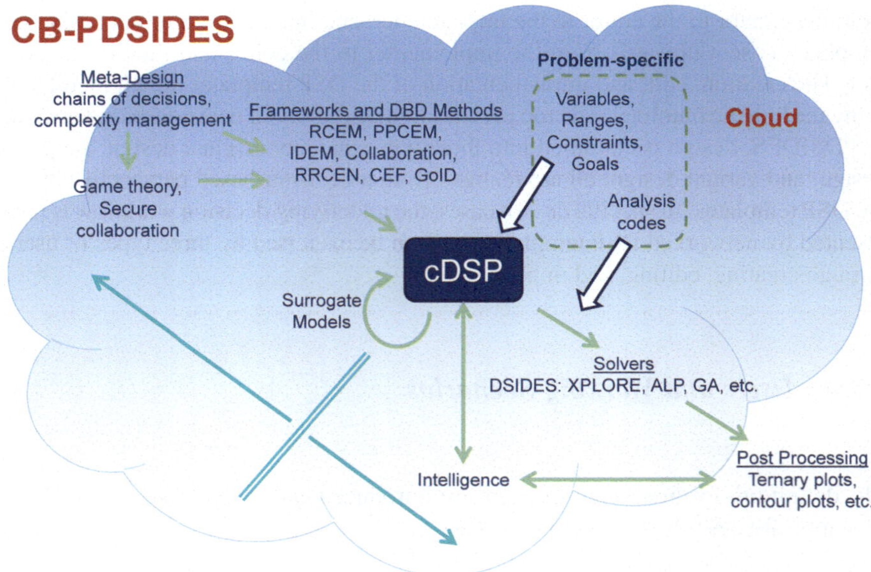

Fig. 8.5 An illustration of the cloud-based platform for decision support in the design of engineering systems concept

service providers and consumers. It is a type of parallel and distributed system consisting of a collection of inter-connected physical and virtualized service pools of design and manufacturing resources (e.g.: parts, assemblies, CAD/CAM tools) as well as intelligent search capabilities for design and manufacturing solutions."

In this chapter, we present PDSIDES Version 2.0, namely, cloud-based PDSIDES (CB-PDSIDES). CB-PDSIDES is currently developed to run in a virtual machine on Google Cloud Computing Engine. The platform PDSIDES integrated with the cloud has in its core a decision support gene. We address two types of decision support genes via CB-PDSIDES: compromise and selection decision support problems (known as cDSP and sDSP, respectively) (Mistree et al. 1993, 1994). The execution of decision support genes on a computer is foundational to the platform to realize complex engineered systems. The developed ontologies for selection decision, compromise decision, and hierarchical coupled decisions are implemented in CB-PDSIDES for representing decision workflows. In Fig. 8.5, we illustrate the concept of CB-PDSIDES with the cDSP construct at its core as the fundamental decision support construct. The problem-specific information (declarative knowledge) is captured using the cDSP via the keywords: *Given*, *Find*, *Satisfy*, and *Minimize*. Procedural knowledge is associated with how the information transformation is carried out and details on how the transformation is executed via a decision workflow or decision network. This is captured via decision-based design templates like *Process Template, sDSP Template, cDSP Template, Surrogate Modeling Template, Design Space Exploration Template, Robust Design Template* (Wang et al. 2018). The analysis codes and simulations associated with the problem framed are also communicated to the decision support construct (see Fig. 8.5). Surrogate modeling techniques and tools are available in the platform via the *Surrogate Modeling Template* to support the designer in managing the complexity and coming up with reduced order models; see (Wang et al. 2018) for details. Post-processing tools, like ternary plots, contour plots, and so on, that are automated with rules are available via the *Design Space Exploration Template* to help the designer easily explore the solution space and identify robust solution regions of interest by managing uncertainty; see (Wang et al. 2018) for details. Frameworks and design methods anchored in the decision-based design paradigm that incorporates the decision support genes (cDSP or sDSP) are incorporated in the cloud-based platform to support the designer to formulate and execute design problems systematically. These include robust concept exploration method (RCEM) (Chen et al. 1997), product platform concept exploration method (PPCEM) (Simpson et al. 1996), inductive design exploration method (IDEM) (Choi et al. 2008), concept exploration framework (CEF) (Nellippallil et al. 2017a), goal-oriented, inverse design (GoID) method (Nellippallil et al. 2018b), and GoID method with robustness (Nellippallil 2018, 2018a). Integrated knowledge, that is, integration of declarative and procedural knowledge is captured via the design methods and frameworks available in the platform. The computational solvers associated with the execution of problem formulated like DSIDES (decision support in the design of engineering systems) (Reddy et al. 1992) for cDSP construct are available to be accessed via the cloud. Collaborating designers can access the cloud-based platform from different parts of the world to share information and formulate design problems worthy of further exploration. The knowledge associated is captured and stored in CB-PDSIDES and can be retrieved and shared instantly via cloud with the collaborators depending on the design requirements. The issues of collaboration and information sharing are also addressed as collaboration, and communication is the key in the cloud-based PDSIDES. This is addressed via paradigms like crowd-sourcing, mass collaboration, and social product development (Thames and Schaefer 2017).

8.3.6 Architecture of Cloud-Based PDSIDES

In Fig. 8.6, we show the architecture of CB-PDSIDES. The computing architecture for CB-PDSIDES follows the architecture of cloud-based design and manufacturing systems proposed by Wu and co-authors (2015). The architecture of CB-PDSIDES includes five layers: (i) user layer, (ii) web portal layer, (iii) logic layer, (iv) virtual layer, and (v) physical layer.

Since CB-PDSIDES is deployed in the cloud, users can gain access to CB-PDSIDES via PCs and smartphones over the internet. The web portal layer of CB-PDSIDES includes the user interaction GUI for accessing the design templates available. The user interaction GUI includes: template searching and browsing GUI which are designed for locating the required DSP templates and presenting them,

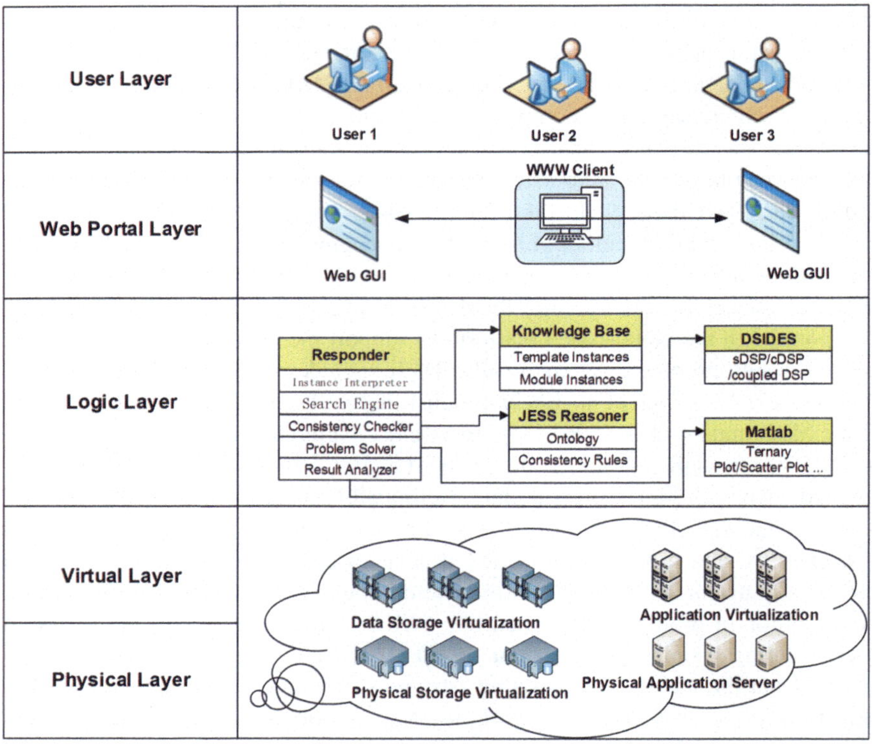

Fig. 8.6 Computing architecture of CB-PDSIDES

template creating and editing GUI which are designed based on the DSP template structures for the purpose of instantiation and modification of the DSP templates, the template execution and analysis GUI which are designed for executing DSP templates and performing post-solution analysis. The GUI is allowed to communicate with the logic layer of CB-PDSIDES by a request-response mode using the hypertext transfer protocol (HTTP). The logic layer of CB-PDSIDES includes five main parts, namely, response server, knowledge base, JESS reasoner, DSIDES, and a link to the commercial software MATLAB. The response server is the central "brain" that integrates the other four parts for responding to requests. The response server itself has five components, including a search engine, an instance interpreter, a consistency checker, and a problem solver. The instance interpreter is for interpreting the data collected from the template creators (or editors) and formatting it into DSP template instances according to the DSP ontologies. The generated template instances and module instances are stored in the knowledge base. The search engine is connected to the knowledge base to provide ontological semantic-based knowledge retrieval. Consistency checking is facilitated through a consistency checker together with the JESS reasoner—the rule engine for the JavaTM platform, which provides rule-based intelligence inference. The problem solver is connected to DSIDES for solving the DSPs. DSIDES (decision support in the design of engineering systems) is a tailored computational environment that supports the execution of the decision support problem templates. DSIDES is invoked when a template executer executes a DSP template. The result analyzer is included to help users, especially template implementers, analyze the results produced by the problem solver, DSIDES. MATLAB and its features are used to carry out data visualization through ternary plots and scatter plots. These plots are used for visualizing the DSP results and further carry out solution space exploration. Therefore, MATLAB and its features are linked to CB-PDSIDES for solution space visualization, solution space exploration, and post-processing. The virtual layer and physical layer in CB-PDSIDES support data storage and retrieval for multiple users who are connected via the cloud, thereby establishing a networked collaborative environment. As noted by Wu and co-authors (Wu et al. 2015), through virtualization the computing resources of CB-PDSIDES like the DSP templates, solvers, post-processing resources, data storage, and so on are separated from physical computing hardware and reallocated dynamically to the different applications based on the needs of users. Through this unified computing architecture of CB-PDSIDES, we are able to support multiple tenants through a single instance of the platform PDSIDES; defined as multitenancy (Wu et al. 2015). These features of the computing architecture of cloud-based PDSIDES differentiate it from the other web-based services. In the next section, we present some of the core functionalities offered by CB-PDSIDES for integrated materials and product realization.

8.4 Core Functionalities Offered by CB-PDSIDES

CF1. ***Modular Reuse of Design Workflows along a Design Process (Knowledge Management)***

Core functionality 1 (CF1) is the capability of CB-PDSIDES to support a designer in designing new design workflows by reusing past knowledge from similar design problems. Specifically,

- Support for reusing and reconfiguring workflows for different conditions and problems;
- Reconfiguring the computational workflow developed for one product to the design of another product, see Fig. 8.7.

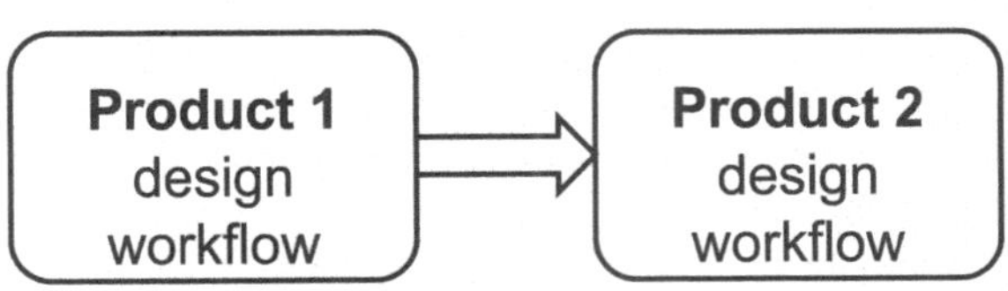

Fig. 8.7 Modular reuse of design workflows along a design process

CF1 is possible using CB-PDSIDES by means of an ontology that provides a common vocabulary for representing domain-specific knowledge. The ontologies for representing the knowledge captured via the decision support genes, cDSP, and sDSP have been developed, and they form the foundations for CB-PDSIDES; see (Ming et al. 2017a) for details. A PEI-X (phase-event-information-X) ontology for meta-design is included in CB-PDSIDES for designing decision workflows. Details regarding the ontologies developed are available in (Ming et al. 2017a, 2018; Wang et al. 2018).

The focus from integrated materials and product design context is on concurrently exploring the design space of both the products and the materials and narrow the set of possible options in the shortest possible time with minimum expense. Hence, instead of exploring the complete design space from first principles using detailed models, the focus is on exploring simplified models that are good enough to compare different design alternatives. Additionally, the notion of designing the workflows by reusing past knowledge from similar design problems is important because of:

(i) evolving simulation models, resulting in multiple fidelities of models at different stages of a design process, and
(ii) significant model development and execution costs, necessitating judicious use of resources.

Moreover, the needs for accurate information depend on whether the goal is to narrow down the alternatives to a specific class of materials (i.e. during the early design phase) and products or to design the composition and structure of a specific material system (i.e. during the later stages of design). To support the need to generate information at variable fidelities during the design process, the following features are offered via CF1 by CB-PDSIDES:

(i) *Support for reusing and reconfiguring workflows for different conditions and problems;*
(ii) *Support for complexity and uncertainty management: Provide computational techniques to measure how complexity and uncertainty changes by replacing different components of the workflow;*
(iii) *Model management and knowledge-based idealizations: Representation of models at different levels of abstraction, along with information about their accuracy.*

CF2. ***Design Workflows in Distributed Collaborative Settings***

CF2 is the capability of CB-PDSIDES to support designers in design collaborations. CF2 is possible using CB-PDSIDES through a secure co-design (SCD) framework that uses information theory (Panchal et al. 2004, 2005) and game theory (Lewis and Mistree 1998; Panchal et al. 2009) protocols to identify co-design solutions while preserving confidentiality of the information shared between the participating design collaborators (see Fig. 8.8).

From the integrated materials and product design context, this could be a scenario where components are designed by one organization and materials are designed by

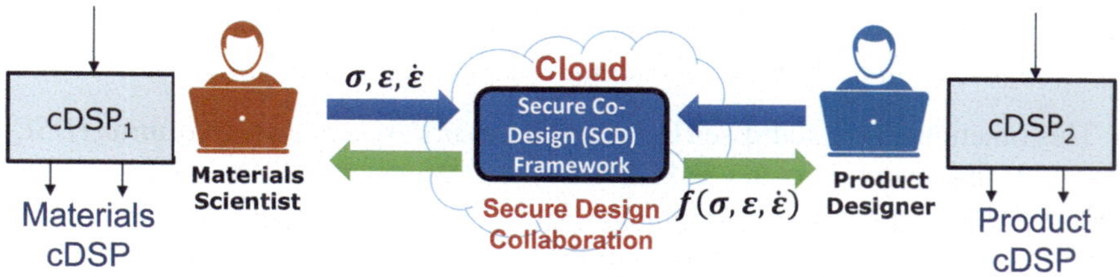

Fig. 8.8 Secure co-design framework for cloud-based secure collaboration

another organization where information and knowledge sharing needs to be managed. An example of this could be a product-level cDSP formulated by product designers and a materials-level cDSP formulated by materials scientists. Let us assume that the materials scientists have proprietary models that they do not wish to share (explicitly or implicitly) with the product designers. In a similar sense, the product designers do not wish to share all the information in the product-level cDSP with material scientists. However, both parties would like to jointly design the product and the material and are connected to each other via the cloud. The collaborative nature of the design process induces additional design issues to be addressed for the management of design workflows, such as:

- Collaborative authoring of workflow templates (check for consistency among collaborating entities)
- Privacy-preserving collaboration in integrated products and materials design

CF3. ***Reduced Cost of Computation and Management of Complexity***

There are several methods for simulating various aspects of materials processing and product design. However, it can be very costly and time-consuming to compute and re-compute these simulations in the process of design, especially in the early stages of design where it is desirable to explore a wide range of options rather than developing detailed designs. For situations like this, there are several ways of developing surrogate models (metamodels) which rapidly provide design information, each of these will give meta-models of different degrees of accuracy at different costs which may be used at different stages of design; see (Simpson et al. 2001; Montgomery 2008) for details. Thus, there is a need to assess the benefits of using different meta-models in different stages of design and compare these with the costs of developing these meta-models. Using meta-models of increasing fidelity in a design process is one way of exploring the design space, and an alternative way is by using robust design with decreasing bands of robustness. The advantages and limitations of each of these approaches have to be considered. To support the need to reduce computational cost and manage complexity, CB-PDSIDES offers the following capabilities:

(i) *Support to develop reduced-order models of various degrees of fidelity using simulations and assess reductions in computational costs when using these models.*

(ii) *Support to combine the use of meta-models with varying degrees of robust design and assess tradeoffs between accuracy and computational costs.*

This functionality is addressed by two modules in *design space exploration (DSE) process template* ontology in CB-PDSIDES; see (Wang et al. 2018) for details. The DSE process template has three subtemplates: *Problem Model* (PM), *compromise Decision Support Problem* (cDSP), and *Post-Solution Analysis* (PSA), see Fig. 8.9. The *Problem Model* (PM) subtemplate in *DSE* process template has two modules: *Theoretical and Empirical Model* and *Surrogate Model*. The *Theoretical and Empirical Model* module is used to capture the information and knowledge associated with already existing and available material and product models. An example of this is capturing the information and knowledge associated with an existing constitutive material model in the literature that establishes stress as a function of strain, strain rate, temperature, and other material internal variables. The *Surrogate Model* module is used when there is a need to develop reduced-order models or meta-models that capture the relationships between responses and the corresponding factors. Template instances of the surrogate models developed are stored as new knowledge in CB-PDSIDES to facilitate reuse. Collaborating designers are able to use the cloud to develop, share, and use models to formulate design problems using CB-PDSIDES.

CF4. ***Cost-efficient Integration of Templates for Product Development—Carry out Meta-Design***

CF4 is the capability of a designer to use CB-PDSIDES in exploring the effects of changing the ways in which templates are integrated on the outcomes of the design workflows. This is achieved in a modular fashion by means of a 3-P information model proposed by Panchal and co-authors (Panchal et al. 2005). The three Ps refer

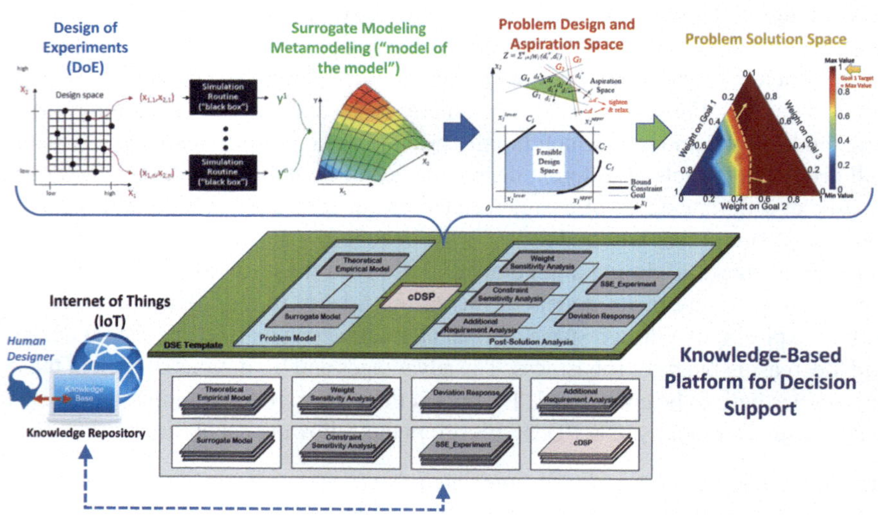

Fig. 8.9 Design space exploration using CB-PDSIDES

to the key design information elements—*Product*, *Problem*, and *Process*. Each of these is defined in an independent modular fashion. Designing a product is possible by using different types of un-instantiated problem and process templates. Examples of different possibilities in problem template include Archimedean cDSP formulation (Mistree et al. 1993), pre-emptive cDSP formulation (Mistree et al. 1993), robust design Type I formulation (Chen et al. 1996), robust design Type II formulation (Chen et al. 1996), Type I, II, and III robust design formulation (Choi et al. 2005), robust design Type IV formulation (Choi et al. 2008), traditional optimization, and so on. Examples of different possibilities in the process template include point-based iterative search, sequential design, set-based design, RCEM or CEF using response surface models (Chen et al. 1997), goal-oriented inverse design of process chains (Nellippallil et al. 2018a), and so on. Examples of the different product templates include design of pressure vessel (Ming et al. 2016), design of hot rod after rolling (Nellippallil et al. 2017b), design of slab after casting (Shukla et al. 2015), design of multiscale materials and products (Sinha et al. 2013), and so on. The benefit associated with this functionality from the integrated materials and product design context is the capability to realize configurations of different problems with different processes and applying these for a variety of product design scenarios. This leads to faster product realization schemes in a cost-efficient manner as needed for ICME. Using the cloud, participating material scientists and designers are able to collaborate in the integration of templates for product development depending on their expertise—in problem, product, and/or process information elements.

CF5. ***Systematic Solution Space Exploration by Managing Uncertainty***

Design and solution space exploration by considering system uncertainty is essential for the model-based realization of complex material/product systems. As discussed in previous chapters, the models that are available are typically incomplete, inaccurate, and not of equal fidelity. Hence, seeking single-point optimum solutions are typically not valid as these optimum solutions no more hold if any variations occur—which is bound to happen with a complex material/product system. This necessitates the need for systematic design and solution space exploration to identify satisficing solutions that perform well and are relatively insensitive to the uncertainty present in the system—defined as robust solutions. Using the cloud-based platform CB-PDSIDES, designers can collaborate from different parts of the world in formulating robust design problems and exploring the design and solution space using rules defined in the platform to identify solutions that are relatively insensitive to these variations. The exploration of the solution space provides designers with the knowledge to refine or improve the model especially at the early stages of design. This functionality is supported by the *Post-Solution Analysis* (PSA) subtemplate within the *Design Space Exploration* (DSE) template of CB-PDSIDES; see (Wang et al. 2018) for details. The three modules *Weight Sensitivity Analysis*, *Constraint Sensitivity Analysis*, and *Additional Requirements Analysis* that form part of PSA subtemplate of CB-PDSIDES are used depending on the requirements of the problem to explore and visualize the solution space and identify solutions that are relatively insensitive to uncertainty (see

Fig. 8.9). The designers are able to formulate a robust design problem by instantiating a Type I, II, III, or IV robust design problem template in CB-PDSIDES.

CF6. ***Cloud-based Design Communication—Instant Feedback Across Design Workflows***

A key goal in design is the proper communication of the design process. Wu and co-authors (2012) identify the key issue of fully understanding a complex design process in order to improve design communication. The key issue includes design tasks that need to be completed, the source for specific information that is needed for design, the individual to be contacted for the right information, the extent of distortion in the information available, and the extent to which the distorted information affects design. As discussed in Sect. 8.2, traditional product design paradigms are limited as design communication is a one-way mapping in a linear sequence across design phases/domains. PDSIDES in the cloud settings improves design communication through multiple information channels facilitated by the cloud.

The implication of CB-PDSIDES in the product realization process is explained using Gero's model of design in Fig. 8.10. Gero's model of design is explained in detail in Chap. 3. Assume that each stage of product realization starting from function to final manufacturable product descriptions involves distributed designers as shown in Fig. 8.10. PDSIDES facilitates a decision network where information is shared in one-way fashion. CB-PDSIDES, however, can facilitate collaboration and can result in a two-way and multi-way network (denoted by the dotted two-way arrows) for product realization where the distributed designers are connected

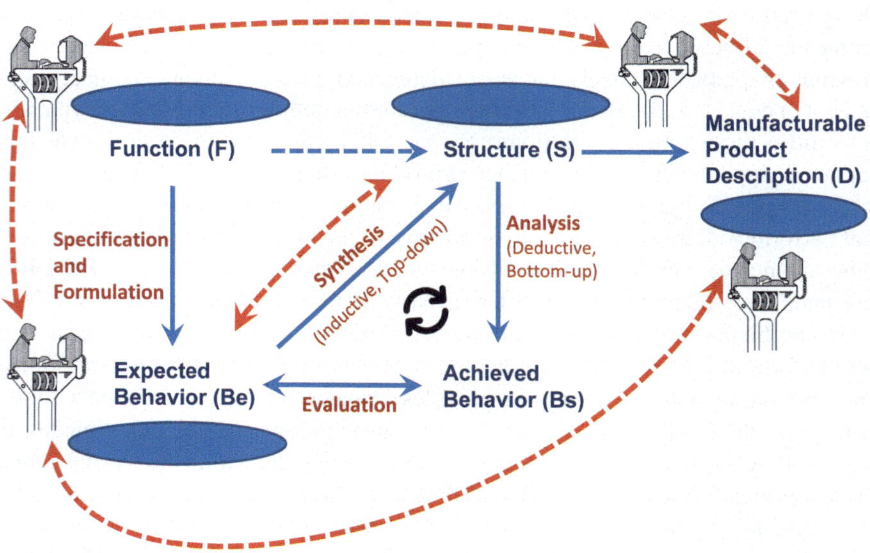

Fig. 8.10 Implication of CB-PDSIDES in product realization process explained using Gero's model of design

through the cloud. This will allow for information flow in multiple directions facilitating dynamic changes during product development and instant communication between the different design domains. In the next section, we demonstrate the core functionalities discussed via an industry-inspired problem.

8.5 Cloud-Based Design of Materials and Products—Design of a Hot Rolled Steel Rod Example

The core functionalities CF1 (modular reuse), CF3 (manage complexity), CF4 (meta-design), and CF5 (solution space exploration) are essential to the platform PDSIDES and have been tested using several example problems, see Chaps. 5, 6, 7 and the following articles (Ming et al. 2016, 2017a, 2018). In this chapter, we focus on demonstrating the core functionalities CF2 (design workflows in distributed collaborative settings) and CF6 (instant feedback across decision workflows) that are prominent to CB-PDSIDES—design workflows in distributed collaborative settings and cloud-based design communication. We define collaboration in this context as the sharing of information, knowledge, and resources between designers either in a sequential (leader-follower) or a concurrent (co-operative) manner. In order to demonstrate the functionality CF2, we view collaboration as the sequential fashion of information sharing where the first designer shares the design decisions with the second, following which the second designer makes their respective design decisions. Later, we discuss the concurrent fashion of collaboration where complete information is shared instantly with all the designers. We demonstrate these two core functionalities of CB-PDSIDES by means of an industry-inspired example problem, namely the design of a hot rolled steel rod; see (Nellippallil et al. 2017a, 2018a) for details regarding the problem.

Problem Description (see Chaps. 1 and 6 for details): Steel manufacturers are focused on developing newer grades of steel with improved mechanical properties and performances at the product level. A steel rod is a product produced after several manufacturing processes like casting, reheating, and hot rolling, and further used to develop automotive components like gears. In order to produce a gear with specified properties, there is a need to design the rod (semi-product) to satisfy certain defined mechanical properties. These mechanical properties, however, are defined by the microstructure, chemical composition, and processing of the steel. Hence there is a need to design the microstructure and processing route of the steel to achieve the target mechanical properties desired for the rod. The mechanical properties that we focus on for the rod are yield strength, tensile strength, and hardness. These mechanical properties depend on the material microstructure like ferrite grain size and phase fractions of ferrite and pearlite which in turn are defined by the initial austenite grain size, carbon and manganese concentrations, and cooling rate during the cooling process after rolling. The problem-specific requirements and goals are detailed by Nellippallil et al. (2018b) and are not repeated here. In this problem, there are three decision-makers—*Customer, Product Designer,* and *Materials Designer*. The customer specifies the requirements from the market side—this could be changing performances for the product depending on the type of applications as specified by industrial experts. Product designers are focused on designing the end product—in this case, the hot rolled and cooled rod. Materials designers are focused on designing the microstructure factors in order to realize the product.

Goal-oriented, Inverse Design Method (see Chaps. 4 and 6 for details): In goal-oriented, inverse design method (Nellippallil et al. 2018b), the designer starts with the end performance requirements for the product and designs the whole system in an inverse manner to meet these end performance goals. In Fig. 8.11, we represent the four design domains of product development on the axis of the quadrant as customer attributes (CA), functional requirements (FR), design parameters (DP), and processing variables (PV). In the lower right quadrant is the *Production Domain* involving *Process Variables* and their relationship to design parameters (Numbered 3 in Fig. 8.11). In the upper right quadrant is the *Engineering Domain* involving *Design Parameters* and *Functional Requirements* (Numbered 2 in Fig. 8.11). In the upper left quadrant is the *Customer Domain* involving *Functional Requirements* and their dependencies to *Customer Requirements* (Numbered 1 in Fig. 8.11).

In this chapter, we adopt a customer-centric perspective toward product realization where the customer attributes/needs (CAs) are the performance desired for the product. The product-level properties are the functional requirements (FRs). The microstructure of the material is the design parameters (DPs) in the physical domain

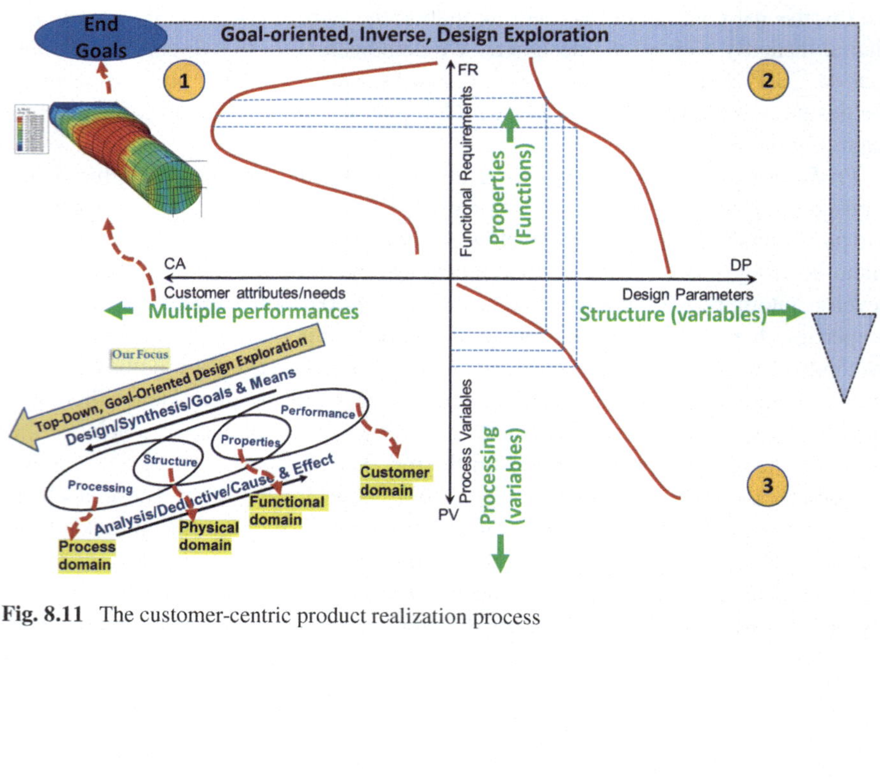

Fig. 8.11 The customer-centric product realization process

that define the product properties. The processing variables are the process variables (PVs) in the process domain that define the microstructure. The mapping of the design domains in terms of the processing–structure–property–performance is shown in the lower left quadrant in Fig. 8.11. Using goal-oriented, inverse design (GoID) method, we start from the product performance (Numbered 1 in Fig. 8.11) and design first the product properties to meet the performance (Numbered 2 in Fig. 8.11) and further the material microstructure to meet the properties (Numbered 3 in Fig. 8.11) in an inverse manner (see Fig. 8.11).

In CB-PDSIDES, using the decision workflow design panel, a decision workflow for the hot rolling problem is created, as shown in Fig. 8.12. Two decisions are defined in terms of cDSPs—a product design cDSP (created by *Product Designer*) and a materials design cDSP (created by *Materials Designer*). A detailed description of the cDSP template instantiated for the product using CB-PDSIDES is shown in Fig. 8.13. Using the template in Fig. 8.13, the product designer captures the information associated with designing the rod, like the problem variables, parameters, constraints, goals, preferences, and so on. More details on the execution of this template are available in (Ming et al. 2018). The execution in this decision workflow is sequential where the product designer first executes the product design DSP and communicates the results via the cloud to the collaborating materials designer; see

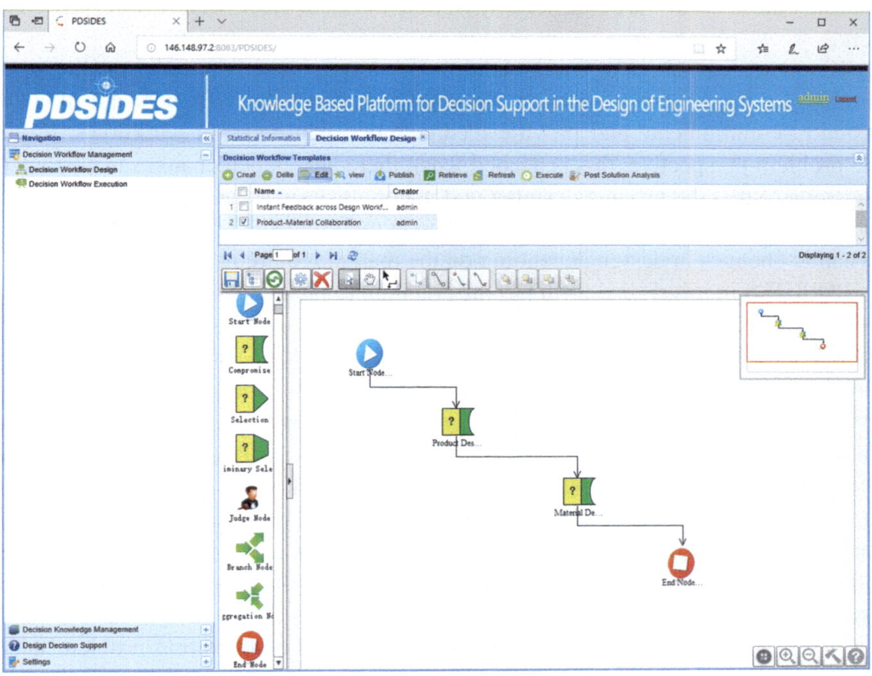

Fig. 8.12 Decision workflow in CB-PDSIDES illustrating collaboration between product design group and materials design group

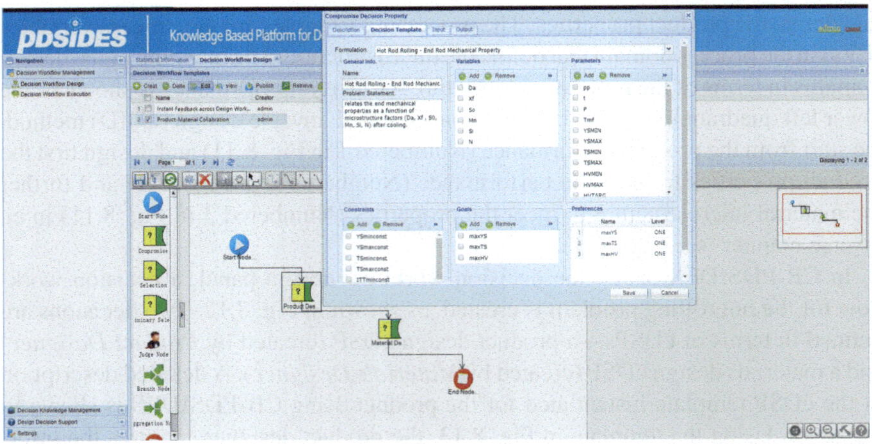

Fig. 8.13 Detailed description of cDSP template instantiated by product designer

(Nellippallil et al. 2018b) for details. After analyzing the results obtained, the materials designer further executes the materials design cDSP to make decisions. The results for this sequential execution of cDSP templates following the GoID method to design the hot rolled are described by Nellippallil et al. (2018b) and are not repeated here.

The major advantage of using cloud-based design is the core functionality CF6 (instant feedback across decision workflows)—where the designers are able to instantly communicate and provide feedback between decision workflows. From the problem perspective, this is needed when there are sudden changes in customer demands so that the product designer and materials designer are required to change their design instantly to meet the new customer requirements. Thus, there need to be dynamic changes in design depending on the market requirements.

PDSIDES in its original form has the potential to facilitate one-way communication via decision workflows. Integrating PDSIDES in the cloud settings will improve design communication through multiple information channels as facilitated by the cloud. This will allow for information flow in multiple directions thereby facilitating dynamic changes during product development and instant communication between the different design domains like the customer and physical domains or functional and process domains; as shown in Fig. 8.14. CB-PDSIDES thus can facilitate collaboration (in concurrent fashion) and can result in a two-way and multiway network for product realization where the distributed designers are connected through the cloud. This facilitates dynamic product updates, design changes, and feedback in the product realization process.

In Fig. 8.15, we show the decision workflow for the hot rolling problem with the three stakeholders—customer, product designer, and materials designer. In this configuration, all the stakeholders are connected via the cloud. The forward arrows (from start to customer to product designer to materials designer) denote the forward

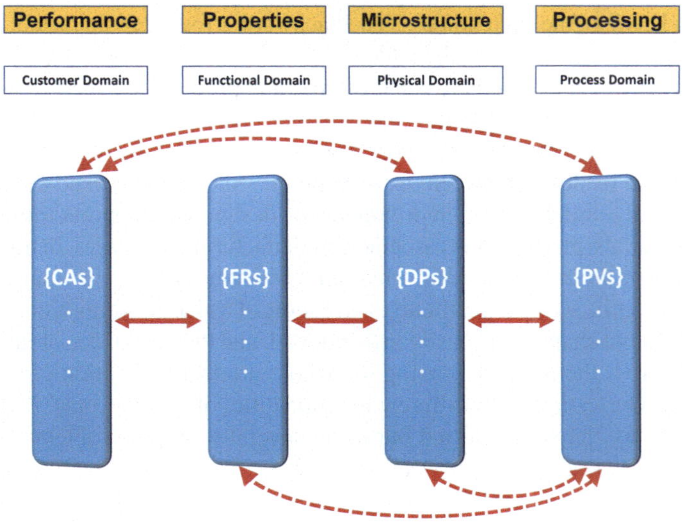

Fig. 8.14 Illustration of communication via information channels using cloud-based design for materials and product realization

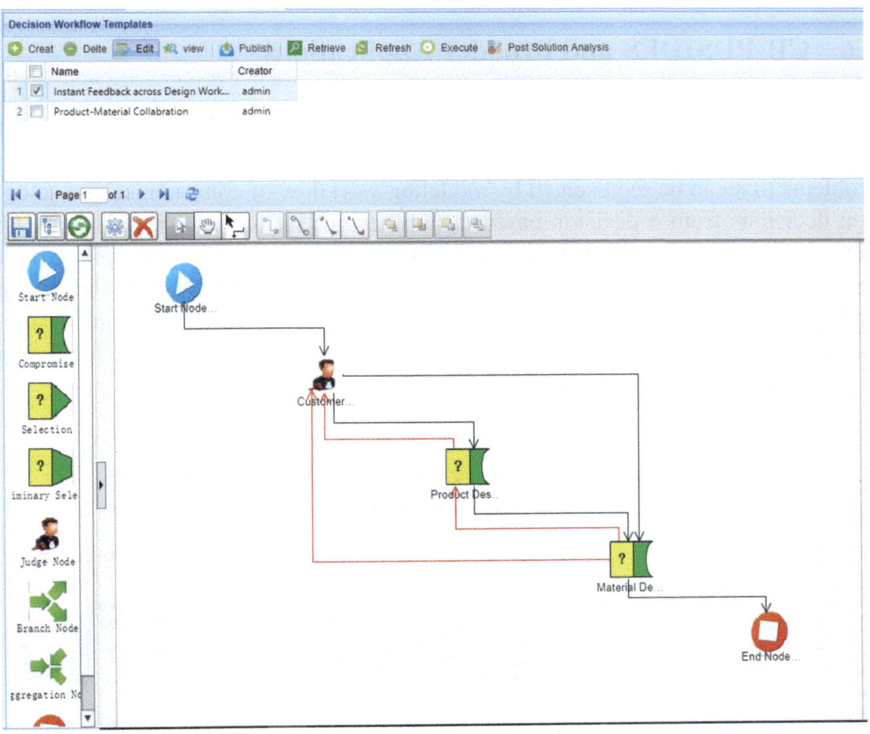

Fig. 8.15 Networked decision workflow in CB-PDSIDES

connection between the stakeholders. The backward arrows denote instant feedback communication between the stakeholders. Thus, as can be seen in Fig. 8.15, there is a direct two-way mapping between customer and product designer; customer and materials designer; and product designer and materials designer—facilitated via the cloud. This is a networked decision workflow in the cloud that facilitates instant communication and feedback between decision points. Such a networked configuration promotes a truly collaborative environment for design and supports the integrated design of materials, products and associated manufacturing processes. In this configuration, designers are able to share information, knowledge, and design/manufacturing resources instantly so as to facilitate co-creation of value in a more cost-effective manner. Dynamic changes are easily incorporated, and the design decisions are communicated across the network making the whole product realization process faster and efficient. The testing of the different functionalities of the platform CB-PDSIDES discussed in this chapter is carried out using several example problems (see Ming et al. 2016, 2017a, 2018; Wang et al. 2017, 2018, 2019; Nellippallil et al. 2019). Interested readers are directed to these articles for details. In Sect. 8.6, we discuss the broader applications of CB-PDSIDES and how CB-PDSIDES integrate with and complement different other projects and products.

8.6 CB-PDSIDES and Broader Applications

CB-PDSIDES is developed as a tool for cloud-based design decision support that supports multiple applications and problem requirements. Complex systems design problems that can be represented by modeling workflow of compromise and selection decisions from a decision-based design perspective are addressed using CB-PDSIDES. In Fig. 8.16, we show some of the applications and examples of focus for CB-PDSIDES. We explore the applications of CB-PDSIDES for the following Cyber-Physical-"X" systems. The "X" here denotes product/material, social, design, or manufacturing systems, respectively, depending on the application of focus. We discuss each of these applications briefly in this section.

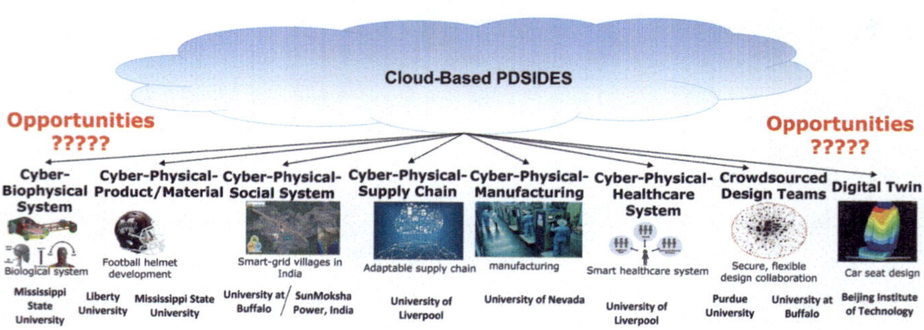

Fig. 8.16 Broader applications of CB-PDSIDES

8.6.1 Applications to Cyber-Biophysical Systems

In this application, our focus is on defining the emerging research frontier that involves the integration of sensing, computational, and communications networks with human biology. The specific application of focus is the head and neck injury-based robust design for vehicular crashworthiness. From an engineering design standpoint, a major design goal is to minimize the effects of road impacts on occupants. This necessitates a need to quantify and manage the injury risks on the human body in terms of the different vehicular impact variables and their associated uncertainties for different crash scenarios. In this application, we use the systems-based design architecture presented to quantify and manage the impact-based injury risks on occupants for different model-based car crash scenarios. The efficacy of the architecture is tested for driver's side-impact scenarios with impact velocity and angle of impact as the critical variables of interest. Two injury criteria, namely head injury criterion (HIC) and neck injury criterion (Lateral Nij), are selected to quantitatively measure the head and neck injury risks in crash simulations. In Fig. 8.17, we show the application of the concept exploration framework in carrying out human safety-centric crashworthiness modeling and design decision support. The functionalities of CB-PDSIDES are used to quantify and manage impact-based injury risks on occupants for different model-based car crash scenarios. This includes performing: i) cloud-based car crash simulations based on a clarified design problem, ii) developing surrogate models capturing injury responses in terms of crash variables, iii) formulating robust design problems and exploring the solution space to manage car crash impact-based injury risks on occupants (see Fig. 8.17).

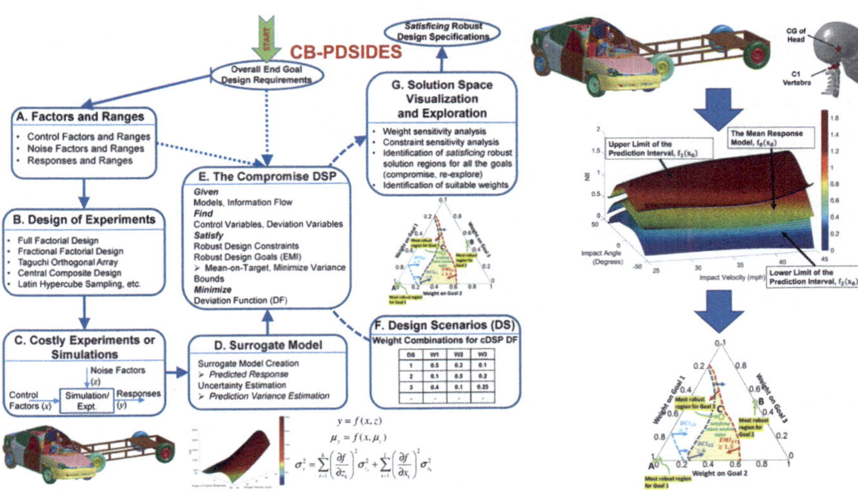

Fig. 8.17 CB-PDSIDES for cyber-biophysical applications (Nellippallil et al. 2020b)

8.6.2 Applications to Cyber-Physical-Product/Material Systems

From the ICME context, we are interested in designing the next generation of cyber-physical-product/material systems. The design of products, subcomponents, and materials with improved system-level performance goals is the major focus here. One application area of focus is the design of American football helmets. The goal-oriented, inverse design method and concept exploration framework (CEF) incorporated into the CB-PDSIDES platform is used to design two components (composite shell and foam liner) of an American football helmet by Fonville and co-authors (Fonville et al. 2019). The performance goals defined for the helmet in this problem include dissipation of impact energy, mitigation of stress waves, and minimization of helmet weight (see Fonville et al. 2019). In Fig. 8.18, the application of the GoID method described in this monograph to the helmet design exploration problem is shown. The end goal is to completely absorb the energy impacting the human head/brain. With this requirement in mind, the system is designed in an inverse manner starting from the foam and TPU liner. The cDSP for foam and TPU liner is formulated and the results obtained for the components are used to formulate the cDSP for the next component, the composite shell. The larger vision here is to design products, subcomponents, and materials using multiscale modeling efforts and systems-based robust design techniques so as to achieve the integrated design exploration of multiscale, multifunctional materials and products.

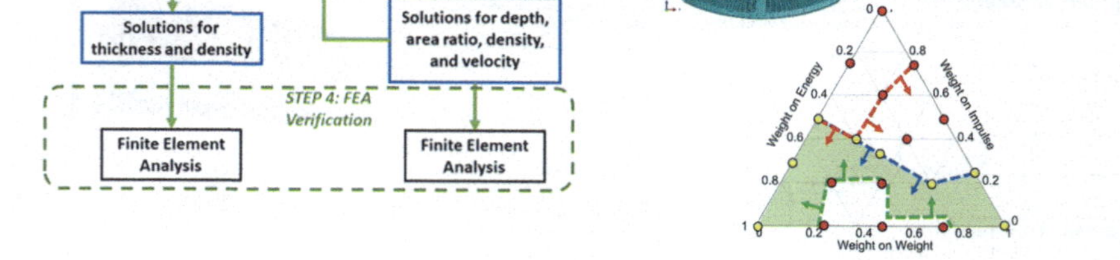

Fig. 8.18 Application of GoID for Cyber-Physical-Product systems (Fonville et al. 2019)

8.6.3 Applications to Cyber-Physical-Design

One other application of focus for CB-PDSIDES from the ICME domain is in cyber-physical design. An application example is the design of the steel manufacturing process chain discussed in this chapter and in Chap. 6. TCS PREMΛP—platform for realization of engineered materials and products is developed by TCS Research, Pune as a comprehensive IT platform that facilitates the integration of models, knowledge, and data for designing both the material and the product (Gautham et al. 2013). The platform TCS PREMΛP is developed for different types of users like expert users, non-expert end-user, and researchers. Gautham et al. (2013) define the domain-independent and domain-dependent components of TCS PREMΛP. The CB-PDSIDES presented in this chapter has the potential to support platforms such as TCS PREMΛP with several of its components. Based on the key functionalities of CBPDSIDES, it is envisioned that CB-PDSIDES can support the platforms from the relevant ecosystems in *Robust design and MDO, Decision Support, Knowledge Engineering; Guided Experimentation, Product Design* and *Product Performance.*

8.6.4 Applications to Cyber-Physical-Manufacturing Systems

Application of CB-PDSIDES to cyber-physical-manufacturing systems is explored from the context of Industry 4.0. Smart manufacturing systems design in the era of Industry 4.0 demands a need to address the distributed and networked nature of manufacturing processes and their associated products. There is a need to design manufacturing systems that facilitate seamless data, information, and knowledge sharing between the different physical, cyber components of the system and the stakeholders (customers, suppliers, manufacturers, etc.) involved. Milisavljevic-Syed et al. (2019) present a computational framework for design of dynamic management of such networked manufacturing systems from the context of Industry 4.0. It is envisioned to incorporate the framework and associated constructs to CB-PDSIDES with possible applications of design for dynamic management in steel manufacturing by expanding on the steel manufacturing example problems presented in (Nellippallil et al. 2016, 2017a, b, 2018a, 2019, 2020a), additive manufacturing, health care systems, and so on. Next, we briefly describe the application of the constructs for additive manufacturing problems and our vision for future.

Cloud-Based Robust Design for Additive Manufacturing (CB-RDfAM)

There have been major transformations in the manufacturing paradigm due to the advent of additive manufacturing (AM) technologies. It is projected that AM will have a market value of more than $15 billion by 2020 and the figure is expected to surge to almost $35 billion by 2024. Some of the major features that distinguish AM from traditional manufacturing are its design freedom and ease of manufacturability. Due to this, in the past several years there have been significant research interests in this domain both from academia and industry. Simulation-based studies focusing on the multiphysics, multiscale nature of AM processes and experiment-based studies focusing on establishing AM processing–structure–property–performance relationships are the major focuses in these research efforts. A major opportunity here is to look at the systems-based robust design of AM process chain to realize AM products. The challenge, however, is in establishing "good enough" models to capture the information workflow in AM starting from processing to the final product performance. Part performance in AM is known to be highly dependent on the melt-pool. The melt-pool defines the local microstructures of the AM product. The presence of internal defects such as pores and cracks results in inferior fatigue and other mechanical properties of the fabricated parts. One of the current research interests in this area is to use data-driven methodologies to predict the size and location of porosities based on in situ process signatures, that is, thermal history (Seifi et al. 2019). Fatigue life and other mechanical properties will be predicted from the porosity features predicted; see (Seifi et al. 2020) for initial study carried out. It is noted that AM-built parts are prone to these porosities and other related effects due to the extreme heating and cooling cycles. These porosity features are randomly distributed over the AM part resulting in features that are stochastic in nature. This necessitates the need for robust design studies taking into account the stochastic nature of the AM part features. Once the forward relationships are defined, we plan to use the cloud-based decision support platform with the robust design constructs presented in this monograph to carry out systems-based robust concept exploration of additive manufacturing process chain. Our vision for this work planned is captured in Fig. 8.19.

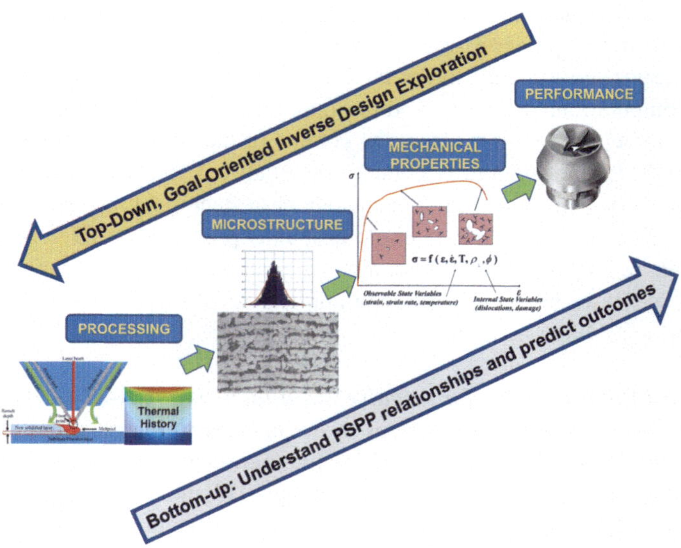

Fig. 8.19 Inverse Robust Design for AM

8.6.5 Applications to Cyber-Physical-Social Systems

CB-PDSIDES for automated micro-enterprise design and analysis for sustainable rural development of villages in India is also explored as an application to cyber-physical-social systems. Three constructs that support this application are: (i) village-level baseline sustainability index, (ii) dilemma triangle, and (iii) village level system dynamics, described by Yadav et al. (2019). Yadav et al. (2019) present a computational framework incorporating these constructs to facilitate dialog between the stakeholders involved: corporate social responsibility (CSR) investors and social entrepreneurs. These stakeholders are geographically dispersed and an increase in the number of stakeholders demands a need for them to effectively communicate and collaborate in order to come up with sustainable solutions (micro-enterprises) worthy of further investment. We envision incorporating the framework and constructs developed by Yadav and co-authors in CB-PDSIDES to facilitate collaboration and open innovation between these stakeholders. Using CB-PDSIDES, the stakeholders will be able to direct attention to issues and challenges that are typically ignored or missed while solving social wicked problems.

All these applications discussed are work in progress in the International Systems Realization Partnership between the Institute for Industrial Engineering @ The Beijing Institute of Technology, The Systems Realization Laboratory @ The University of Oklahoma, and the Design Engineering Laboratory @ Purdue. Other collaborating institutions include Center for Advanced Vehicular Systems at Mississippi State University, Liberty University, University at Buffalo, SunMoksha India, TCS Research, and the University of Liverpool.

8.7 Closing Remarks: Advancing the Vision for the Co-design of Materials, Products, and Manufacturing Processes and Opportunities Worthy of Investigation

To fully realize integrated materials, products, and manufacturing process design as required by ICME and its capabilities in this digital era, there is a need to facilitate a network of participants, which includes material scientists, systems designers, software developers, end service customers to come together and share material/product/manufacturing process/market data, information, knowledge, and resources instantly and in an integrated fashion, and thereby collaborate so as to facilitate a cost-effective co-creation of value supporting open innovation, defined in this monograph as the capability to facilitate co-design. In this chapter, we identify the requirements for a cloud-based computational platform (CB-PDSIDES) to support the integrated design of materials and products as required by ICME. The core competencies needed for a designer using the cloud-based platform for implementing integrated materials, product, and manufacturing process realization in the industry are identified. Based on these core competencies, we present a cloud-based computational platform anchored decision-based design paradigm and the core functionalities offered by the platform in keeping with integrated materials and product design through cloud-based decision support. A hot rolling example problem for the production of a steel rod is used to demonstrate two core functionalities associated with CB-PDSIDES and some preliminary results are presented. The domain-independent cloud-based platform presented supports designers in seamless, yet controllable, information, knowledge, and resource sharing. Further work and development are needed to successfully institutionalize the integrated design of materials, product, and manufacturing processes in the industry using the technologies associated with digital manufacturing and design innovation paradigm. Addressing the impacts of other digital technologies like big data, internet of things (IoT), cyber-physical-product/material (CPP/M) systems, and so on on the goals defined for ICME needs to be carried out in the future to fully realize the capabilities of ICME in the digital era. One area of focus based on work presented in this monograph is the development of an intelligent concept exploration framework by bringing in machine learning techniques to carry out design problem formulation and solution space exploration. We discuss this briefly next.

An Intelligent (Machine Learning Enhanced) Concept Exploration Framework
Future work is planned to deliver an intelligent (machine learning enhanced) concept exploration framework (iCEF) that is generalizable to the realization of complex systems. iCEF will be an end-goal driven, "top-down" design strategy by taking a goal-oriented, inverse approach by using machine learning techniques. Three key features of the iCEF are described in the context traversing the process–structure–performance scales in hot rod rolling discussed in this monograph to generate solutions (hypotheses) that are relatively insensitive to uncertainties:

Step 1: The capability to capture forward relationships using supervised learning techniques (e.g. artificial neural networks). Both the relationship between A and B (i.e. processing and microstructure), and the relationship between B and C (microstructure and property/performance) are established by training based on the data from simulations or physical experiments.

Step 2: The capability to capture the inverse decision workflow using the compromise decision support problem (cDSP), construct, and the associated template. The forward relationships obtained by supervised learning techniques are used to formulate multiple cDSPs for inverse decision-making in the design of the system. Decision-related knowledge is captured using multiple cDSP templates.

Step 3: The capability to integrate forward relationships and inverse decision workflows to generate multiple solutions that are relatively insensitive to uncertainties, achieved using the intelligent concept exploration framework, see Fig. 8.20.

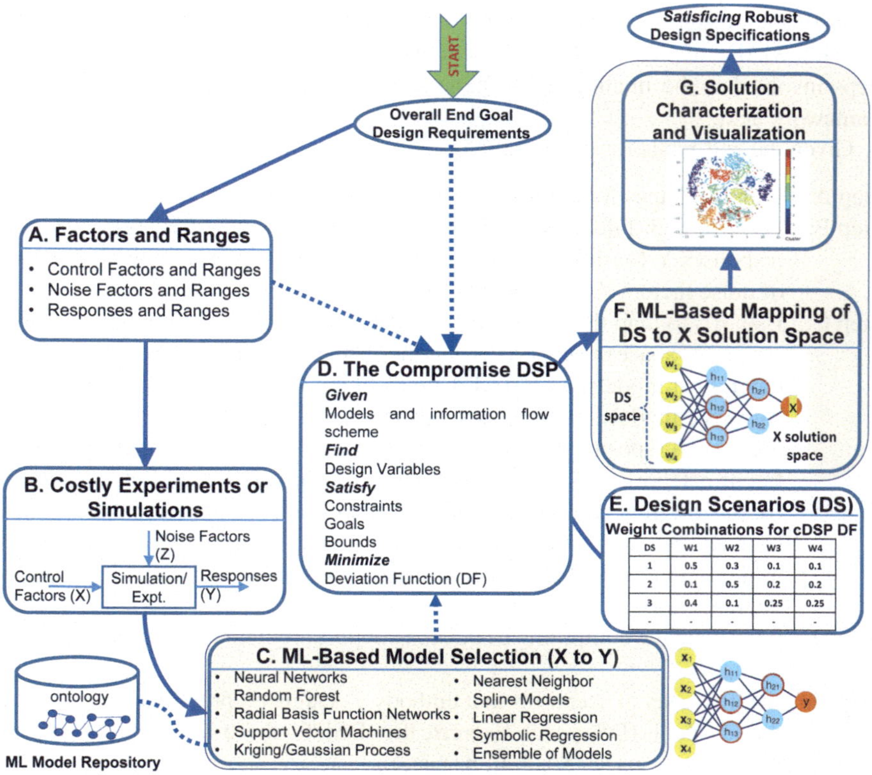

Fig. 8.20 Intelligent (ML-enhanced) concept exploration framework

We propose to evaluate these solutions using unsupervised learning techniques (e.g. using clustering algorithms to map the design solution space to lower dimensions and annotating the reduced dimensions as feasible and infeasible, or sensitive and insensitive). A key challenge is to invoke the right machine learning algorithms for the limited availability of data. In Fig. 8.20, we show the steps of the proposed iCEF for systematic problem formulation, and design and solution space exploration of satisficing, robust design solutions using machine learning (ML). We hypothesize that by constructing machine learning-based surrogate models using data derived from high-fidelity simulation models, fast approximations of the responses at new design points can be made, thereby providing opportunities for design sensitivity and exploration studies in a feasible manner. Our rationale for using ML-based surrogate models is that they can be used to predict the varying responses within the entire design space in a computationally efficient manner while also providing information on the uncertainty measure of the prediction, thereby supporting analysis, design exploration, and uncertainty quantification, as well as management. Supervised machine learning-based surrogate models, such as the Gaussian process (GP) models, are used to model highly nonlinear behavior of the responses and also to quantify the epistemic uncertainty, the uncertainty due to lack of data or incomplete knowledge of physics of the problem in the model predictions. We describe the steps involved in the intelligent (machine learning enhanced) concept exploration framework next.

Given the end-goals for the design:

Step A: Identify ranges for control factors X, noise factors Z, and the responses Y.
Step B: Perform high-fidelity simulations and/or physical experiments to generate responses Y for different ranges of the control factors X. Note the impact of noise factors Z on the responses.
Step C: Use ML to map the control factor/design space (X space) to the response space (Y space). Supervised learning ingests design space points from theoretical models, empirical data, and high-fidelity simulations to create surrogate models mapping design space criteria to physical outcomes. An ML model repository involving ontology to manage the knowledge of different categories of ML models is used to facilitate the model selection for the different data sets and problem requirements. ML model properties such as accuracy, time-efficiency, and parameter-complexity are captured in the ontology to distinguish different ML models as available assets in the repository.
Step D: Formulate the compromise decision support problem (cDSP) as illustrated in Fig. 8.20.
Step E: Identify a set of scenarios that embody a designer's preference for each of the end-goals (DS space). Exercise the cDSP for these scenarios to generate the solution space (X solution space).
Step F: Use supervised learning to facilitate solution space exploration by mapping cDSP weights and constraint sensitivity to satisficing solution regions in X solution space.
Step G: Use unsupervised learning to map the design solution space to lower dimensions for visualization. This information is used by a designer to support his/her decision that may incorporate judgment and inductive speculation.

In Fig. 8.21, we capture the key enablers in terms of tools, technologies, concepts, and a digital knowledge platform as proposed and envisioned in this monograph for enabling the co-design of materials, manufacturing processes, and products leveraging physics-based simulations, data-driven reasoning, decision-support tools, and knowledge engineering systems. The technologies associated with the digital manufacturing and design innovation paradigm are creating disruptive transformations in the design, and manufacturing domains and hence are paving new paths in the way we realize materials and products. Having discussed some of the tools and technologies that are part of this shift, we see several opportunities for improvements in the integrated design of materials, products, and manufacturing processes. These opportunities are identified as:

- Methods and approaches to prioritize the material phenomena to be modeled for a given material design problem and ensuring the proper vertical and horizontal

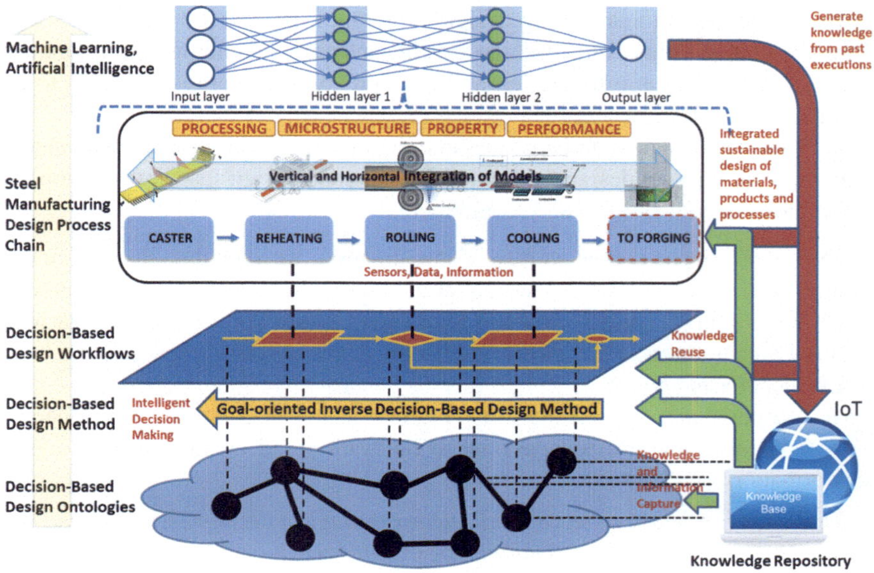

Fig. 8.21 Key enablers in terms of tools, technologies, concepts, and a digital knowledge platform for enabling the co-design of materials, manufacturing processes as proposed and envisioned in this monograph

integration of models to facilitate the physics-based, history-dependent material information flow.

- Rapid concept exploration approaches to predict the value/utility of investing in a new material/product system and their processing routes.
- Approaches to address beyond concept generation and early stages of materials and product design need to be explored, however, ensuring that the approaches maintain the characteristics of design exploration and not detailed design.
- Digital design catalogs for material/product systems that keep track of any new material models developed and deployed for applications of interest.
- Approaches to establishing the protocols and rules for coming up with **digital surrogate models** replacing the physics-based engineering models for specific application domains.
- Digital threads that capture knowledge transition by means of the digital surrogate models across the entire material/product lifecycle.
- Digital tools and constructs to quantify and manage the uncertainty in material process–structure–property digital surrogate models in order to support **decision-based digital robust design** such that robust, resilient, and affordable material solutions can be sought.
- **Microstructure-mediated digital design** using approaches like the goal-oriented, inverse design method proposed using advances in high-performance computing (HPC) to enable high-resolution simulations of complex material systems across multiscale.

- Approaches to automating the design exploration process by using rules and improved algorithms using machine learning and artificial intelligence techniques are required to ensure the design and solution space exploration process to be practical and efficient.
- Improved visualization and post-processing tools need to be developed to handle multiple numbers of design goals while carrying out design space exploration for complex material systems and requirements.

Given these opportunities worthy of investigation, we speculate that the future academic and industrial research and collaborations in this area should be adaptive to these transformative changes to remain competitive. In the words of Charles Darwin, *"It is not the strongest of the species that survive, nor the most intelligent but ones most responsive to change."*

We recognize and suggest the following non-technical competencies as necessary for learning that are foundational to innovation and creation in this digital transformation:

1. *The ability to learn, unlearn, and relearn through reflection on doing and the associated creation and articulation of knowledge.*
2. *The ability to speculate and to identify gaps in state-of-the-art or state of practice that may lead to innovation.*
3. *The ability to ask questions, actively listen, reflect, and identify opportunities worthy of further investigation.*
4. *The ability to make decisions to move forward using incomplete information.*
5. *The ability to assess and think critically (deductive reasoning and inductive speculation) and identify a way forward.*

Recognizing that there are transformations taking place in the materials/product design and manufacturing domains and a need to adapt, we look forward to these accelerating disruptions and trends creating peace, hope, and happiness in the lives of the people across the globe.

References

Bascaran, E., Bannerot, R., & Mistree, F. (1987). The conceptual development of a method for solving multi-objective hierarchical thermal design problems. In *ASME 1987 national heat transfer conference*. Pennsylvania: Pittsburgh. Paper Number: 87-HT-62.

Beitz, W., Pahl, G., & Grote, K. (1996). Engineering design: A systematic approach. *MRS Bulletin 71*.

Chen, W., Allen, J. K., Tsui, K.-L., & Mistree, F. (1996). A procedure for robust design: minimizing variations caused by noise factors and control factors. *Journal of Mechanical Design, 118*(4), 478–485.

Chen, W., Allen, J. K., & Mistree, F. (1997). A robust concept exploration method for enhancing productivity in concurrent systems design. *Concurrent Engineering, 5*(3), 203–217.

Choi, H.-J., Austin, R., Allen, J. K., McDowell, D. L., Mistree, F., & Benson, D. J. (2005). An approach for robust design of reactive power metal mixtures based on non-deterministic micro-scale shock simulation. *Journal of Computer-Aided Materials Design, 12*(1), 57–85.

Choi, H., McDowell, D. L., Allen, J. K., Rosen, D., & Mistree, F. (2008). An inductive design exploration method for robust multiscale materials design. *Journal of Mechanical Design, 130*(3), 031402.

Pollock, T. M., Allison J. E., et al. (2008). Integrated computational materials engineering: a transformational discipline for improved competitiveness and national security. National Academies Press.

Fonville, T. F., Nellippallil, A. B., Horstemeyer, M. F., Allen, J. K., & Mistree, F. (2019). A goal-oriented, inverse decision-based design method for multi-component product design. *ASME Design Automation Conference*, Paper Number: IDETC2019-97388.

Franke, N., Von Hippel, E., & Schreier, M. (2006). Finding commercially attractive user innovations: A test of lead-user theory. *Journal of Product Innovation Management, 23*(4), 301–315.

Gautham, B., Singh, A. K., Ghaisas, S. S., Reddy, S. S., & Mistree, F. (2013). PREMΛP: A platform for the realization of engineered materials and products. In *ICoRD'13* (pp. 1301–1313). Berlin: Springer.

Gero, J. S. (1990). Design prototypes: A knowledge representation schema for design. *AI magazine, 11*(4), 26.

Gruber, T. R. (1993). A translation approach to portable ontology specifications. *Knowledge Acquisition, 5*(2), 199–220.

Hazelrigg, G. A. (1996). *Systems engineering: An approach to information-based design*. Upper Saddle River, NJ: Prentice Hall.

Hazelrigg, G. A. (1998). A framework for decision-based engineering design. *Journal of Mechanical Design, 120*(4), 653–658.

Hehenberger, P., & Bradley, D. (2016). Mechatronic futures: challenges and solutions for mechatronic systems and their designers. Springer.

Horstemeyer, M. F. (2012). *Integrated Computational Materials Engineering (ICME) for metals: using multiscale modeling to invigorate engineering design with science*. New York: Wiley.

Kraft, E. (2013). Expanding the digital thread to impact total ownership cost. In *NIST MBE Summit*.

Lewis, K., & Mistree, F. (1998). Collaborative, sequential, and isolated decisions in design. *Journal of Mechanical Design, 120*(4), 643–652.

McDowell, D. L. (2018). Microstructure-sensitive computational structure-property relations in materials design. In *Computational materials system design* (pp. 1–25). Berlin: Springer.

Milisavljevic-Syed, J., Allen, J. K., Commuri, S., & Mistree, F. (2019). Design of networked manufacturing systems for indusrty 4.0. In *CIRP manufacturing systems conference*.

Ming, Z., Yan, Y., Wang, G., Panchal, J. H., Goh, C. H., Allen, J. K., & Mistree, F. (2016). Ontology-Based Executable Design Decision Template Representation and Reuse. *Artificial Intelligence for Engineering Design, Analysis and Manufacturing, 30*, 390–405.

Ming, Z., Wang, G., Yan, Y., Dal Santo, J., Allen, J. K., & Mistree, F. (2017a). An ontology for reusable and executable decision templates. *Journal of Computing and Information Science in Engineering, 17*(3), 031008-031013.

Ming, Z., Wang, G., Yan, Y., Panchal, J. H., Goh, D., Allen, J. K. & Mistree, F. (2017b). Ontology-based representation of design decision hierarchies. *Journal of Computing and Information Science in Engineering, 18*(1), 011001-12.

Ming, Z., Nellippallil, A. B., Yan, Y., Wang, G., Goh, C. H., Allen, J. K., et al. (2018). PDSIDES—a knowledge-based platform for decision support in the design of engineering systems. *Journal of Computing and Information Science in Engineering, 18*(4), 041001.

Mistree, F., Smith, W. F., Bras, B. A., Allen, J. K., & Muster, D. (1990). Decision-based design: a contemporary paradigm for ship design. *Transactions, Society of Naval Architects and Marine Engineers. 98*, 565–597.

Mistree, F., Hughes, O. F., & Bras, B. (1993). Compromise decision support problem and the adaptive linear programming algorithm. In *Progress in Astronautics and Aeronautics, 150*, 251–251.

Mistree, F., Lewis, K., & Stonis, L. (1994). Selection in the Conceptual Design of Aircraft. *AIAA Journal* 1153–1166.

Mocko, G. M., Rosen, D. W., & Mistree, F. (2007). Decision retrieval and storage enabled through description logic. In *ASME Computers and Information in Engineering Conference*. Las Vegas, Nevada. Paper Number: DETC2007-35644.

Montgomery, D. C. (2008). *Design and analysis of experiments*. New York: Wiley.

Musen, M. A. (2015). The protégé project: a look back and a look forward. *AI matters, 1*(4), 4.

Muster, D., & Mistree, F. (1988). The decision support problem technique in engineering design. *International Journal of Applied Engineering Education, 4*(1), 23–33.

Nellippallil, A. B. (2018). The Integrated Realization of Materials, Products and Associated Manufacturing Processes. Ph.D. Dissertation, School of Aerospace and Mechanical Engineering, University of Oklahoma, Norman.

Nellippallil, A. B., Song, K. N., Goh, C.-H., Zagade, P., Gautham, B., Allen, J. K., & Mistree, F. (2016). A goal oriented, sequential process design of a multi-stage hot rod rolling system. *ASME Design Automation Conference*, Paper Number: DETC2016-59402.

Nellippallil, A. B., Rangaraj, V., Gautham, B., Singh, A. K., Allen, J. K., & Mistree, F. (2017a). A goal-oriented, inverse decision-based design method to achieve the vertical and horizontal integration of models in a hot rod rolling process chain. *ASME Design Automation Conference*, Paper Number: DETC2017-67570.

Nellippallil, A. B., Song, K. N., Goh, C.-H., Zagade, P., Gautham, B., Allen, J. K., & Mistree, F. (2017b). A goal-oriented, sequential, inverse design method for the horizontal integration of a multistage hot rod rolling system. *ASME Journal of Mechanical Design, 139*(3), 031403.

Nellippallil, A. B., Mohan, P., Allen, J. K., & Mistree, F. (2018a). Robust Concept Exploration of Materials, Products and Associated Manufacturing Processes. *ASME Design Automation Conference*, Paper Number: DETC2018-85913.

Nellippallil, A. B., Rangaraj, V., Gautham, B., Singh, A. K., Allen, J. K., & Mistree, F. (2018b). An inverse, decision-based design method for integrated design exploration of materials, products, and manufacturing processes. *Journal of Mechanical Design, 140*(11), 111403-111403-17.

Nellippallil, A. B., Ming, Z., Allen, J. K., & Mistree, F. (2019a). Cloud-Based Materials and Products Realization–Fostering ICME via Industry 4.0. *Integrating Materials and Manufacturing Innovation, 8*(2), 107–121.

Nellippallil, A. B., Mohan, P., Allen, J. K., & Mistree, F. (2019b). Inverse Thermo-Mechanical Processing (ITMP) design of a steel rod during hot rolling process. *ASME Design Automation Conference*, Paper Number: IDETC2019-97390.

Nellippallil, A. B., Mohan, P., Allen, J. K., & Mistree, F. (2020a). An Inverse, decision-based design method for robust concept exploration. *Journal of Mechanical Design*, https://doi.org/10.1115/1.4045877.

Nellippallil, A. B., Berthelson, P. R., Peterson, L., & Prabhu, R. (2020b). Head and neck injury risk criteria-based robust design for vehicular crashworthiness. *ASME Design Automation Conference*, Paper Number: IDETC2020-22539.

Panchal, J. H., Fernández, M. G., Paredis, C. J. J., & Mistree, F. (2004). Reusable design processes via modular, executable, decision-centric templates. In *AIAA/ISSMO multidisciplinary analysis and optimization conference*. Albany, NY. Paper Number AIAA-2004-4601.

Panchal, J. H., Fernández, M. G., Allen, J. K., Paredis, C. J., & Mistree, F. (2005). Facilitating meta-design via separation of problem, product, and process information. *ASME 2005 International Mechanical Engineering Congress and Exposition*, Paper Number: IMECE2005-80013.

Panchal, J. H., Gero Fernández, M., Paredis, C. J., Allen, J. K., & Mistree, F. (2009). A modular decision-centric approach for reusable design processes. *Concurrent Engineering, 17*(1), 5–19.

Reddy, R., Smith, W., Mistree, F., Bras, B., Chen, W., Malhotra, A., Badhrinath, K., Lautenschlager, U., Pakala, R., & Vadde, S. (1992). DSIDES user manual. In *Systems design laboratory, department of mechanical engineering*. Texas, no.: University of Houston, Houston.

Reddy, R., Smith, W., Mistree, F., Bras, B., Chen, W., Malhotra, A., Badhrinath, K., Lautenschlager, U., Pakala, R., & Vadde, S. (1996). DSIDES User Manual. In *Systems realization laboratory, woodruff school of mechanical engineering*. Atlanta, Georgia, no.: Georgia Institue of Technology.

Reddy, S., Gautham, B., Das, P., Yeddula, R. R., Vale, S., & Malhotra, C. (2017). An ontological framework for integrated computational materials engineering. In *Proceedings of the 4th world congress on integrated computational materials engineering (ICME 2017)*. Berlin: Springer.

Rockwell, J. A., Grosse, I. R., Krishnamurty, S., & Wileden, J. C. (2010). A semantic information model for capturing and communicating design decisions. *Journal of Computing and Information Science in Engineering, 10*(3), 031008.

Seifi, S. H., Tian, W., Yadollahi, A., Doude, H., & Bian, L. (2019). In-Situ Fatigue Prediction of Direct Laser Deposition Parts Based on Thermal Profile. In *ASME 2019 international mechanical engineering congress and exposition*. American Society of Mechanical Engineers Digital Collection.

Seifi, S. H., Nellippallil, A. B., Tian, W., Prabhu, R., & Bian, L. (2020). Multi-objective inverse design exploration of process-defect-property relationships in additive manufacturing. In *2020 annual IISE conference*. New Orleans.

Shukla, R., Goyal, S., Singh, A. K., Panchal, J. H., Allen, J. K., & Mistree, F. (2015). Design exploration for determining the set points of continuous casting operation: An industrial application. *Journal of Manufacturing Science and Engineering, 137*(3), 034503.

Shupe, J. A. (1988). Decision-based design: Taxonomy and implementation. Ph.D. Dissertation, Department of Mechanical Engineering, University of Houston, Houston, Texas.

Simpson, T. W., Chen, W., Allen, J. K., & Mistree, F. (1996). Conceptual design of a family of products through the use of the robust concept exploration method. In *6th. AIAA/USAF/NASA/ISSMO symposium on multidisciplinary analysis and optimization*.

Simpson, T. W., Poplinski, J., Koch, P. N., & Allen, J. K. (2001). Metamodels for computer-based engineering design: survey and recommendations. *Engineering with Computers, 17*(2), 129–150.

Sinha, A., Bera, N., Allen, J. K., Panchal, J. H., & Mistree, F. (2013). Uncertainty management in the design of multiscale systems. *Journal of Mechanical Design, 135*(1), 011008.

Smith, W. F. (1985). The Development of AUSEVAL: An Automated Ship Evaluation System. *Department of Mechanical Engineering*, University of Houston, Houston, Texas. M.S.

Suh, N. P. (1990). *The principles of design* Oxford: Oxford University Press on Demand.

Thames, L., & Schaefer, D. (2017). Cybersecurity for Industry 4.0, Berlin: Springer.

US_Air_Force. (2013). Global horizons: United States air force global science and technology vision AF/ST TR 13-0, July 2013.

Wang, H., Noy, N., Rector, A., Musen, M., Redmond, T., Rubin, D., Tu, S., Tudorache, T., Drummond, N., & Horridge, M. (2006). Frames and OWL side by side. *Presentation Abstracts*. Citeseer.

Wang, R., Wang, G., Yan, Y., Sabeghi, M., Ming, Z., Allen, J. K., & Mistree, F. (2017). Ontology-based representation of meta-design in designing decision workflows. *ASME Design Automation Conference*, Paper Number: DETC2017-67817.

Wang, R., Nellippallil, A. B., Wang, G., Yan, Y., Allen, J. K., & Mistree, F. (2018). Systematic design space exploration using a template-based ontological method. *Advanced Engineering Informatics, 36*, 163–177.

Wang, R., Nellippallil, A. B., Wang, G., Yan, Y., Allen, J. K., & Mistree, F. (2019). Ontology-based uncertainty management approach in designing of robust decision workflows. *Journal of Engineering Design, 30*(10–12), 726–757.

Wu, D., Thames, J. L., Rosen, D. W., & Schaefer, D. (2012). Towards a cloud-based design and manufacturing paradigm: looking backward, looking forward. *ASME Design Automation Conference*, Paper Number: DETC2012-70780.

Wu, D., Rosen, D. W., Wang, L., & Schaefer, D. (2015). Cloud-based design and manufacturing: A new paradigm in digital manufacturing and design innovation, *Computer-Aided Design 59*, 1–14.

Yadav, A., Das, A. K., Allen, J. K., & Mistree, F. (2019). A computational framework to support social entrepreneurs in creating value for rural communities in India. *ASME Design Automation Conference*, Paper Number: IDETC2019-97375.

Yeddula, R. R., Vale, S., Reddy, S., Malhotra, C. P., Gautham, B., & Zagade, P. (2016). A knowledge modeling framework for computational materials engineering. *SEKE*.

Zhan, P., Jayaram, U., Kim, O., & Zhu, L. (2010). Knowledge representation and ontology mapping methods for product data in engineering applications, *Journal of Computing and Information Science in Engineering, 10*(2), 021004–021004–021011.

Index

A. B. Nellippallil et al., *Architecting Robust Co-Design of Materials, Products, and Manufacturing Processes*, https://doi.org/10.1007/978-3-030-45324-4

The manufacturer's authorised representative in the EU is Springer Nature Customer Service Centre GmbH, Europaplatz 3, 69115 Heidelberg, Germany. If you have any concerns regarding our products, please contact ProductSafety@springernature.com

Printed and bound by CPI Group (UK) Ltd, Croydon, CR0 4YY
22/09/2026
02221060-0017